Advances in Robotics and Optimization Approach in Engineering

Advances in Robotics and Optimization Approach in Engineering

Editors
Dilip Kumar Bagal
Abhishek Barua

CWP
Central West Publishing

This edition has been published by Central West Publishing, Australia

For more information about the books published by Central West Publishing, please visit https://centralwestpublishing.com

Disclaimer
Every effort has been made by the publisher, editors and authors while preparing this book, however, no warranties are made regarding the accuracy and completeness of the content. The publisher, editors and authors disclaim without any limitation all warranties as well as any implied warranties about sales, along with fitness of the content for a particular purpose. Citation of any website and other information sources does not mean any endorsement from the publisher and authors. For ascertaining the suitability of the contents contained herein for a particular lab or commercial use, consultation with the subject expert is needed. In addition, while using the information and methods contained herein, the practitioners and researchers need to be mindful for their own safety, along with the safety of others, including the professional parties and premises for whom they have professional responsibility. To the fullest extent of law, the publisher, editors and authors are not liable in all circumstances (special, incidental, and consequential) for any injury and/or damage to persons and property, along with any potential loss of profit and other commercial damages due to the use of any methods, products, guidelines, procedures contained in the material herein.

NATIONAL LIBRARY OF AUSTRALIA

A catalogue record for this book is available from the National Library of Australia

ISBN (print): 978-1-922617-18-7

Contents

Preface

This book focuses on questions of importance from the point of view of robotics and optimization for interested beginners, scholars and researchers. It consists of thirteen chapters under three main themes. This book examines the actual implementation of service automation technologies via robotic process automation and cognitive automation techniques. This newest and extensive research demonstrates that automation is developing, maturing and rising in global businesses.

The first chapter focuses on robotics foundations, describing the basic concepts and techniques necessary to build a robotic system. The robot is a non-linear coupling system owing to the complicated dynamics. Besides movement control, management of the interactive forces between the robot and the environment is required for the handling or interaction with people. Recent progress has been made in cloud computing, big data and artificial intelligence technology during the second phase of robotic process automation. The last section of the book also covers sophisticated methods for optimization in different mechanical applications.

We appreciate the support of institutes like the Government College of Engineering, Kalahandi, National Institute of Technology, Rourkela, and Centre of Post Graduate Studies, Rourkela, in the creation of this book. All the authors who have contributed for this book are likewise honored for their outstanding research in the field of robotics, automation and optimization.

List of Contributors

Abhishek Barua
Centre for Advanced Post Graduate Studies, BPUT, Rourkela, Odisha, India

Abhishek Samantaray
Scan Steels Limited, Rajgangpur, Odisha, India, India

Ajay Sudhir Bale
CMR University, Bengaluru, Karnataka, India

Ajit Kumar Pattanaik
Government College of Engineering, Kalahandi, Bhawanipatna, Odisha, India

Ankita Barua
College of Agricultural Engineering and Technology, OUAT, Bhubaneswar, Odisha, India

Antarjyami Giri
Government College of Engineering, Kalahandi, Bhawanipatna, Odisha, India

Arati Rath
Kalinga Institute of Industrial Technology, Bhubaneswar, Odisha India

Ashok Kumar Bhoi
Government College of Engineering, Kalahandi, Bhawanipatna, Odisha, India

Ashutosh Khuntia
College of Engineering and Technology, Bhubaneswar, Odisha, India

Asini Kumar Baliarsingh
Government College of Engineering, Kalahandi, Bhawanipatna, Odisha, India

Asish Kumar Bagal
College of Engineering and Technology, Bhubaneswar, Odisha, India

B S Parinitha
CMR University, Bengaluru, Karnataka, India

Baby Chithra R
CMR University, Bengaluru, Karnataka, India

Bibekananda Naik
CSIR - Institute of Minerals & Materials Technology, Bhubaneswar, Odisha, India

Biswajit Parida
Centre for Advanced Post Graduate Studies, BPUT, Rourkela, Odisha, India

Deepak Kumar Bagal
Veer Surendra Sai University of Technology, Burla, Odisha, India

Deepak Kumar Pradhan
Government College of Engineering, Kalahandi, Bhawanipatna, Odisha, India

Dilip Kumar Bagal
Government College of Engineering, Kalahandi, Bhawanipatna, Odisha, India

Dillip Ranjan Nayak
Government College of Engineering, Kalahandi, Bhawanipatna, Odisha, India

Dipti Acharjya
College of Engineering and Technology, Bhubaneswar, Odisha, India

Dulu Patnaik
Government College of Engineering, Kalahandi, Bhawanipatna, Odisha, India

Gopal Charan Behera
Government College of Engineering, Kalahandi, Bhawanipatna, Odisha, India

Kanchan Kumari
Parala Maharaja Engineering College, Berhampur, Odisha

Kaushik M
Bannari Amman Institute of technology, Sathyamangalam, Tamil Nadu, India

Kishor Chandra Panda
Government College of Engineering, Kalahandi, Bhawanipatna, Odisha, India

Kishore P
CMR University, Bengaluru, Karnataka, India

Manjunath V
CMR University, Bengaluru, Karnataka, India

Mohamed Lounis
Ziane Achour University of Djelfa, Djelfa, Algeria

Naveen Ghorpade
CMR University, Bengaluru, Karnataka, India

Nirmalya Parhi
Government College of Engineering, Kalahandi, Bhawanipatna, Odisha, India

Pradyut Kumar Muduli
Government College of Engineering, Kalahandi, Bhawanipatna, Odisha, India

Pravesh P
CMR University, Bengaluru, Karnataka, India

Pulkit Kumar Agrawal
Government College of Engineering, Kalahandi, Bhawanipatna, Odisha, India

Rajendra Prasad Nayak
Government College of Engineering, Kalahandi, Bhawanipatna, Odisha, India

Rajesh Kumar N
CMR University, Bengaluru, Karnataka, India

Ram Krishan
Mata Sundri University Girls College, Mansa, India

Ritesh Barkey
Government College of Engineering, Kalahandi, Bhawanipatna, Odisha, India

Sandeep Bhoi
Parala Maharaja Engineering College, Berhampur, Odisha

Sangam Kumar Singh
Delhi Technical Campus, Bahadurgarh, Haryana, India

Sasmita Nayak
Government College of Engineering, Kalahandi, Bhawanipatna, Odisha, India

Siba Sankar Mahapatra
National institute of Technology, Rourkela, Odisha India

Siddharth Jeet
Centre for Advanced Post Graduate Studies, BPUT, Rourkela, Odisha, India

Soumya Ranjan Mangaraj
Government College of Engineering, Kalahandi, Bhawanipatna, Odisha, India

Sriparna Paul
Council of Scientific and Industrial Research–Advanced Materials and Process Research, Bhopal, Madhya Pradesh, India

Surya Narayan Panda
Birsa Institute of Technology, Sindri, Dhanbad, Jharkhand, India

Swarnima Subhadarsini
Veer Surendra Sai University of Technology, Burla, Odisha, India

Swastik Pradhan
Lovely Professional University, Phagwara, Punjab, India

Varun Yogi S
CMR University, Bengaluru, Karnataka, India

Vinayak N
CMR University, Bengaluru, Karnataka, India

1

Robotics Terminology and its Advancement

Dulu Patnaik[1], Deepak Kumar Pradhan[2], Siddharth Jeet[3], Mohamed Lounis[4], and Antarjyami Giri[2]

[1]*Department of Electrical Engineering, Government College of Engineering, Kalahandi, Bhawanipatna, Odisha, India*
[2]*Department of Mechanical Engineering, Government College of Engineering, Kalahandi, Bhawanipatna, Odisha, India*
[3]*Department of Mechanical Engineering, Centre for Advanced Post Graduate Studies, BPUT, Rourkela, Odisha, India*
[4]*Department of Agro-veterinary Science, Faculty of Natural and Life Sciences, Ziane Achour University of Djelfa, Djelfa, Algeria*

1.1 Introduction

In today's culture, robots can be found everywhere; they come in a variety of forms and sizes, and they are programmed to do a broad range of tasks. Typically, when we think about robots, we think of huge humanoid robots from science fiction films like "Transformers" or giant robotic arms utilised in automated manufacturing lines [1-5]. What is a robot, exactly? As defined by the International Robotics and Automation Society, a robot is an artificially intelligent device that can move on its own. Its mobility may be modelled, planned, detected, and controlled, and its motion and behaviour may be influenced by programming. A robot is a programmable manipulator that can be used to perform a wide range of activities in various environments. In reality, it's typically an electromechanical system that communicates its own purpose via the motions and appearance of its components. Humanoid robots are now extensively employed in commercial and industrial settings, and they are capable of performing jobs at a lower cost or with better precision and dependability than people. They are also used for tasks that are too dirty, dangerous, or boring for humans to do. Automation is used in a variety of applications including manufacturing, assembly and packaging, earth and space exploration, surgery, weaponry, laboratory research, and large-scale production of consumer and commercial products. As robotic technology develops, domestic robots for cleaning and mowing the lawn, as well as robotic toys for children of all ages, are becoming more accessible for consumers. When we speak about robots, we're really referring about machines that will do jobs that would normally be performed by people. These devices are intended to mimic the functioning of the human body, or at least some parts of it. The vast majority of shop floor robots have a human-like look; their motors resemble muscles, their general-purpose computer resembles the human brain, and their robotic vision system resembles the eye of the human. The robot is built by combining these different components in a variety of ways. Karel Capek, a Czech playwright, is credited with coining the word "robot" in a play titled Rossum's Universal Robots, which was performed in 1920. (RUR). [6] The term "robot" in Czech implies "worker" or "servant."

The robot arm is the most common kind of robot now in use in industry. These arms are used for a range of tasks such as welding, packing, painting, placing, and assembling a variety of products that we use on a daily basis [7]. A robot arm is simply a collection of linkages that are connected together in such a way that each joint may be controlled by a servo motor, as seen in the illustration. The controlling computer, or brain, of the robot is programmed to drive the robot's many motors in such a manner that it can carry out certain tasks. It is possible to build the robot arm in a number of ways, but its size and form are critical to the robot's robotic architecture. Using the robot's arm, it is possible to position the final grabber arm or spray head in order for them to do their pre-programmed duties. A big or small design of the arm may prevent this position from being achieved. Many arms resemble human arms in that they feature shoulders, elbows, wrists, and hands. The human arm is very well-designed, enabling for precise and complicated motions to be performed. For each degree of freedom needed, one motor is often required. A single joint movement is often used to describe a degree of freedom. For example, a robot with three degrees of freedom may move in three directions: up and down; left and right; forward and back. This simple pan and tilt robot, which has two degrees of freedom, is powered by two servo motors. It has the capacity to rotate from left to right as well as elevate and lower. Many robots can now be constructed with seven degrees of freedom, which is a significant advance [8].

The Robot Institute of America is the term given to any machine built by one of our members and is comprised of our members. A robot is a reprogrammable, flexible manipulator that can perform a variety of tasks by moving material, components, tools, or specialised devices via a variety of programmed movements. Robots are used in a variety of industries to accomplish a variety of activities. Founded in 1979 [9], the Robot Institute of America is a non-profit organisation dedicated to the advancement of robotics.

1.2 Laws of Robotics

Asimov proposed three "Laws of Robotics," with the "zeroth law" being added later on to complete the set.

Law 0: A robot may not injure humanity or through inaction, allow humanity to come to harm.
Law 1: A robot may not injure a human being or through inaction, allow a human being to come to harm, unless this would violate a higher order law.
Law 2: A robot must obey orders given to it by human beings, except where such orders would conflict with a higher order law.
Law 3: A robot must protect its own existence as long as such protection does not conflict with a higher order law [4].

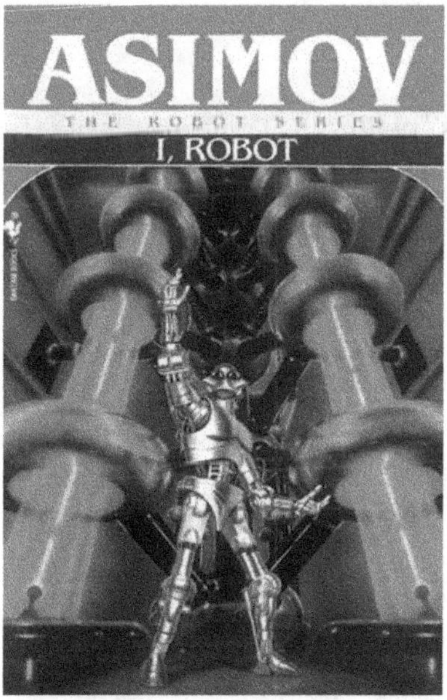

Figure 1.1 Asimov, I, Robot.

1.3 History of Robotics

The "UNIMATE" robot was the world's first industrial robot. It was in 1954 when George Devol, the man who invented the term "Universal Automation," designed the world's first programmable robot. In the year 1962, he abbreviates this to "Unimation," which goes on to become the name of the first robotics company. [7].

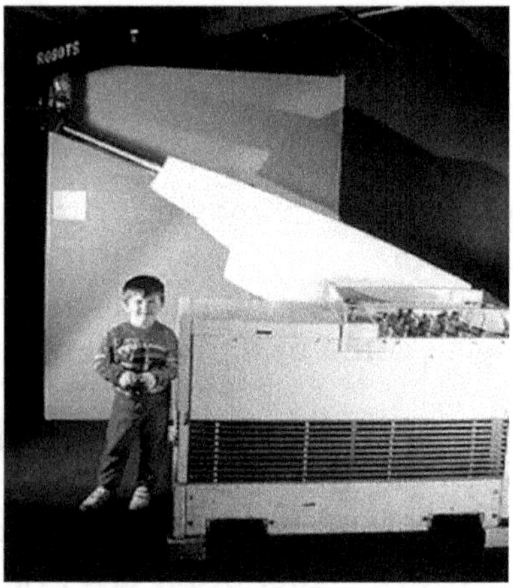

Figure 1.2 UNIMATE originally automated the manufacture of TV picture tubes.

Unimation develops the Puma (Programmable Universal Machine for Assembly) robot in 1978, with design assistance from General Motors.

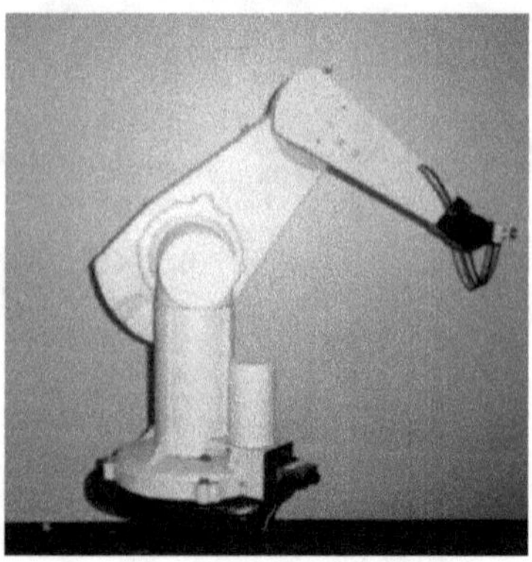

Figure 1.3 PUMA 560 Manipulator.

The robot industry starts a period of tremendous expansion in the 1980s. Many institutions offer robotics programmes and courses. Mechanical engineering, electrical engineering, and computer science departments all offer robotics courses [10].

Figure 1.4 (a) Adept's SCARA Robots, (b) Cognex In-Sight Robot and (c) Berrett Technology Manipulator.

Emerging applications in tiny robotics and mobile robots have fueled a second wave of start-up businesses and research since 1995.

Figure 1.5 In 2003, NASA's Mars Exploration rovers will launch toward Mars in search of answers about the history of water on Mars [2].

1.4 Knowledgebase for Robotics

A typical knowledgebase for robotics system design and operation.

- Modeling and analysis of dynamic systems
- Computer programming

- Feedback control
- Sensors and signal conditioning
- Actuators (muscles) and power electronics
- Hardware/computer interfacing

Mathematics, physics, biology, mechanical engineering, electrical engineering, computer engineering, and computer science are some of the disciplines studied [11].

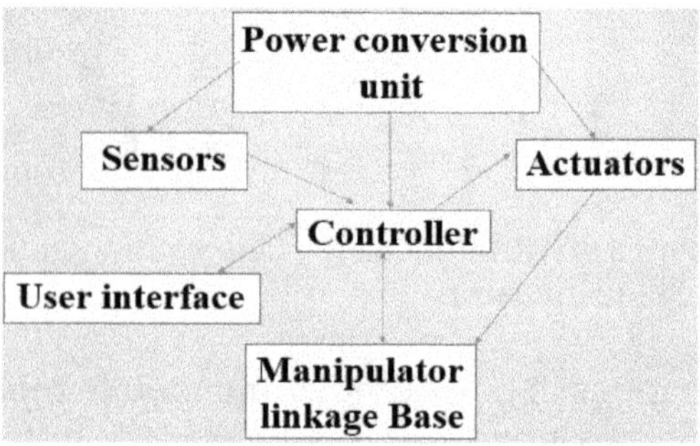

Figure 1.6 Key Components.

Robot Base: Fixed V/S Mobile

Fixed robots include robotic manipulators used in manufacturing. They are unable to relocate their base away from the construction under progress.

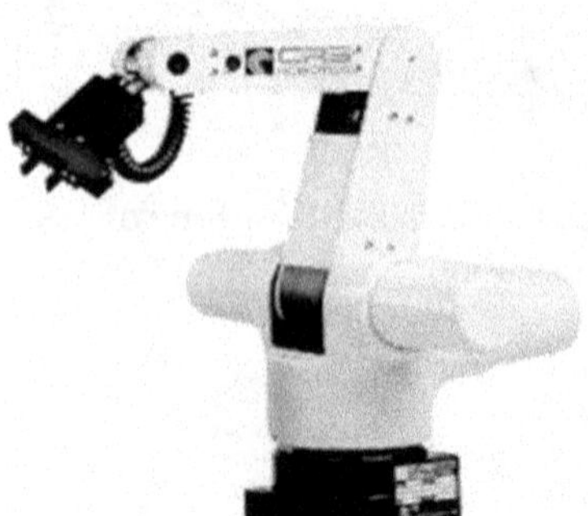

Figure 1.7 CRS-A255 robot.

Platforms with wheels or tracks are commonly used as mobile bases. Some robots move about by using their legs rather than wheels or tracks.

Figure 1.8 Robots having legs.

Robot Mechanism

a) Sensors: Our senses of sight, hearing, touch, taste, and smell provide us with critical data that allows us to function and survive. Robot sensors gather information about the robot's configuration and state, as well as its surroundings, and provide it to the robot controller through electrical impulses (e.g., arm position, presence of toxic gas). Robots frequently require information that goes beyond the five human senses (e.g., ability to: see in the dark, detect tiny amounts of invisible radiation, measure movement that is too small or fast for the human eye to see) [12].

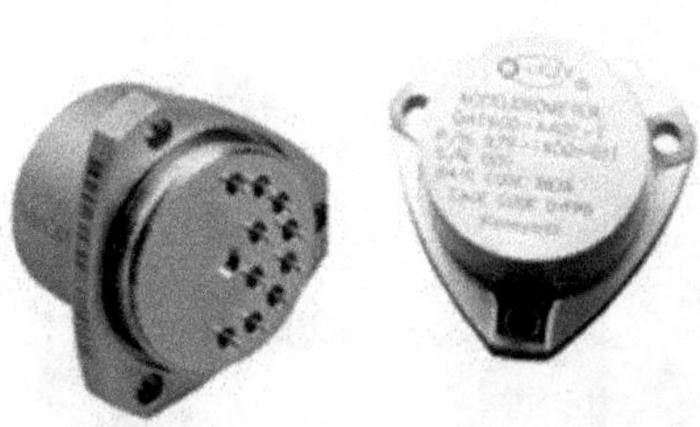

Figure 1.9 Accelerometer using piezoelectric effect.

Figure 1.10 Flexi force sensor.

b) Vision Sensors: Picking bins, performing inspections, and other tasks need the use of a vision sensor. Part-Picking: Using a 3-D vision sensor, the robot can handle work components that are haphazardly stacked. An automated system may be built at a cheap cost since alignment operations, a specific parts feeder, and an alignment pallete are not required.

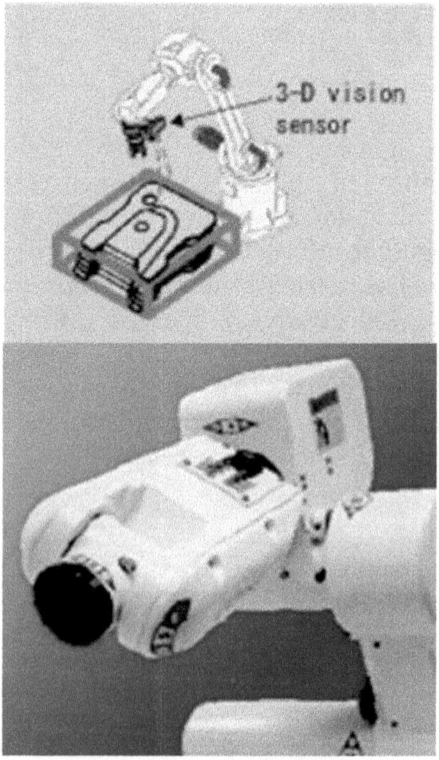

Figure 1.11 (top) 3D vision sensor and (bottom,) in-sight vision sensors.

c) Force Sensors*:* Force Sensor: e.g., part fitting and insertion, robotic surgery force feedback Fitting and insertion of machine components: Using force sensors, robots can conduct accurate fitting and insertion of machine parts. In addition to merely inserting components that have the phases, a robot can insert parts that have the phases after matching their phases. It has the ability to automate occupations that need a high level of competence [13].

Figure 1.12 Force sensors.

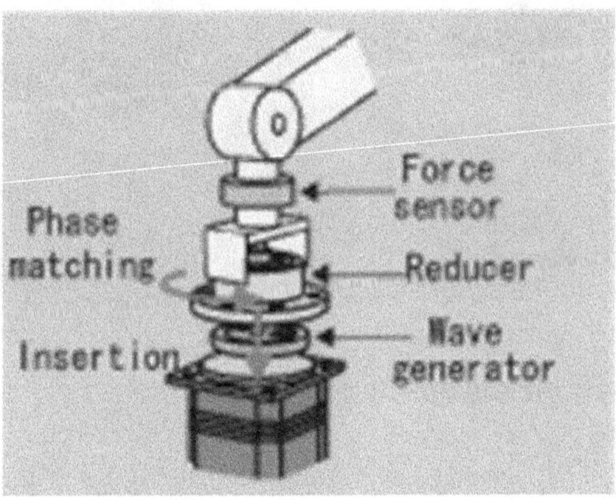

Figure 1.13 Force sensors.

d) Proximity Sensors

- 6 ultrasonic sonar transducers for wide-area exploration
- 16 built-in infrared proximity sensors (range 5-20cm)
- Obstacle detection over a wide range of 15cm to 3m
- Infrared sensors function as a "virtual bumper," allowing you to manoeuvre through tight areas.

Figure 1.14 Infrared ranging sensor.

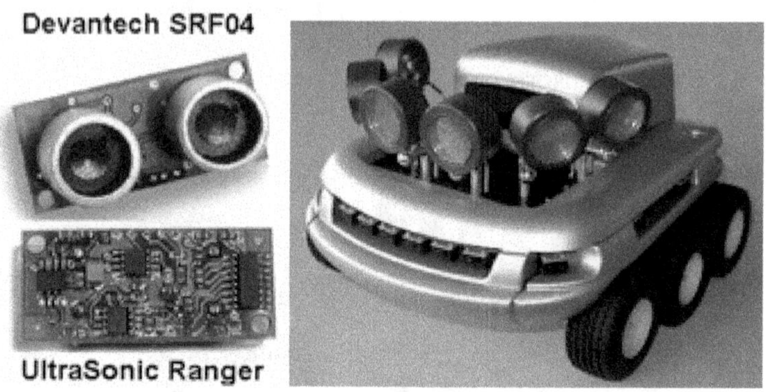

Figure 1.15 Ultrasonic sensors and KOALA robot.

e) Tilt Sensors: Tilt sensors are used to balance a robot.

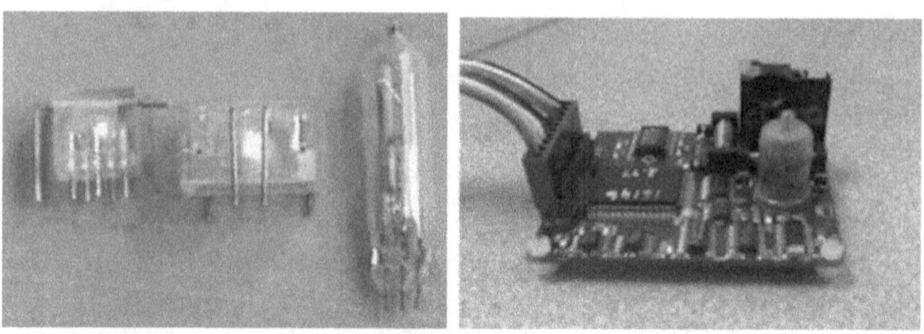

Figure 1.16 Tilt sensor.

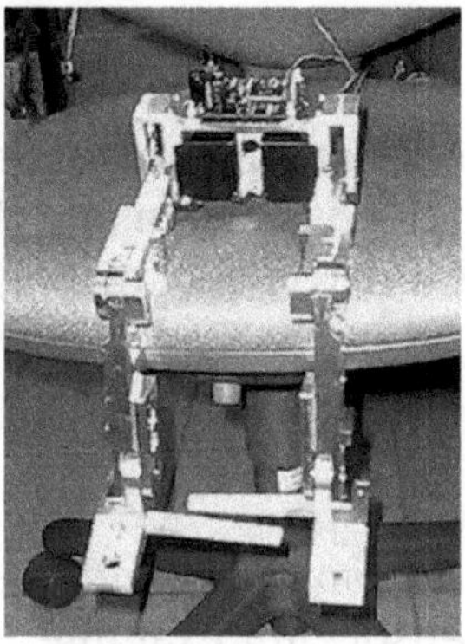

Figure 1.17 Planar bipedal robot.

f) Actuators/Muscles: Combinations of several electro-mechanical devices are commonly used in robotic actuators.

- Motor with synchronisation
- stepper motor
- Servo motor (AC)
- DC servo motor with no brushes
- DC servo motor with brushed finish [14].

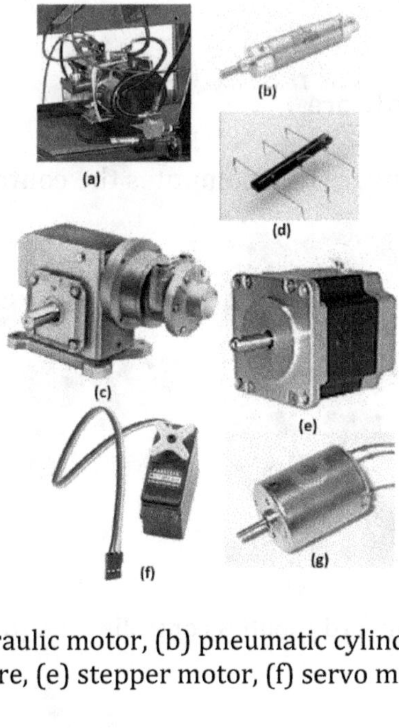

Figure 1.18 (a) Hydraulic motor, (b) pneumatic cylinder, (c) pneumatic motor, (d) muscle wire, (e) stepper motor, (f) servo motor, (g) DC motor.

g) Controller

Controller process sensory data and calculate control orders for the actuators to carry out defined tasks. In addition, it also provides essential intelligence so as to effectively manage the manipulator/mobile robot.

h) Storage Hardware

Memory to store the control software and the status of the robot system acquired from the sensors, for example.

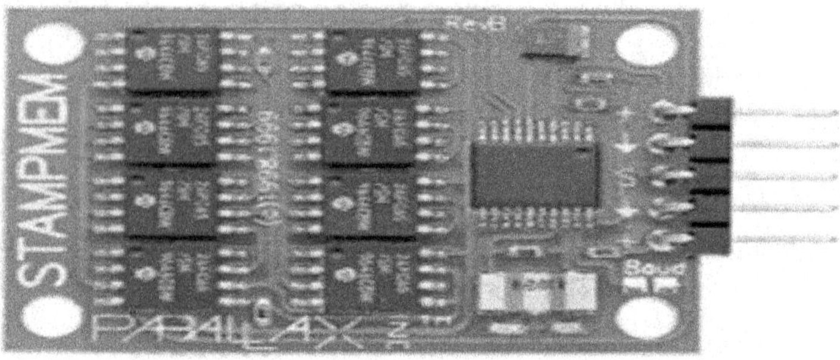

Figure 1.19 Storage hardware.

i) Computation Hardware

Computational Hardware that computes the control commands.

Figure 1.20 (a) Robo-board robotics controller and (b) BASIC Stamp 2 module.

j) Interface Hardware

Interface units are pieces of hardware that connect a digital controller to the outside world (sensors and actuators)

Figure 1.21 (a) Analog to digital converter, (b) LM358 amplifier, (c) LM358 amplifier and (d) LM1458 dual operational amplifier.

1.5 Robots in Industry

Robots are usually have following applications which are helpful to humankind.

- Agriculture
- Automobile
- Construction
- Entertainment
- Health care (hospitals, patient care, surgery, research, and so on)
- Science, engineering, and other laboratories
- Surveillance, patrol, and other aspects of law enforcement
- Manufacturing
- Military: demining, monitoring, and attack, among other things
- Exploration, mining, and excavation
- Air, land, rail, space, and other modes of transportation
- Utilities: electricity, gas, and water
- Warehouses [15].

Industrial Applications of Robots

- Spot welding
- Continuous arc welding
- Spray coating
- Assembly
- Inspection
- Material handling
- Material transfer
- Machine loading and/or unloading [1-16]

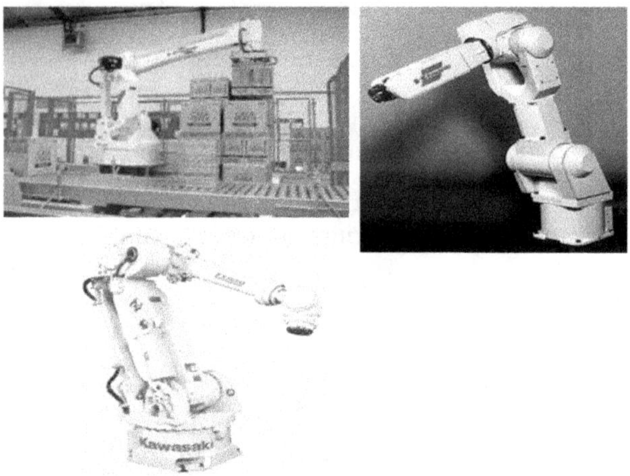

Figure 1.22 (a) Material handling manipulator, (b) assembly manipulator and (c) spot welding manipulator [6].

Robots in Space

Figure 1.23 NASA space station robots.

Robots in Hazardous Environments

Figure 1.24 :(a) TROV in Antarctica operating under water and (b) HAZ-BOT operating in atmospheres containing combustible gases.

Medical Robots

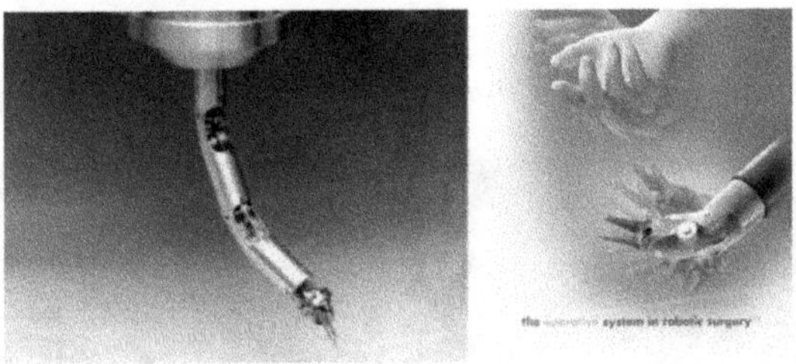

Figure 1.25 (a) Robotic assistant for microsurgery and (b) operative system of robotic surgery.

Robots in Military

Manta might launch small mine-seeking AUVs or engage in more destructive warfare if launched from the hull of a submarine.

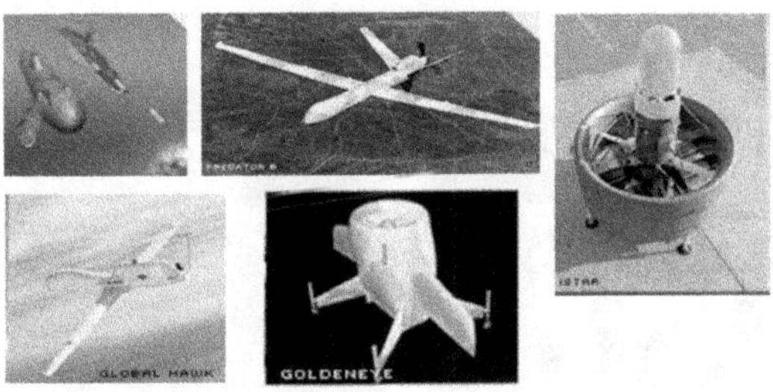

Figure 1.26 (a) Split Strike, (b) Predator, (c) ISTAR, (d) Global Hawk and (e) Goldeneye.

Robots at Home

Figure 1.27 (a) Sony SDR-3X Entertainment Robot and (b) Sony Aido.

Future of Robots

Artificial intelligence-based robots:

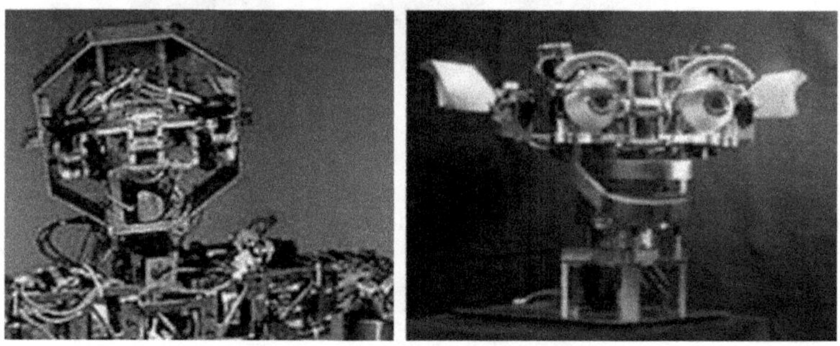

Figure 1.28 (a) Cog and (b) Kismet.

Autonomy based robots:

Figure 1.29 (a) Robot Work Crews and (b) Garbage Collection Cart.

Human-based robots:

Figure 1.30 (a) Humanoids and (b) Honda Humanoid robot.

1.6 Degrees of Freedom (DOF) of the Robot

The following will demonstrate the degrees of freedom using the human arm.

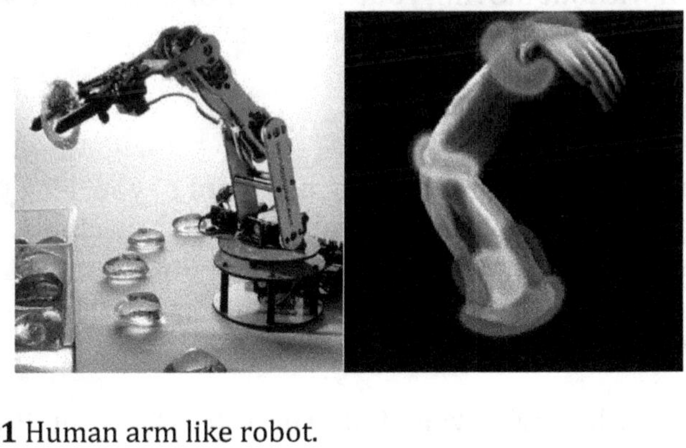

Figure 1.31 Human arm like robot.

First Degree: Shoulder Pitch

Make a straight line with your full arm in front of you. Raise and lower your shoulder. The shoulder pitch is the upward and downward movement of the shoulder.

Second Degree: Arm Yaw

Make a straight line with your full arm in front of you. Your entire arm should be moved from side to side. The arm yaw is the term for this side-to-side movement.

Third Degree: Shoulder roll

Make a straight line with your full arm in front of you. Now, as if you were putting in a light bulb, roll your entire arm from the shoulder. A shoulder roll is the name for this spinning motion.

Fourth Degree: Elbow Pitch

Make a straight line with your full arm in front of you. Keep your arm steady and just bend your elbow. Your elbow has the ability to move up and down. The shoulder pitch is the up and down movement of the shoulder.

Fifth Degree: Wrist Pitch

Make a straight line with your full arm in front of you. Flex your wrist up and down without moving your shoulder or elbow. The wrist pitch is the up and down movement of the wrist.

Sixth Degree: Wrist Yaw

Make a straight line with your full arm in front of you. Flex your wrist from side to side without moving your shoulder or elbow. The wrist yaw is the side-to-side movement [5].

Seventh Degree: Wrist Roll

Make a straight line with your full arm in front of you. Rotate your wrist as if you were twisting a doorknob without moving your shoulder or elbow. The wrist roll is the rotation of the wrist. The Crust Crawler Robot is a Revolute Robot with 5 degrees of freedom, as illustrated in the diagram below [17].

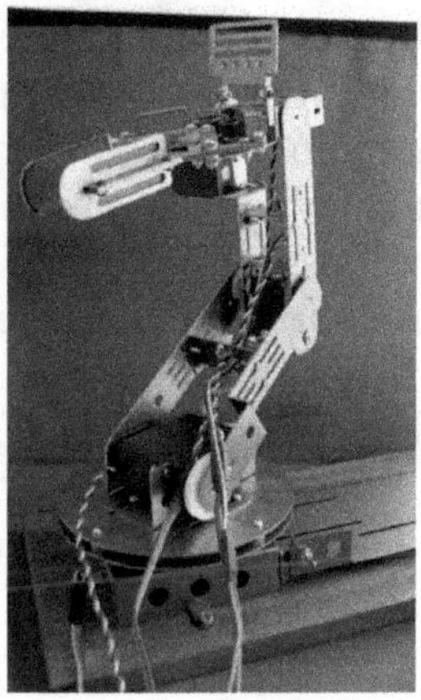

Figure 1.32 DoF Crust Crawler robot.

1.7 Robotic Joints

It goes without saying that different robotic joints are required to accomplish varying degrees of freedom. Unlike human joints, which have three degrees of flexibility in the shoulder, robot joints typically have only one degree of freedom to simplify the manipulator's mechanics and control. Rotary joints and prismatic joints are the two most prevalent types of joints found in robots.

Rotary or Revolute Joints

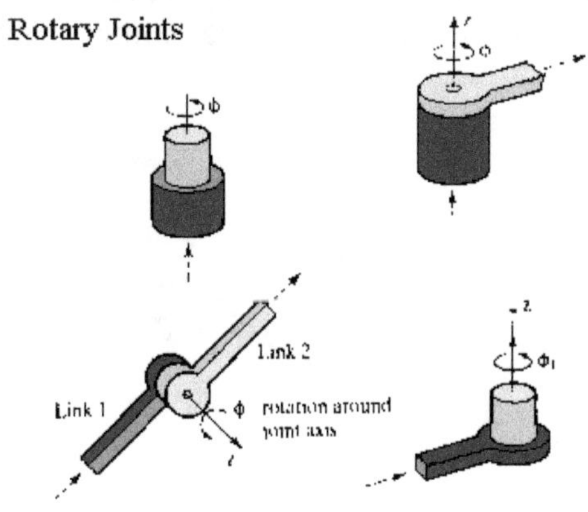

Figure 1.33 Rotary joints.

In the design above, the top two connections are commonly utilised as a waist joint. The waist joint of the Crust Crawler robot is an example of a waist joint; it is the joint that pivots the entire robot. The blue portion remains stationary, while the green spins around it. Elbow joints may be seen in the bottom connections. Because each of the four joints may only rotate one degree, the joint variable is the angle at which the joint moves. Because the arm structure and the servo motor are mechanically limited, most rotary joints cannot rotate 360 degrees.

Linear Joints

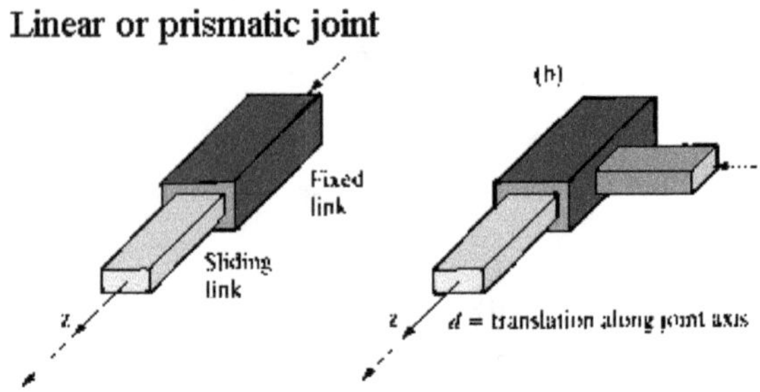

Figure 1.34 Linear joints.

A sliding joint is a prismatic joint. It may be utilised in just about any axial direction. Although linear joints are less frequent than rotary joints, they are quite helpful. Both types of joints are seen on the cylindrical robot below. The grabber may revolve around the axis thanks to the rotary joint, while the linear joints Y&Z govern height and reach [18].

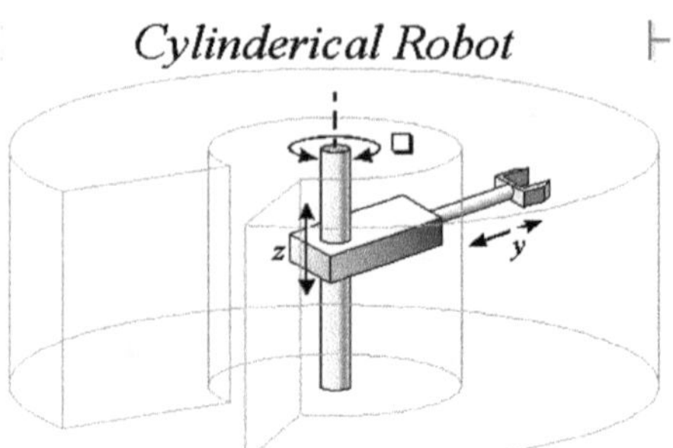

Figure 1.35 Cylindrical robot.

1.8 Work Envelope of Robot

The Work Envelope is the volume of space in which a robot operates. The work envelope is the area around a robot that the end effector or gripper can reach.

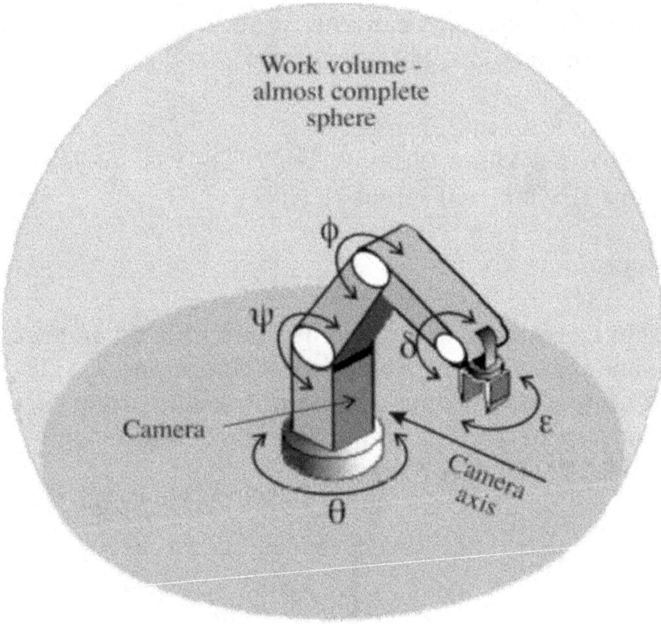

Figure 1.36 Work envelope of robot.

As a robot moves around the limits of its reaches it traces out a specific shape. The Cylindrical robot in the last image has a visible work area of a cylinder. A Cartesian robot sweeps out a rectangular volume, a polar a partial sphere.

Classification of Robots

A variety of variables assist to determine the kind of robot that is needed for a certain job. The primary element is the kind of movement required for the intended robotic movement. Some applications need just one robot to move a product along a specified axis, while other robotic arms must manoeuvre an item over many axes at once. The placement of joints fulfils various coordination systems in different ways.

Coordinate Systems/Frames

Coordinate system types establish a point's location using distance or angle measurements or their combination. A space point needs three measurements for each of these kinds of coordinates. It should be emphasized that in any system the same point may be discovered. Different co-ordinating systems can only provide for another scenario. Three main robotics co-ordinate systems utilized are:

1) Cartesian Systems/Frames
2) Cylindrical Systems/Frames
3) Spherical Systems/Frames

Cartesian Coordinate Frame

The Cartesian coordinate system is the most utilized and recognizable coordinate system. Most would know this as the X, Y; the axis is 90° to each other. The distance of a point from its origin (0, 0) along each axis may be situated in a plane. This is a two-dimensional depiction, and hence the two X & Y axes.

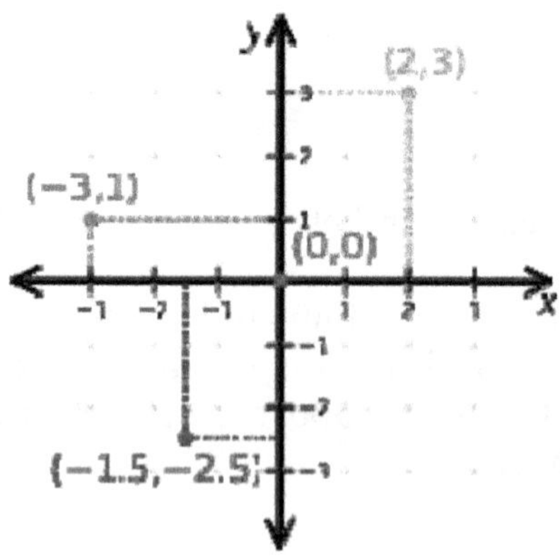

Figure 1.37 Cartesian coordinate frame.

To locate a point in space, a third axis must be added (Z). This third axis forms a three-dimensional grid that matches a collection of coorders to one single space point.

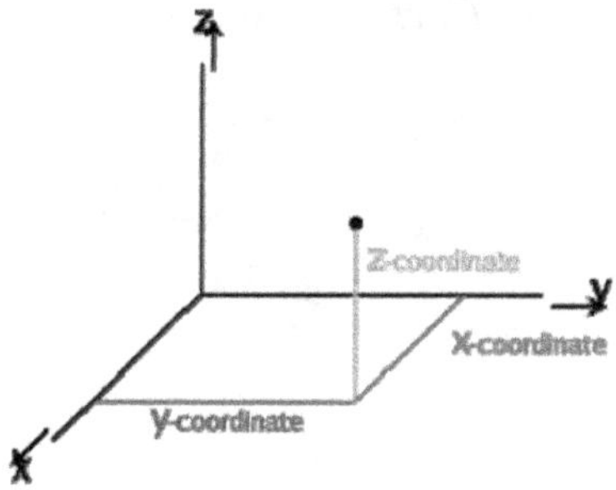

Figure 1.38 Different axes of coordinate frame.

The machine axis is always determined by what is called the right-hand rule. Taking the thumb to indicate the positive X-Axis, the second finger points to the positive Y-Axis and the middle finger to the positive Z-Axis. As illustrated in the 'Cartesian robot,' the Z axis is always directed toward the spindle or grab arm.

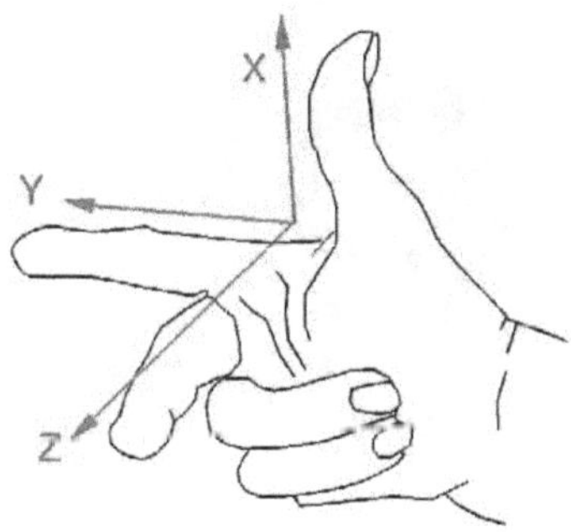

Figure 1.39 Right-hand rule.

Cartesian Robots

For pick and place work, sealant application, assembly activities, equipment for handling machines and arc welding cartesian robots are utilized. It is a robot with three prismic joints whose axes correspond with the Cartesian coordinators.

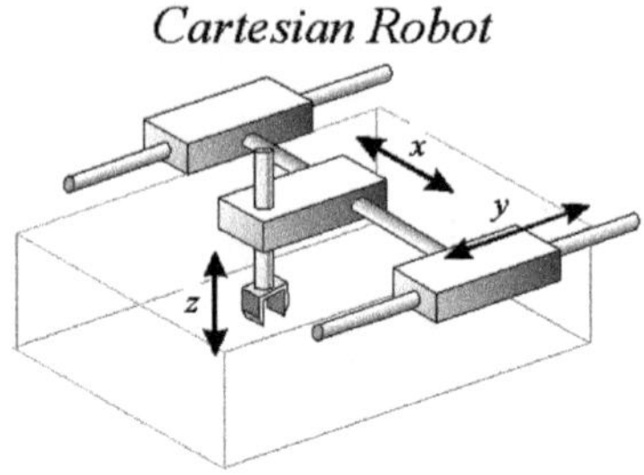

Figure 1.40 Cartesian robot system.

Application of Cartesian Robots

Applying adhesive: This robot is used to adhere to a glass panel. This robot can handle big amounts of labor. The concept comparable to this robot is used by most laser cutters and CNC machines.

Figure 1.41 Adhering robot.

Pallet transfer: This orientation of a Cartesian robot moves integrated circuits (IC) from a pallet to a particular location.

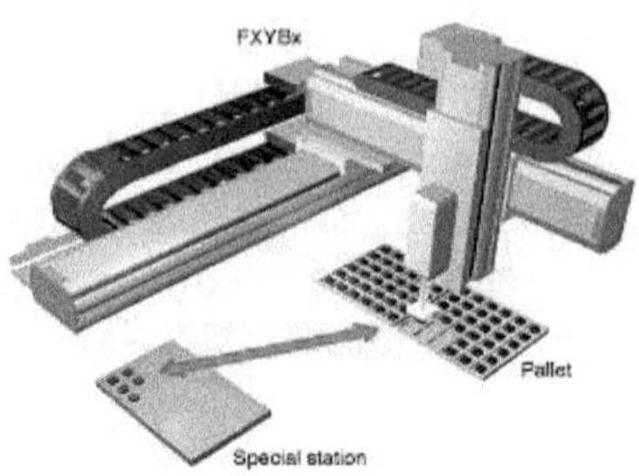

Figure 1.42 Pallet transfer robot.

Product inspection: Companies must check their goods to guarantee their quality is excellent. Cameras placed on top of the Cartesian robot monitor the in exactitude of the passing components. Because of its design, the robot can travel with the moving conveyor and concentrate on a product immediately.

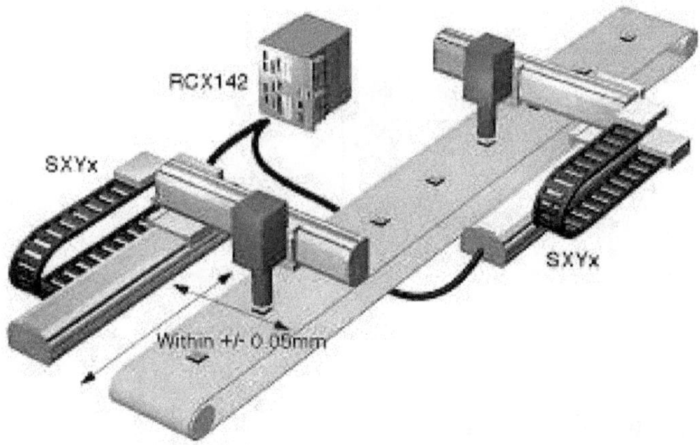

Figure 1.43 Inspection robot.

Transfer and stacking: The Cartesian robot is excellent for the transfer and stacking of sheets of metal or wood sheets because of its linear mobility. It may feed sheets into machinery or pull them as completed goods.

Figure 1.44 Transfer and stacking robot.

Cylindrical Coordinate Frame

A three-dimensional "A" point is assumed to be situated on a cylinder "R" with a height of "Z" in a cylindrical coordinate frame. The third piece of data needed to define the point comes from an angle on the XY plane.

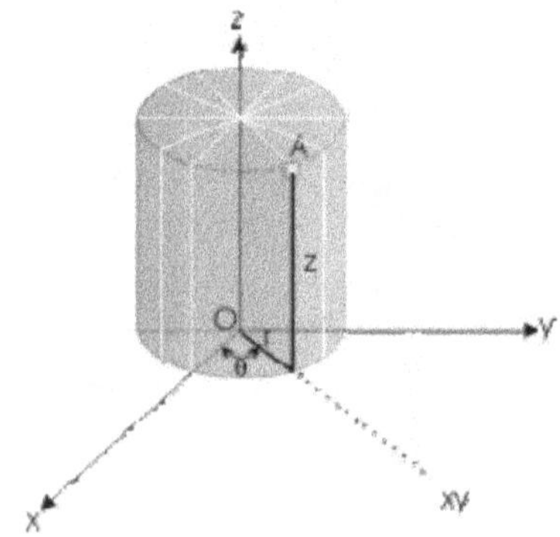

Fig 1.45 Cylindrical coordinate system.

Cylindrical Robots

Cylinder robots are utilized in assembly, machine tools handling, spot welding and die cast processing. They are also used in various medical tests. The following sample includes two prismatous joints and a rotating joint. A cylindrical robot may spin along its primary axis and create a cylindrical form.

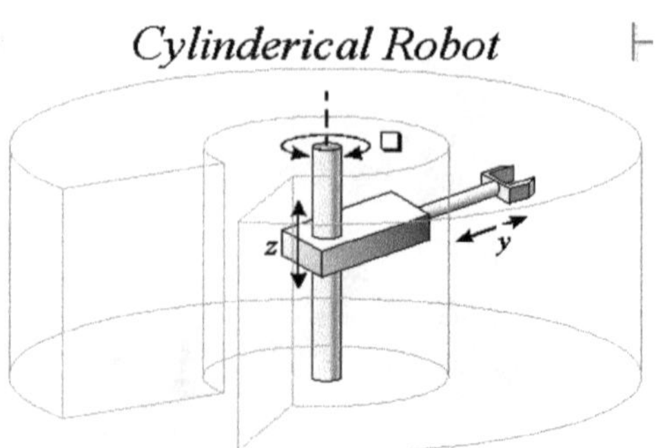

Figure 1.46 Cylindrical robot.

Application of the Cylindrical Robot

In a wide range of medical applications, the robot is utilized for DNA testing, forensics, drug development and toxicity. These robots are ideal for medical research, where hundreds of samples have to be examined and repeated tasks repeatedly done. The robot removes human mistake, which provides repeatable outcomes and consistent results. One common example would be to pull a drawer out for access to a test plate, raise a sample plate, shut the drawer and then transfer it to another tool to be tested.

Figure 1.47 Basic cylindrical robots.

Figure 1.48 Medical robots.

Spherical / Polar Coordinate Frame

A three-dimensional point A may be located in a spherical coordinate system on a "R" sphere. This point is on a certain cross section of the sphere; like the cylindrical frame, the cross section is an angle "relating" to the ZX plane. Once the plane has reached the final point where point A is attached to the source, the angle (α) created with the Z axis identifies the point.

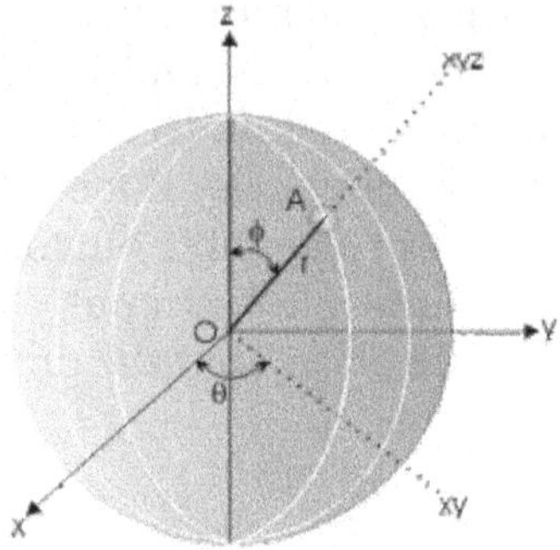

Figure 1.49 Spherical/Polar coordinate system.

Let's see to simplify how a location is found on the earth's surface. Like the spherical coordinate system, the earth's surface becomes "R." Maps utilize longitude (axis) and latitude (ć) lines to identify a certain location on the surface of the globe. We now have three pieces of information that enable us to understand our situation.

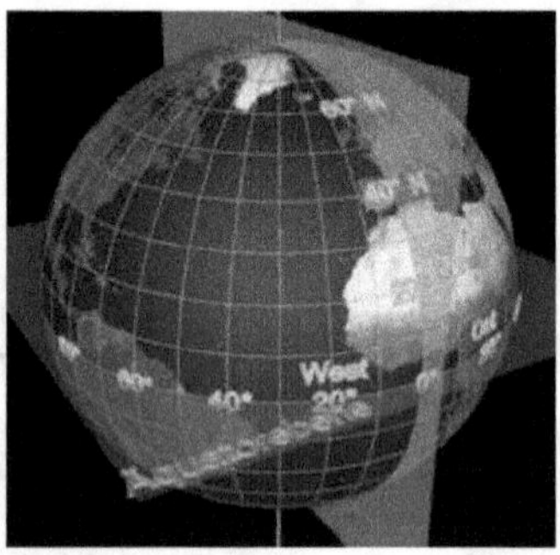

Figure 1.50 Earth's surface.

Spherical or Polar Robots

Spherical or Polar robots combine rotation with single linear arm motions. Sometimes the polar robot is called the gun turret arrangement. They are usually utilized mainly for spot, gas and arc in various welding applications. Polar robots are ideal to access horizontal or sloped tunnels.

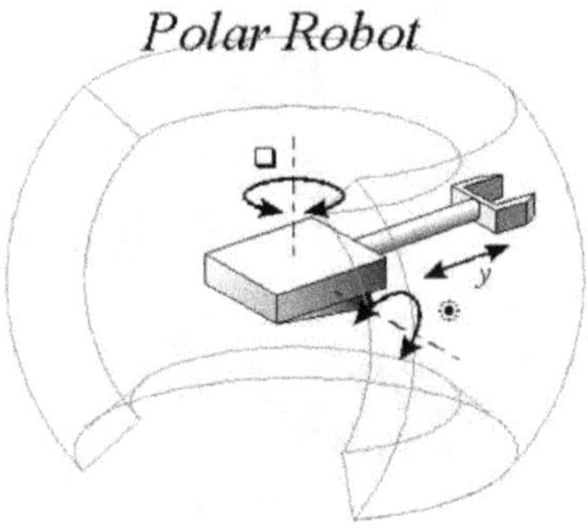

Figure 1.51 Spherical/polar robots.

Applications of a Polar/ Spherical Robot

Welding is the primary use for these kinds of robots. It may be very big and weigh more than 1000 kg. Polar robots are extensively employed in the automotive sector.

Figure 1.52 Numerous polar robots at work.

The SCARA Robot

This arrangement was designed to suit the requirements of contemporary assembly operations where rapid mobility is needed with low payloads. Electronic components are placed quickly on PCB's is a clear use. The SCARA robot is a combination of two horizontal and vertical rotating axes.

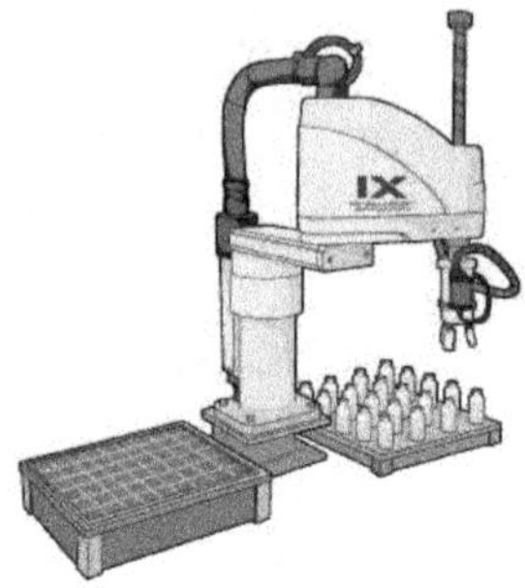

Figure 1.53 SCARA robot.

Applications of a SCARA Robot

The SCARA robot tests a newly developed calculator to make sure it works before packing. The camera monitors the screen to determine whether the robot's action achieves the intended outcome.

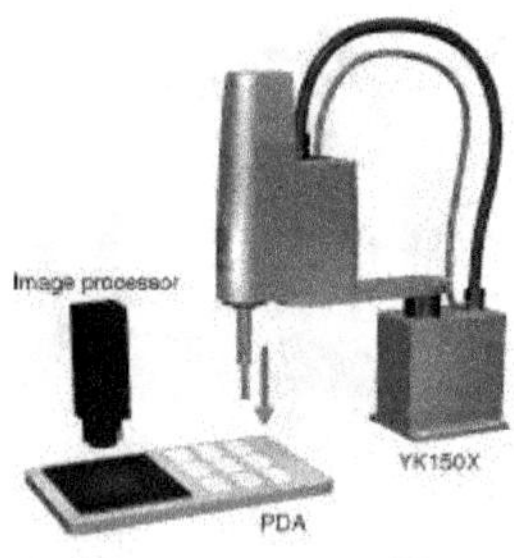

Figure 1.54 Testing a finished product.

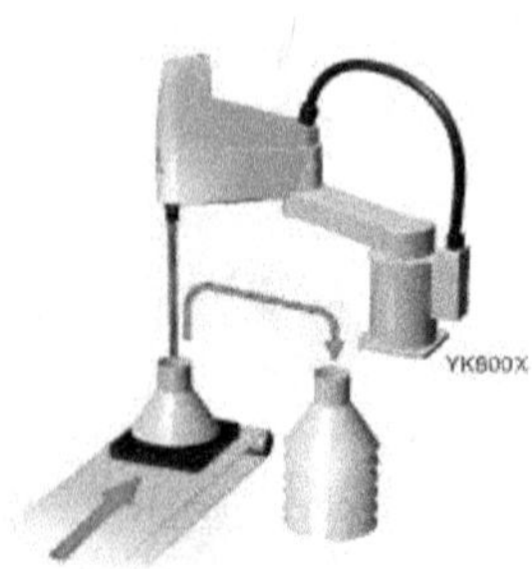

Figure 1.55 Stacking lightweight components.

The SCARA configuration can easily remove and stack components from a mounting line. The SCARA robot is ideal for precision positioning and is ideal for component assembly. The picture above shows the robot taking and assembling parts from the supply unit. The robot can rotate to the left and change its grip to a screw or boiling head. These various robot heads are known as end-effectors [15-18].

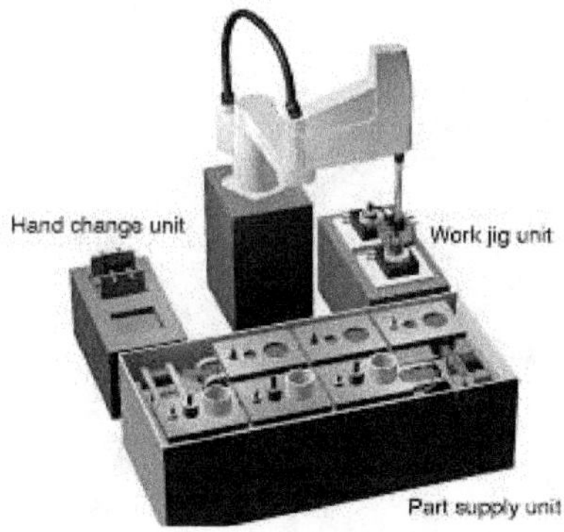

Figure 1.56 Part assembly.

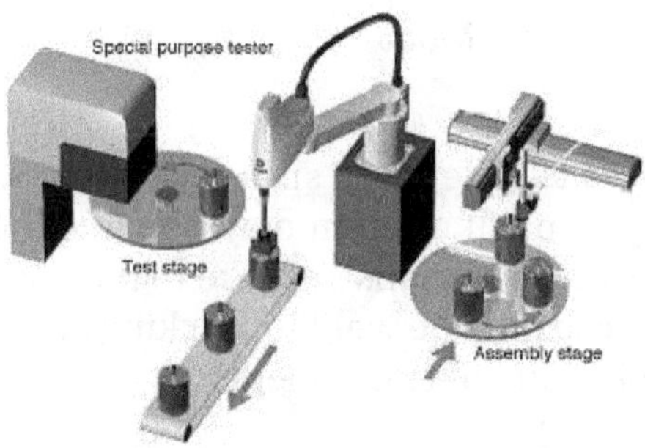

Figure 1.57 Multi-functional system.

Here so the SCARA can be viewed from the selection phase to the test stage. Once the component is quality controlled, the robot arm places it on the conveyor belt for transfer to the next stage.

Revolute Robot

The revolute robot or Puma, as it is most often known, is similar to the human arm with three major freedom rotations. The manipulator revolves like the human waist on the base. The other two axes of rotation are similar to the shoulder and the elbow. The extra wrist motion provides two more degrees of freedom, up and down (pitch) and rotation (roll). A last movement is Yaw, which is the wrist movement from one side to the other.

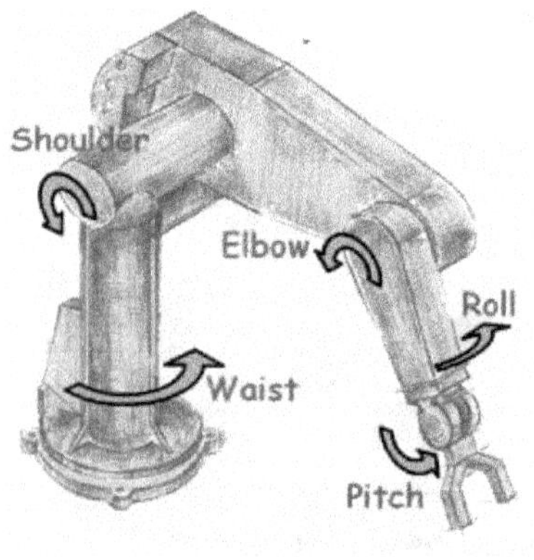

Figure 1.58 The revolute robot.

Applications of a Revolute Robot

This kind of robot is excellent for spray painting, where the human motions necessary to paint an item may be taught. The Revolute robot is ideal for many applications because to its manufacturing based on human hand. These include welding, picking and placing activities, assembling and soldering of components.

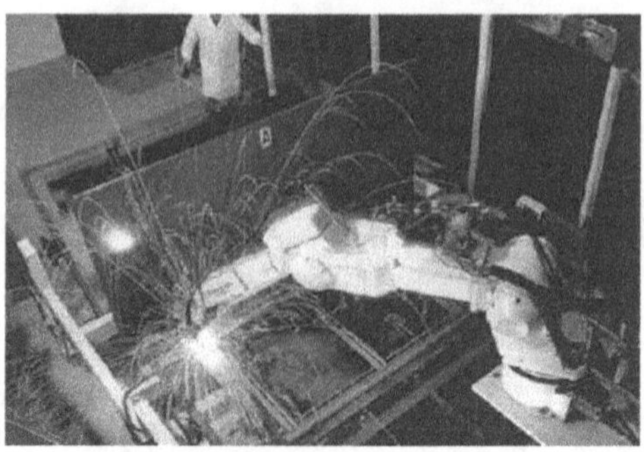

Figure 1.59 Revolute robot MIG welding.

Humanoid Robot

During the past several years, the creation of robots similar to the human body has been of more amusement than usefulness. However, with considerable study on how robots may fuse into our lives and carry out human tasks, this tendency is shifting. Films like as ET, Transformers, AI and I have all played on the notion of machines as human beings with feelings and emotions.

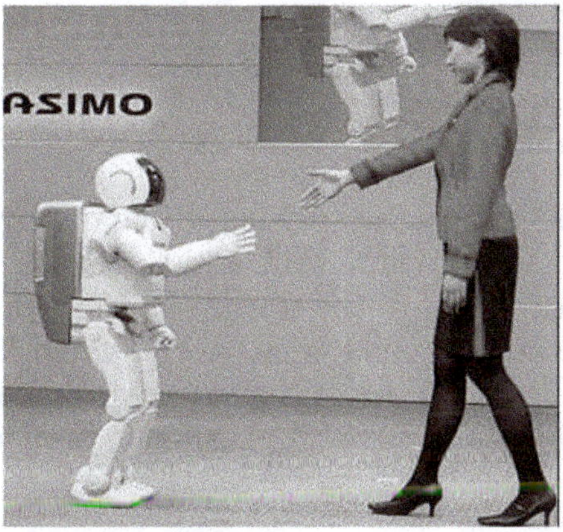

Figure 1.60 Honda humanoid robot.

The four-foot ASIMO robot went up a pair of stairs at the New York stock market event, shook hands with the chairman, placed the opener and fluttered, clapped with centuries of traders on the Exchange floor. HONDA has started to explore possible methods of benefiting its humanoid robots worldwide. Maybe ASIMO will aid the elderly and help the homework in the future [10-18].

1.9 Forces and Moments

For an effective and successful robot, it is essential to examine the different factors that both hinder and aid robot activities. It is essential to choose the appropriate servo motor for a certain joint since the life of the servo may be reduced by loading. The first step is to determine the torque of the servo by the force (moment) applied on it.

Moment = Force (Newton's) x Distance (Meters).

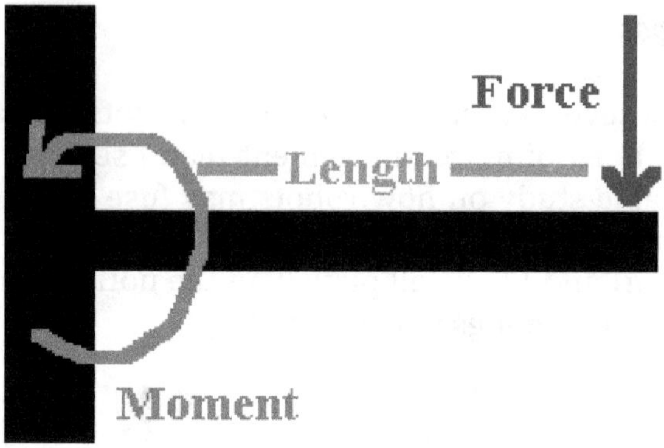

Figure 1.61 Relationship among moment, force and length.

The strength of the arm on the servo is influenced by a variety of variables. The weight of the arm to raise is the primary concern. Second, two significant variables are the weight and length of the arm. The weight of the arm and the load must be combined. The length of the arm is also significant since it affects the distance from the pivot (pivot point). As Moment = Force x Distance, the moment the servo acts, the longer the arm.

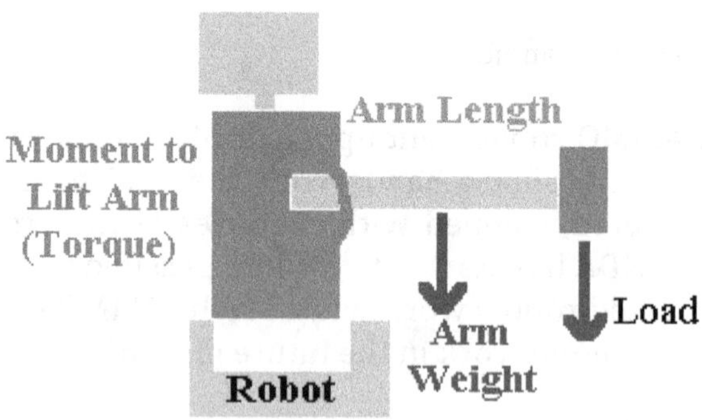

Figure 1.62: Relationship among moment, load and length.

To compute the overall moments of the servo, combine the arm and load moments together.

*Moment = (Load weight x arm length) + (arm weight * ½ arm length)*

Note that we only utilize half the number for the arm length. This is because the weight of the arm is on the whole arm and not at the very end. The center of gravity of the arm is where the arm is completely balanced while hanging. In this example, the center of gravity of the arm is estimated to be the center of the arm. The real torque is applied after this computation is complete. This is the minimum torque that an engine requires. A greater torque value motor should be utilized to enable easy overloading and smooth operation; this is known as the safety factor. If a 100Nm motor is required, it is suggested to search for a 20 percent greater torque engine.

Actuators

Motors are the most frequent method of controlling robotic movement and are called actuators. They may be linked to gears and wheels and provide movement. A variety of various motor types, dc, stepper and servo motors may be utilized.

Figure 1.63 An actuator.

DC Motors

These are the most frequently available motors, linked by two wires to the power source. By changing the polarity of the motor supply voltage, the direction of a DC motor may be altered. DC motors consume a high quantity of current, which means that a control system such as PIC cannot be directly connected. DC motors do not provide precise, controlled rotation.

Stepper Motors

Stepper motors operate similarly to dc motors, however, the stepper has several while a dc motor has only one magnetic. The step is controlled by consecutive switching of these spins on and off. Each time a fresh belt is powered, the engine revolves a few degrees. The step angle is termed the number of degrees to which a motor rotates with each pulse. The sequence repeat leads it to travel a few more degrees until a complete rotation is accomplished. The following graphic illustrates how the basic engine operates. The magnet in the center is attached to the engine shaft. The four magnets on the exterior symbolize each step motor spindle. When the coils are activated, the center magnet is pulled in various directions, and thus the proper pulse sequence is motorized.

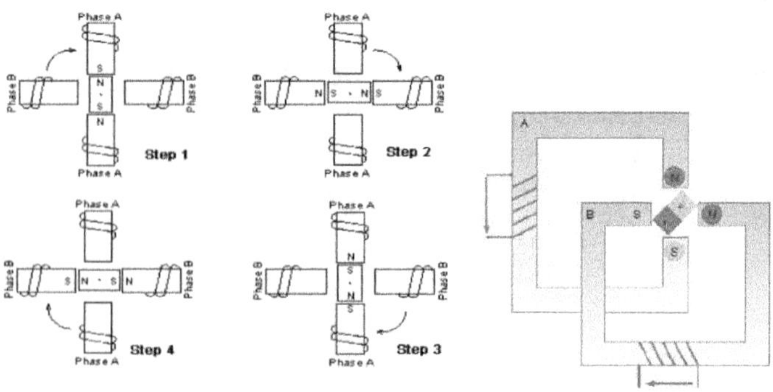

Figure 1.64 Various steps of stepper motors.

The typical stepper motors used in projects work on the same concept as the previous examples. The primary advantage of the stepper engine is regulated movement. The aforementioned engines have 90° step angles. This wide angle does not make the motions smooth. This is overcome by a dented steel disk formed by the gear that substitutes the center spinning magnet.

Figure 1.65: Step 1 of stepper motor.

The top electromagnet (1) is activated and the disc's closest teeth are attracted. The teeth are oriented to the top electromagnet and are somewhat inconsistent with electromagnet (2).

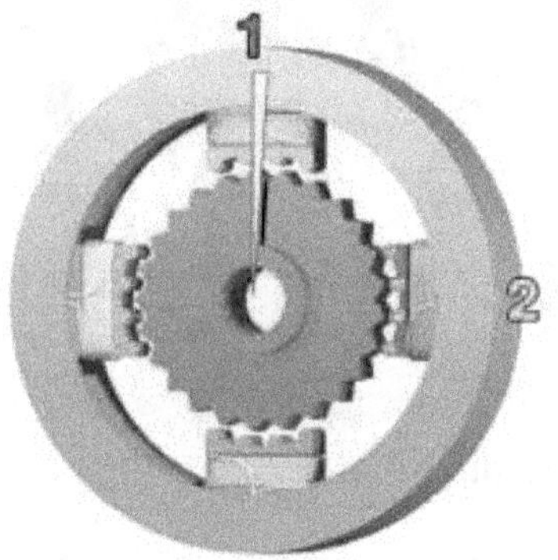

Figure 1.66 Step 2 of stepper motor.

The top electromagnet (1) is switched off and the right electromagnet (2) is activated, pushing the teeth slightly right. The amount of movement is 3.6° for this motor.

Figure 1.67 Step 3 of stepper motor.

Continuing the process, the electromagnet (3) is energized and another step happens

Figure 1.68 Step 4 of stepper motor.

The last electromagnet (4) is initiated, and the rotor is pulled by another 3.6°. When you re-energize the first electromagnet, the teeth in the sprocket will rotate by a tooth. As 25 teeth are on the disk, 100 steps will be taken to complete one rotation. 25teeth x 4 steps = 100 steps per tooth. Stepper motors transform electrical pulses into discreet mechanical movements of rotation. They offer excellent torque holding, speed control and precise movement. Unlike the DC engine, the wear parts in a step motor are limited, but they do not contain brushed parts that can wear and result in malfunction. Steppers are used in printers, belt drives, photo or film processors as well as in scanners.

Figure 1.69 Standard stepper motor.

To operate the stepper engines, a controller is necessary. Logicator provides step control when the correct sequence table for the particular engine is available. The engine has six wires connecting to the

PIC board. Two are used for constant 0v and 5v, and four are direct PIC outputs. The following is a typical table for relationship between various steps of stepper motor and its outputs [15-18].

Table 1.1 Relationship between various steps of stepper motor and its outputs

	Output0	Output1	Output2	Output3
Step 1	0	0	1	1
Step 2	1	0	1	0
Step 3	1	1	0	0
Step 4	0	1	0	1

This table is input in Logicator as output blocks. The speed of the engine is governed by the length of a waiting block between the output blocks. The number of steps may be adjusted by the number of loops.

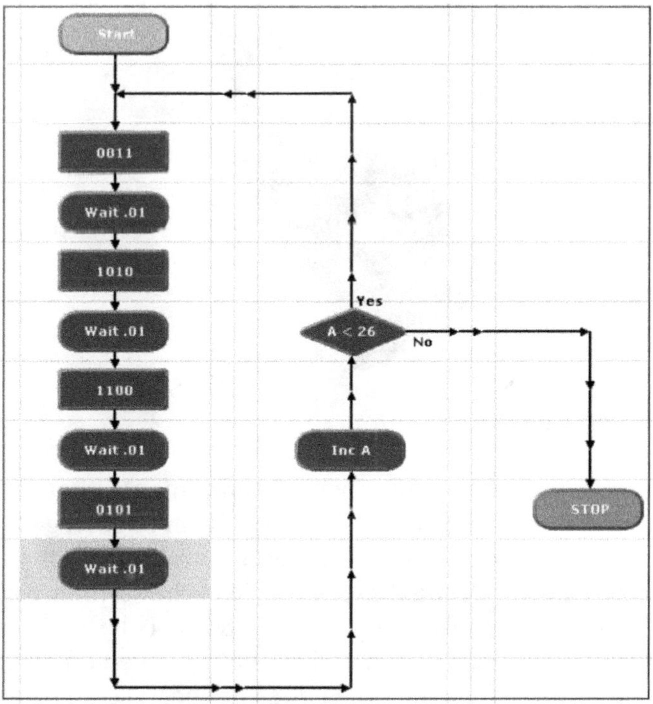

Figure 1.70 Algorithm of robotic program.

The "Inc" command is used in this program to record the number of times the sequence has executed. The program will end after it reaches 26 iterations. By reversing the order of the table steps, the engine may be rotated in the other direction, thus it can be used in open loop systems because it does not feedback the controller. Once the pulse step has been delivered, the control does not know if this point has been reached.

Servo Motor

The servo motor provides the smoothest and most control of the three motors. You may rotate to a certain point, making them suitable for applications requiring precision movement. The servo rotation is restricted, most of which revolve between 90° and 180°, although others can complete a complete revolution. Their construction makes them unsuitable for driving wheels, but their torque and control make them ideal to power robotic arms and others.

Figure 1.71 Servo motor.

Servo motors consist of a number of components that are neatly housed in an autonomous unit. They include an engine, gearbox and driver controller which may be directly operated from a microcontroller. Three wires are linked to Servos. Like the stepper motor, two wires are available for power supply - 0v & 5v. The third wire feeds directly from the PIC output pin and cannot be received by a transistor chip to guarantee that the servo does not get a straight current at frequency. The potentiometer is linked through a gear train to the engine. The motor is given a signal to spin to a certain location while the motor rotates it also moves the potentiometer to alter its resistance. The control circuit monitors this resistance to ensure that the servo is in the right position. The control circuit will recognize and adjust its placement if the servo is knocked out of place unintentionally. This is an example of a closed loop control that is subsequently discussed. Servos are positioned using the Pulse Width modulation method. Serves usually need 0.75ms to 2.25ms every 20ms pulse width. This pulse width has to be repeated continuously every 20ms when the pulse is lost. Pulse widths between 0.75 and 2.25ms are between 0° and 150°, pulse widths may be chosen between these to get the desired position. It's the pulse width that determines the servo position, not the number of times per second is repeated. Logicator is a range of pulse widths from 0.75m to 2.25m as 75 and 225m [14-18].

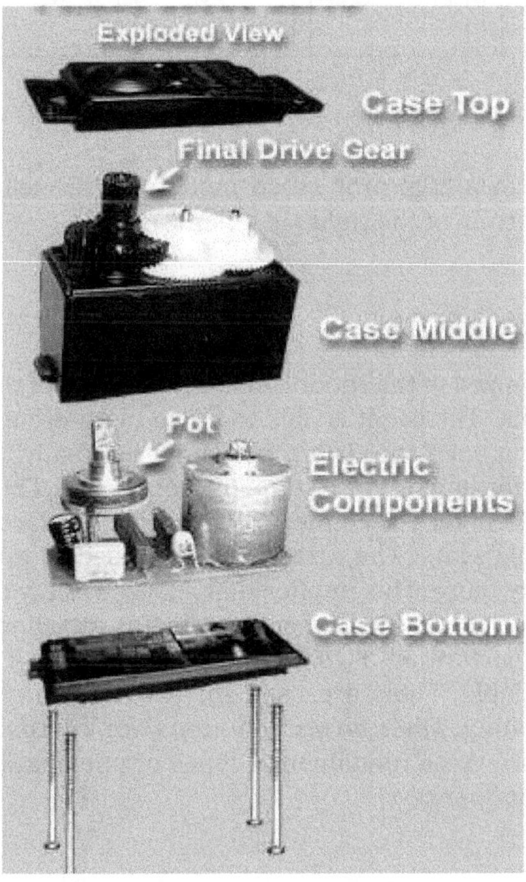

Figure 1.72 Various structural levels of servo motors.

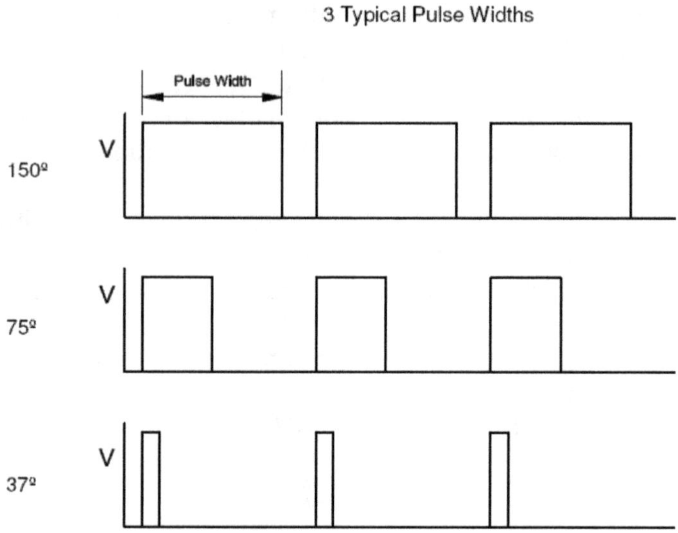

Figure 1.73 Various pulse widths of Servo motors.

Programming of a Logicator servo motor is done using the Servo Block at the bottom of the right side menu.

End Effectors

It is well recognized in the robotic field that the end of the wrist is the end of the robot. The robot is able to move to different locations inside its working envelope. The robot is not yet ready for the procedure to be performed; it has not the proper "hand." The end effector is the proper term for the connection that may be placed on a wrist-attached bolting plate. The fittings may be used to grab, lift, weld, paint and many more. This implies that a large variety of various applications based on the final effector which is attached to the basic robot may be performed. Two kinds of end effectors – grippers and tools are available. Tools are used for activities such as welding, painting or boiling. Their forms and kinds are multifaceted and diverse. There are three fundamental types of pneumatic, mechanical and magnetic influencers.

Figure 1.74 Mechanical gripper.

Figure 1.75 Drilling end effector with brush attached.

Mechanical Gripper

Figure 1.76 Multiple fingered gripper.

The two finger types as shown in the image are the most popular kind of gripper. Multiple finger kinds are also offered for more complicated jobs. The mechanics of a gripper are that the fingers of the gripper shut the item with the mechanical force necessary to keep the object securely against gravity and motion forces. However, the force should not be too strong and harm the component. Servos, pneumatic or hydraulic power may power the grippers. Due to their size or delicacy, mechanical grippers may not always be appropriate for handling certain components. Magnetic or pneumatic grippers provide alternate options for component management [15-18].

Magnetic Gripper

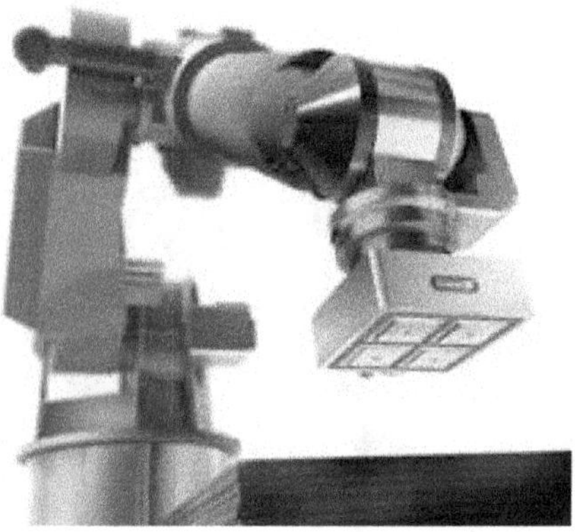

Figure 1.77 Magnetic gripper.

These grippers are used for the treatment of ferrous material. Electromagnetic or permanent magnets are the gripper. It can be picked up and released by switching on and off the magnet. Using a permanent means that the component cannot simply be taken off from the magnet, the pneumatic piston must be eased off. When an electromagnet may be utilized, this might appear unnecessary. The risk of sparking is minimized since there is no electric power, making them more suitable in dangerous situations.

Pneumatic or Vacuum Grippers

Circular vacuum or pneumatically formed vacuum cups made of plastic or rubber. The cups push against the lifting material and pull the air via a pump to create a dip effect. This suction force lifts the component. The components' weight and center of gravity affects the number of suction cups needed.

Figure 1.78 Vacuum gripper.

The item to be lifted must have a reasonably smooth level and clean surface to enable efficient suction. Releasing the component after it reaches its destination merely means vacuum neutralization and air in the suction pads.

1.10 Open and Closed Loop Control

Principles of a Practical Control system

There are three components in all control systems:

- The control
- Current Amplifiers
- Servo Motors

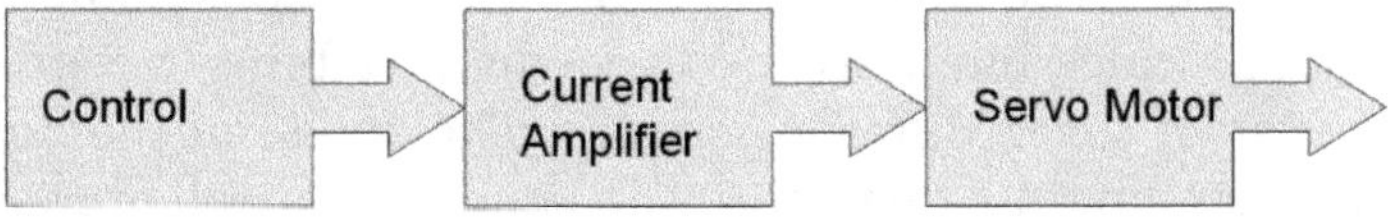

Figure 1.79 Principles of a practical control system.

The control is the system's 'brain.' The control reads the instructions of the program and conducts all measurements and computations that decide where the robot tool should be instantly throughout the program. This may be a complicated procedure requiring sophisticated math. This enables the control to transmit signals to the servo motors for axis position control up to several hundred times per second. Whether open loop or closed loop control is employed – see below – depends on the kind of signal delivered. The control output is generally a modest voltage (typically plus or minus five volts). A current amplifier is required to operate the stepper motors. The voltage signal from the control (referred to as the demand voltage) is used to regulate the current powering stepper motors. The second kind of engines are servo motors AC or DC. These are comparable to the standard AC or DC engine but are intended to provide more consistent operating characteristics. These motors exercise a torque in proportion to the current. They are exclusively utilized with Closed Loop [1-18].

Open Loop

This is the simplest kind of control and includes the three fundamental components mentioned above.

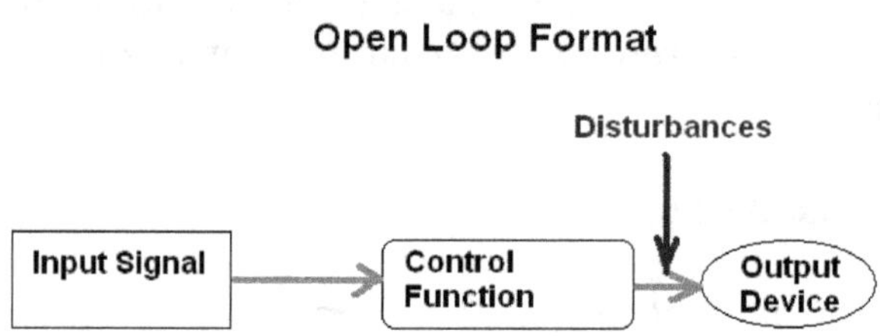

Figure 1.80 Elements of an open loop control.

The open loop system operates as follows:

A new conveyor is built in a plant to transport products from one location to another. The conveyor will operate at a consistent pace for a particular product; but there are three distinct items that may be on the conveyor at three different speeds. The speed of the conveyor may be adjusted by the operator. A DC motor is installed to operate the conveyor belt, a controller will be utilized which will be able to give a variable voltage output. A control dial is provided to adjust the input to the controller and determine the speed. When setting the speed of the conveyor the operators will need to utilize a measuring instrument to record the speed of the belt and manually adjust it to the appropriate speed for a particular operation. Each speed setting may be indicated on the dial and enable the operator to select the appropriate setting whenever required.

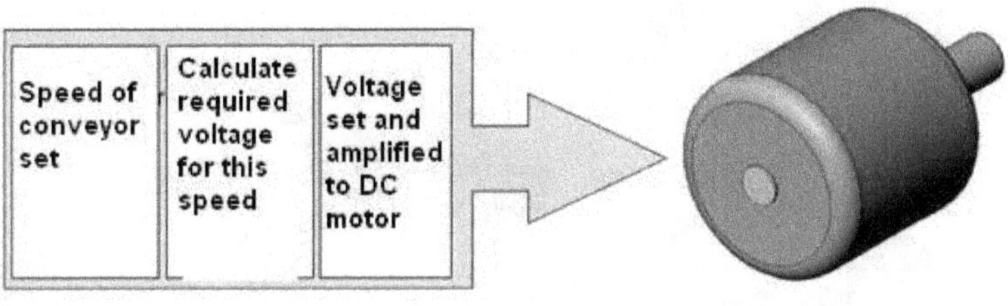

Figure 1.81 Open loop control system.

However, some circumstances are likely to arise which may prohibit the conveyor from operating at the specified speed. If the weight of the item being carried changes the load on the motor will rise and decrease, forcing the conveyor to speed up or slow down. Similarly, if the conveyor is not maintained higher forces due to friction in bearings etc would create larger load and delay the motor. The Open Loop system is not self-monitoring. Once the controller sends out the power to the motor it does not know whether the motor is running at that required speed since there is no sensor on the motor.

Advantages of Open Loop

Open Loop is quite cheap to install, and in many situations high precision requiring constant monitoring may not be necessary. In the conveyor system, a system with an accuracy of only 10 per cent can be acceptable. The Open Loop Control allows the conveyor to run at various speeds at a moderate cost.

Closed Loop Control

This control method has one additional feedback. A closed loop control is a more precise control system. It monitors its movements and actions constantly with sensors. Temperature, speed, pressure etc. can be controlled.

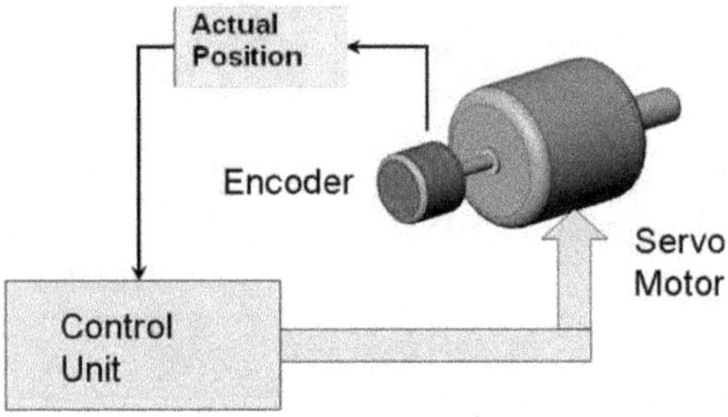

Figure 1.82 Elements of a closed loop control system.

Encoders

It is necessary for a closed loop control that the control system can monitor the motor it controls. We have seen before how some servos use a potenciometer to monitor their position. Most systems use encoders to provide the controller with feedback on the actual position of the engine. The above image shows the basic structure of a progressive optical encoder. This encoder is connected to the engine shaft. A circular disk spins between light beams. A light sensor called a photosensitive transistor monitors and receives a series of pulses of light on the rotation of the disk. The number of slots on the disk determines the encoder's resolution and accuracy. This disk has two sets of slots that give more precision than a single disk. The light sensor transmits light pulse information back to the controller so that the motor speed, position and direction can be calculated [11-16].

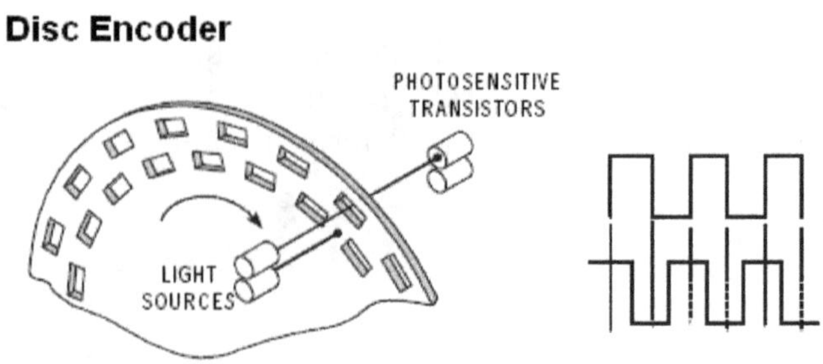

Figure 1.83 Disc encoder.

The value we aim to attain is the input value the users set and is frequently referred to as the 'desired value' or 'set point.' The sensor provides the real value of the condition we control and is frequently referred to as "measured value" or "actual value." A comparator is used for the closed loop control. It compares two values by removing the real value from the intended value. Any discrepancy between the two results in an error that is sent to the controller. The controller produces a process control action to attempt to remove this mistake.

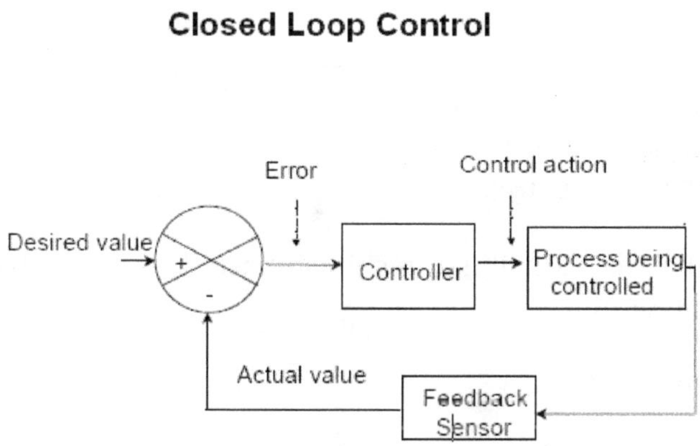

Figure 1.84 Closed loop control.

On / Off Control

This is the easiest way to regulate the closed loop. When the system notices a mistake between the intended value and the actual value, it is completely corrected. Take a basic example of regulating the temperature of water heating in a house. When the temperature of the water is low, the thermostat shuts and the element that begins heating the water rapidly receives full power. When the required heat is reached, it switches off again, but the elements are still heated such that the temperature slightly exceeds the intended temperature. The water refreshes anew, and the cycle begins again. This is not so much a water issue, but the start impact on engines and servos may create inaccurate difficulties.

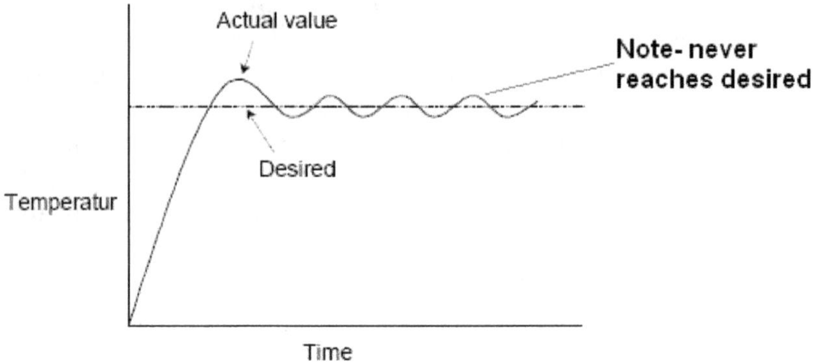

Figure 1.85 On-off control.

The unstable diagram illustrates the impact of on-off control and the output is on or off.

Proportional Control

Proportional control enhances the control on-off with an action proportionate to the mistake. Using the example of household water heating, the heater will be switched on and the control will then provide the element full power to heat the water quickly. However, when the controller notices that the water is getting closer to the temperature, the closer the water gets to the intended temp, the smaller the current flow to the element and the less risk of shooting beyond the desired value of the water temperature.

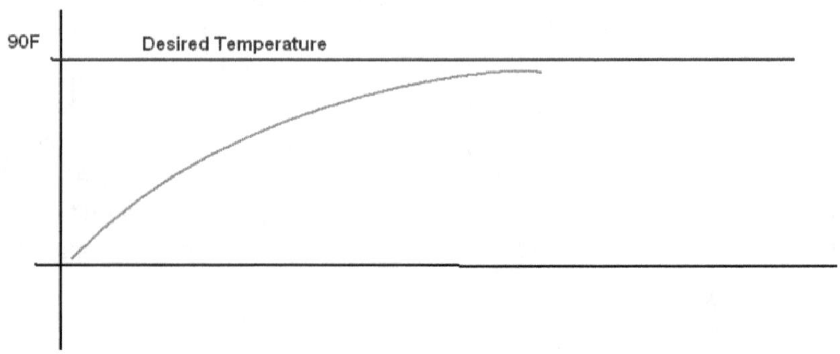

Figure 1.86 Proportional control.

The proportional control removes the fluctuating values that the on-off system may observe. It promotes a smooth ascent towards the desired location and minimizes the excess shot. The restriction of proportional control is that the closer to the goal, the slower the movement becomes, the more delayed the reaction to change is. The graph above shows that the current level has almost disappeared and is not at the intended value. The difference between the intended and the actual assessments is sometimes called the offset [5-15].

Increasing the Gain of a Proportional Controller

The usual way to enhance the proportional control reaction is to raise the gain factor K. The process seems to have returned to an oscillating state. After careful examination of the following graphs, the output finally stabilizes. In the proportional control graph above, the gain was too low, and the reaction was extremely slow. Imagine the temperature of the water heater has slightly decreased. The use of conventional proportional control without a gain factor would never achieve the required offset value. The system calculates a gain K that stabilizes the temperature at the optimal level.

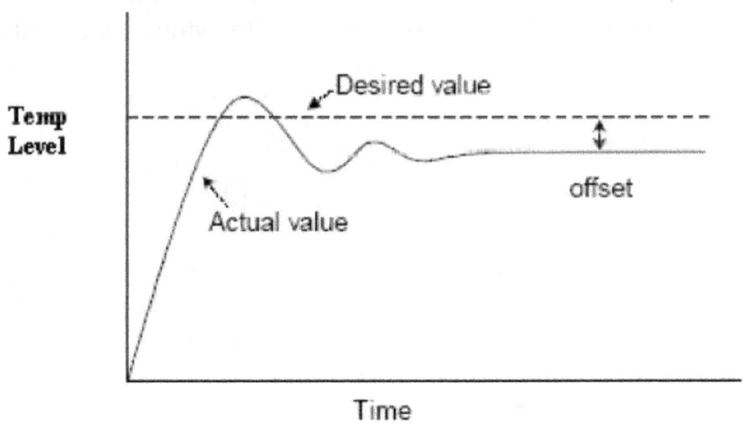

Figure 1.87 Proportional controller values with more offset.

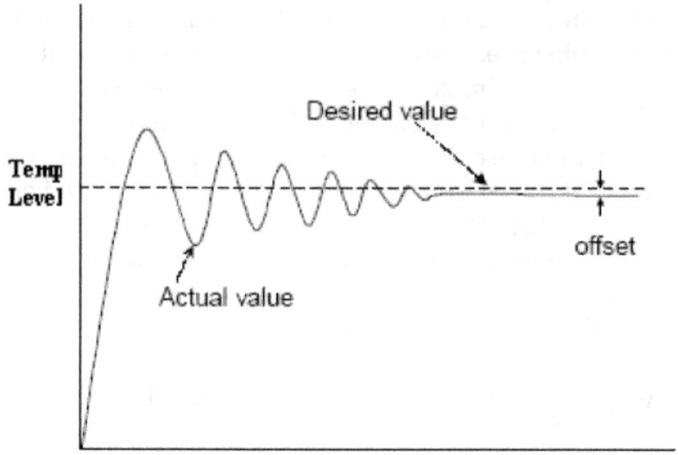

Figure 1.88 Proportional controller values with less offset.

A/D and D/A Conversion

Within control systems, the analog value of an input must be taken and stored as a digital number. First, the distinction between digital and analog and the difficulties posed by analog values is essential to grasp.

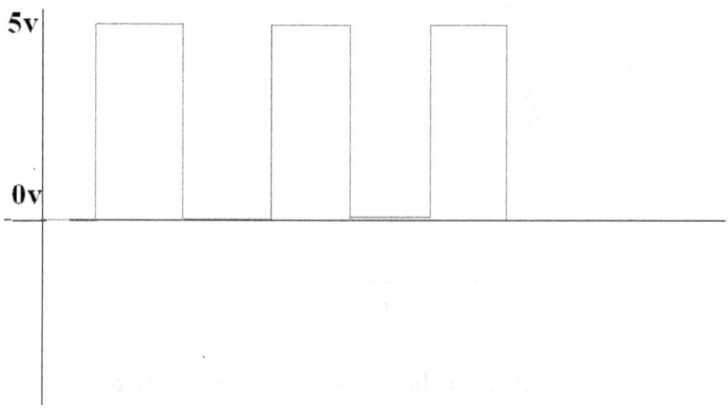

Figure 1.89 Digital wave.

Digital is like a simple switch, on or off. All computers work digitally and store 1 & 0 series. We all know the decimal system with 10 digits, 0, 1,2, ... 8,9. Infinity may be counted by simply organizing these 9 numbers into tens, hundreds, thousands and so on. This system is at the 10th base.

- 0,1,2,3....8.9
- 10,11,12......20,21...30......98....99
- 100.....200....999
- 1000

The digital method counts similarly, except it is based on two digits.0, and 1. This is called the binary system.

Table 1.2 Binary counting- 4bit

$2^3 = 8$	$2^2 = 4$	$2^1 = 2$	$2^0 = 1$	Decimal Equivalent
0	0	0	0	0
0	0	0	1	1
0	0	1	0	2
0	0	1	1	3
0	1	0	0	4
0	1	0	1	5
0	1	1	0	6
0	1	1	1	7
1	0	0	0	8
1	0	0	1	9
1	0	1	0	10
1	0	1	1	11
1	1	0	0	12
1	1	0	1	13
1	1	1	0	14
1	1	1	1	15

This table displays the numbers in binary from 0 to 15 (16 numbers, 0 included). You cannot count more than 15 since all four columns were filled with 1's. Because of four columns, this is a 4bit number. The table below illustrates how the columns may be increased to double numbers. Each column adds a bit; 8 bits Equals 1 bite (256).

Table 1.3 8 bit binary counting

Binary Nember	Decimal Equivalent	Number of Bits	Maximum Number of Values
1	1	1	2
10	2	2	4
11	3		
100	4	3	8
101	5		
110	6		
111	7		
1000	8	4	16
1001	9		
1010	10		
1011	11		
1100	12		
1101	13		
1110	14		
1111	15		
10000	16	5	32
11111	31		
100000	32	6	64
111111	63		
1000000	64	7	128
1111111	127		
10000000	128	8	256
11111111	255		

In an 8 bit system there are 256 max values.

For programming with Pic, this 8-bit system with 256 values is essential. In the logicator the number range may be chosen from 0 to 255 in a comparison block or while monitoring the analog input (256 numbers). This is no accident but because the chip may store these integers with 8 bits or 1 bit. Since 1 bite only contains 256 digits, all this may be recorded in the digital system.

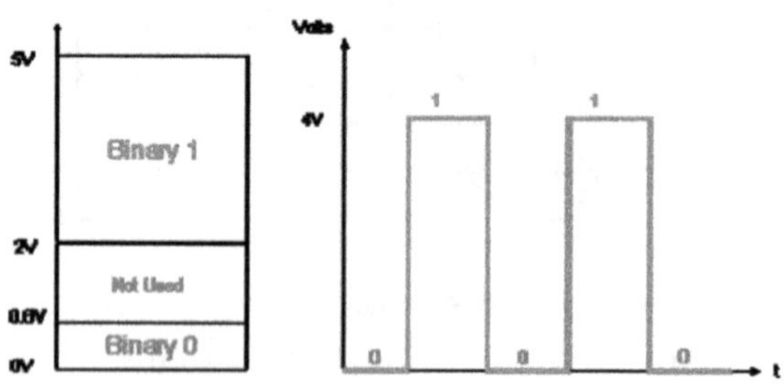

Figure 1.90 Digital reading.

As previously mentioned, digital is on or off. When a control system reads whether something is 1 or 0, it has to determine which voltage means a change between 0 and 1. In a perfect system 0 v is off and 5v is on. However, the voltage may not decrease to 0v or up to 5v. Therefore, the following is regarded as standard:

- Binary 1: Any tension from 2V to 5V
- Binary 0: Any voltage from 0V to 0.8V

Unused: voltage from 0.8V to 2V, this may cause a digital circuit fault. The computer system doesn't care about the voltage levels between 2v and 5v, only that it's intermediate and digitally connected. The precise voltage value is essential in analogue. This isn't fixed to 0 or 1 but may range from 0v to 5v.

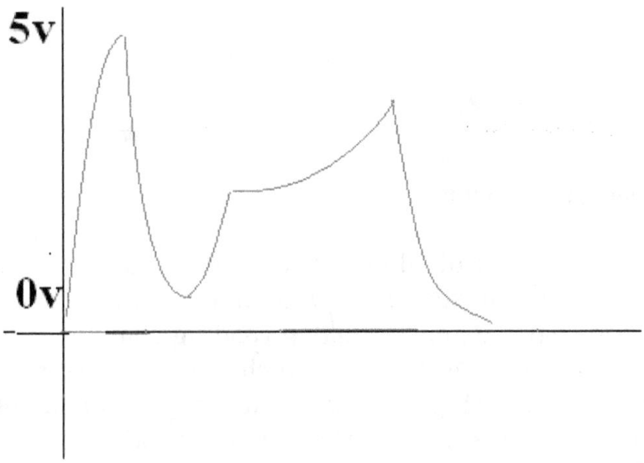

Figure 1.91 Analogue values.

In the T4 training boards we modify the analog inputs to shift the read to the Pic chip. Really all we do is alter the resistance of the potentiometer or the LDR, the voltage across the input increases and falls between 0v and 5v. The image reads the various voltage levels. Next, the analog readings are converted into digital [16-17].

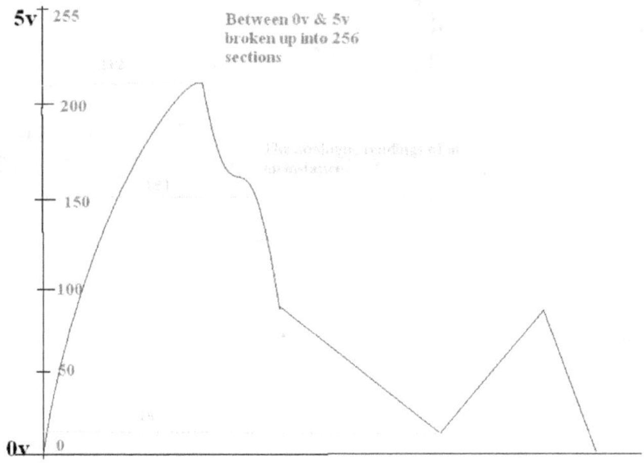

Fig 1.92 Analogue to digital.

The digital values were divided into three strips, and in analog the 0v -5v is divided into 256 segments as seen above in the red. It travels between these parts when an analog reading changes. The Pic chip notes and may store the value in the chip as a binary number. The number of bits in the chip defines the quantity segments into which 0v-5v may be split. The smaller the segments and the more accurate the bits. This is called the resolve.

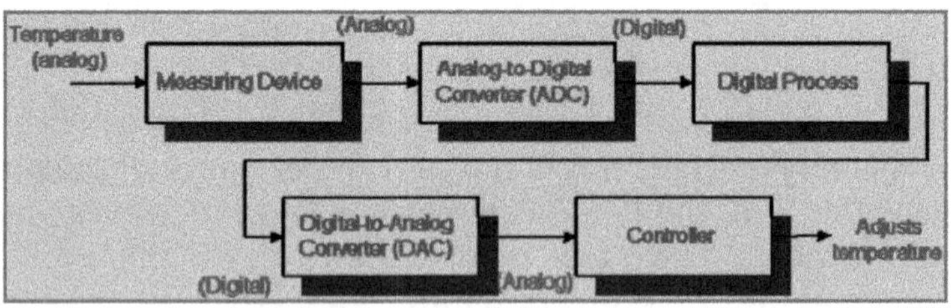

Figure 1.93 Flow diagram of A/D D/A.

This flow diagram illustrates how a temperature control system works. The thermometer is a thermist with a change of temperature resistance. This produces a comparable reading of the potentiometer. This reading is split down into the previously displayed parts and transformed to a binary number readable by the control system. The binary read from the thermometer is monitored and the control system sends a digital signal to change the temperature when it is too warm or cold [15-18]. The digital value must be converted back to analog at this step. This binary number is compared to the segments and the appropriate voltage is determined for the analog controller. This voltage is the control voltage that allows the temperature adjustment by the analog controller [19-22].

References

1. Bailey, S. A., Cham, J. G., Cutkosky, M. R., & Full, R. J. (2000). Biomimetic robotic mechanisms via shape deposition manufacturing. In *Robotics Research* (pp. 403-410). Springer, London.
2. Botelho, S. C., & Alami, R. (1999, May). M+: a scheme for multi-robot cooperation through negotiated task allocation and achievement. In *Proceedings 1999 IEEE International Conference on Robotics and Automation (Cat. No. 99CH36288C)* (Vol. 2, pp. 1234-1239). IEEE.
3. Cid, F., Moreno, J., Bustos, P., & Núnez, P. (2014). Muecas: A multi-sensor robotic head for affective human robot interaction and imitation. *Sensors, 14*(5), 7711-7737.
4. Hafez, M., Lichter, M. D., & Dubowsky, S. (2003). Optimized binary modular reconfigurable robotic devices. *IEEE/ASME Transactions on Mechatronics, 8*(1), 18-25.
5. Petrescu, R. V., Aversa, R., Akash, B., Berto, F., Apicella, A., & Petrescu, F. I. (2017). Forces of a 3R robot. *Journal of Mechatronics and Robotics, 1*(1).
6. Fiore, M., Clodic, A., & Alami, R. (2016). On planning and task achievement modalities for human-robot collaboration. In *Experimental Robotics* (pp. 293-306). Springer, Cham.
7. Lussier, B., Gallien, M., Guiochet, J., Ingrand, F., Killijian, M. O., & Powell, D. (2007, June). Fault tolerant planning for critical robots. In *37th Annual IEEE/IFIP International Conference on Dependable Systems and Networks (DSN'07)* (pp. 144-153). IEEE.
8. Bensalem, S., Gallien, M., Ingrand, F., Kahloul, I., & Thanh-Hung, N. (2009). Designing autonomous robots. *IEEE Robotics & Automation Magazine, 16*(1), 67-77.
9. Wang, N., Pynadath, D. V., & Hill, S. G. (2016, March). Trust calibration within a human-robot team: Comparing automatically generated explanations. In *2016 11th ACM/IEEE International Conference on Human-Robot Interaction (HRI)* (pp. 109-116). IEEE.
10. Kube, C. R. (1997). Task modelling in collective robotics. *Autonomous robots, 4*(1), 53-72.
11. Bruyneel, M., & Granville, D. (2007). Mechatronic analysis of a flexible mechanism using SAMCEF: application to robotics. *Proceedings of the NAFEMS, Copenhagen*, 24-25.

12. Wang, L., & Meng, M. Q. H. (2012, July). A game theoretical bandwidth allocation mechanism for cloud robotics. In *Proceedings of the 10th World Congress on Intelligent Control and Automation* (pp. 3828-3833). IEEE.
13. Stojanov, G. (2001). Petitagé: A case study in developmental robotics. *Proceedings of Epigenetic Robotics, 1.*
14. Moubarak, P. M., Ben-Tzvi, P., Ma, Z., & Alvarez, E. J. (2013, May). An active coupling mechanism with three modes of operation for modular mobile robotics. In *2013 IEEE International Conference on Robotics and Automation* (pp. 5489-5494). IEEE.
15. Wu, S., Ze, Q., Zhang, R., Hu, N., Cheng, Y., Yang, F., & Zhao, R. (2019). Symmetry-breaking actuation mechanism for soft robotics and active metamaterials. *ACS Applied Materials & Interfaces, 11*(44), 41649-41658.
16. Siciliano, B., Khatib, O., & Kröger, T. (Eds.). (2008). *Springer handbook of robotics* (Vol. 200). Berlin: Springer.
17. Nof, S. Y. (Ed.). (1999). *Handbook of Industrial Robotics.* John Wiley & Sons.
18. Lynch, K. M., & Park, F. C. (2017). *Modern Robotics.* Cambridge University Press.
19. Garcia, E., Jimenez, M. A., De Santos, P. G., & Armada, M. (2007). The evolution of robotics research. *IEEE Robotics & Automation Magazine, 14*(1), 90-103.
20. Craig, J. J. (2009). *Introduction to Robotics: Mechanics and Control, 3/E.* Pearson Education India.
21. Ghosal, A. (2006). *Robotics: Fundamental Concepts and Analysis.* Oxford university press.
22. Bajd, T., Mihelj, M., Lenarčič, J., Stanovnik, A., & Munih, M. (2010). *Robotics* (Vol. 43). Springer Science & Business Media.

2

Study of Different Robotic Arm Features and Economic, Social and Ethical Issues

Surya Narayan Panda[1], Swastik Pradhan[2], Arati Rath[3], Dilip Kumar Bagal[4], and Siba Sankar Mahapatra[5]

[1]*Department of Production Engineering, Birsa Institute of Technology, Sindri, Dhanbad, Jharkhand, India*
[2]*School of Mechanical Engineering, Lovely Professional University, Phagwara, Punjab 144411, India*
[3]*School of Computer Engineering, Kalinga Institute of Industrial Technology, Bhubaneswar, Odisha India*
[4]*Department of Mechanical Engineering, Government College of Engineering, Kalahandi, Bhawanipatna, Odisha, India*
[5]*Department of Mechanical Engineering, National Institute of Technology, Rourkela, Odisha, India*

2.1 What is Robotic Arm?

The word robot is derived from the Czech word ROBOTA, which is usually referred to as "forced labor." This is a good description of the bulk of robots. Most robots worldwide are intended for heavy, repetitive production. They do jobs that are tough, risky or uninteresting for people.

Figure 2.1 Robotic arm.

2.2 Concept of Trigonometry in Robotics

Trigonometry was originally believed to describe relationships between triangular components. There are six fundamental components in a triangle: 3 sides and 3 angles. Not three segments can be the sides of a triangle. They do not fulfill the inequality of the triangle, or rather three triangles. The angles of a triangle cannot be any three angles. The three angles of a triangle add up to a straight angle in Euclidean geometry. These constraints restrict the way the relationships between the components are specified. These relationships are expanded to arbitrary angles in contemporary trigonometry. For example, this may be done by monitoring the projections of a spinning cycle radius and of a tangent at the end of the radius.

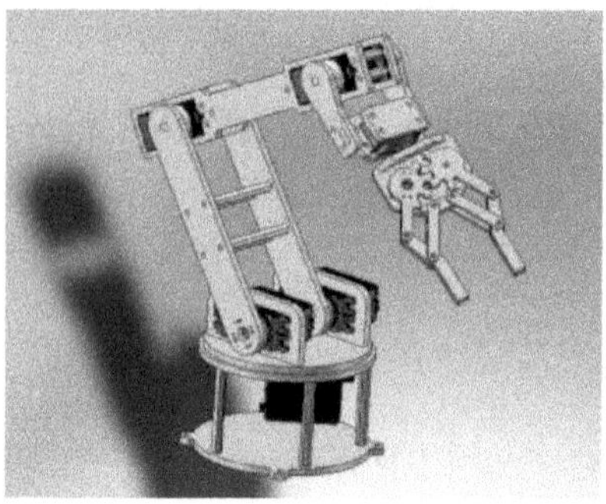

Figure 2.2 Robotic arm with skeleton structure.

2.3 Rules of Trigonometry used in Robotics Law

Law of sines

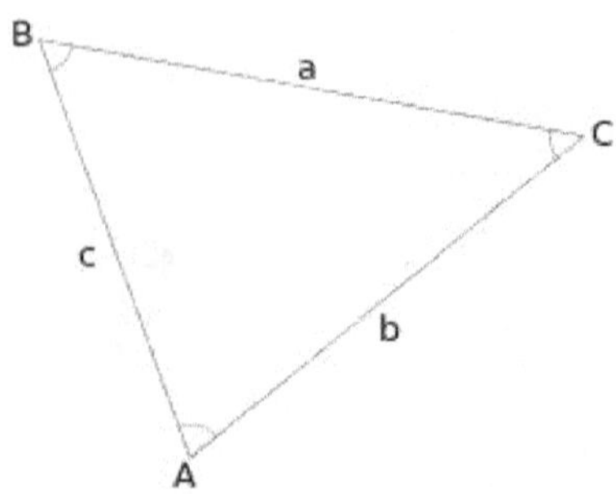

Figure 2.3 Triangle with angles A, B and C.

According to the law,

$$\frac{a}{\sin A} = \frac{b}{\sin B} = \frac{c}{\sin C} = d,$$

Where a, b, and c are triangle sides' lengths, and A, B, and C are the opposing angles, while d is the circumcircle diameter of the triangle [1-10].

Law of cosine rule

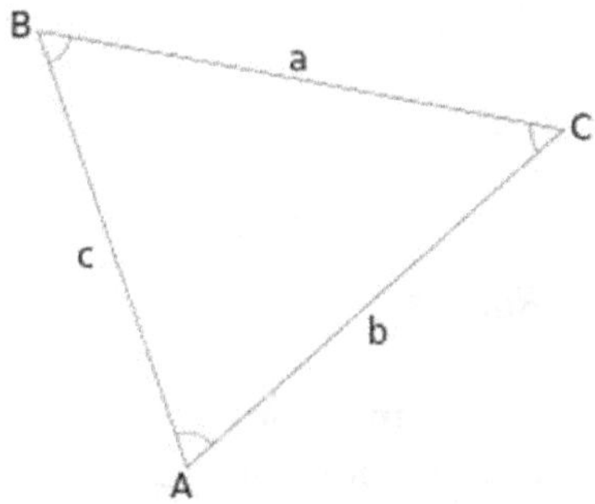

Figure 2.4 Triangle with angles A, B and C.

The law of cosines (also known as the cosine formula or cosine rule) in trigonometry links the sides of a triangle to the cosine of one of its angles. The Law of Cosines is as follows:

$$c^2 = a^2 + b^2 - 2ab\cos\gamma,$$

where γ indicates the angle between length sides a and b and the length sides c, opposite [9-10].

2.4 History of Robotic Arm

The Shuttle's robotic arm has conducted extensive types of jobs throughout the years. It has launched satellites into orbit and recovered others for repair. The first time Canadarm (space shuttle manipulator System) was used in one of several International Space Station

(ISS) construction flights was during Mission STS-88, December 1998 [11-15].

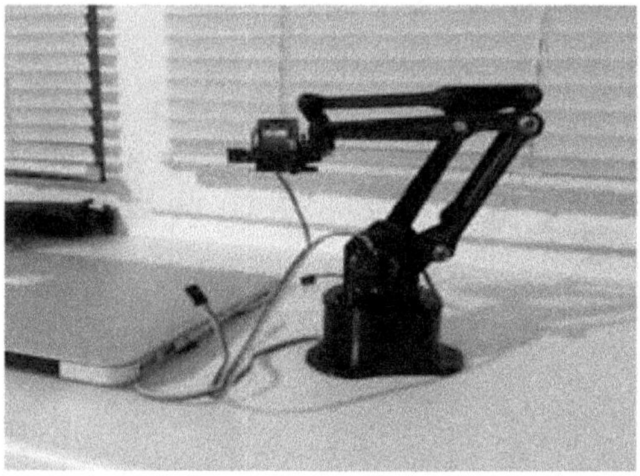

Figure 2.5 Shuttle's robotic arm.

Definition of Industrial Robots

"A robot is a programmable, multipurpose manipulator intended to move materials, components, tools, or special devices via varied programmed movements for the execution of a range of tasks." - By Robot Institute of America (RIA) [14-17].

Applications of Arm Robots

- Material Handling/Palletizing
- Machine Loading/Unloading
- Arc/Spot Welding
- Water jet/Laser cutting
- Spray Coating
- Gluing/Sealing
- Investment casting
- Assembly
- Inspection [16,17]

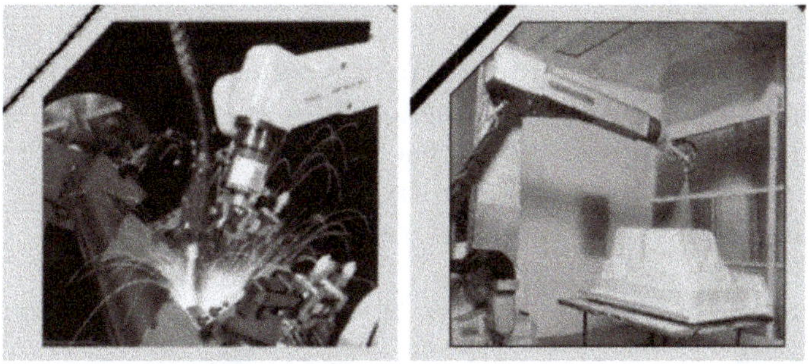

Figure 2.6 Arc/spot welding and spray coating in industry.

Overview of Industrial Robots

- The robots are being used more and more in manufacturing. They perform tasks that are too dangerous or boring for people to address.
- Before robots, people were manufactured in factories. People don't make things very quickly, and sometimes they make mistakes.
- Modern factories use automated robots to more efficiently manufacture products.
- Robots produce products far faster than people and never make mistakes.
- Some people need to keep robots going smoothly.

The Structure of Robotic Arm

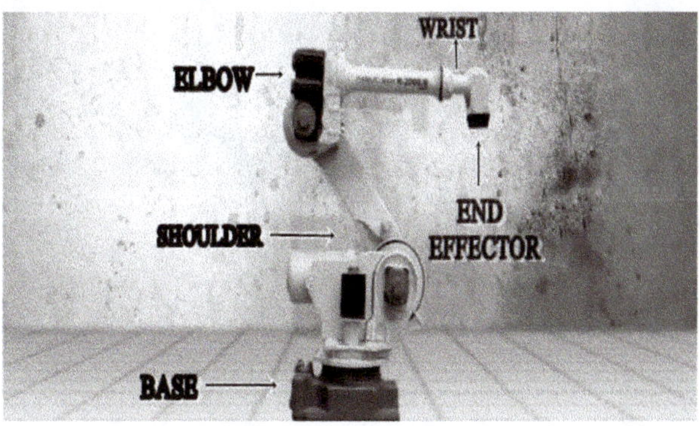

Figure 2.7 Representation of robotic arm.

A typical robotic arm consists of seven sections of metal joined by five joints. This enables the computer to move the arm very precisely and repeat the same movement repeatedly. A six-joint industrial robot closely resembles a human arm. The robot has six degrees of freedom, which means that it can swing in six different ways. In comparison, a human arm has seven degrees of freedom [15-17].

Characteristics and Uses of Robotic Arm

Industrial robots share the following characteristics:

- Roberts are extremely powerful.
- They have powerful motor-powered joints.
- They may have different tool fasteners at the end of the arm (such as grippers, welding / spraying devices).
- They are managed by microprocessors.

Some tasks for which robots are utilized:

- Soldering automobiles in automotive plants.
- Heavy vehicle lifting components.
- Paint vehicles in automobile plants.
- Goods are packaged in boxes.
- Production of microchips.

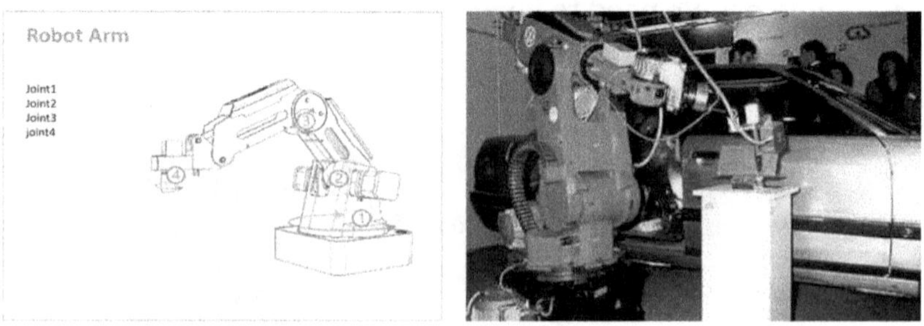

Figure 2.8 Representation of robotic arm with joints.

Advantages and Disadvantages of Robotic Arm

Table 2.1 Advantages and disadvantages of robotic arm

Advantages	Disadvantages
Robots can work non-stop for 24 hours a day, 7 days a week.	Very expensive to set up at first. They can cost 100's of thousands of dollars.
Robots can work in environments that would be too dangerous for humans.	They can take jobs off people. This leads to unemployment.
Robots work faster than humans and can produce more.	Robots cannot make decisions. E.g. if a car arrived at a welding station in the wrong position, the robot would not adjust its own position to compensate.
Robots are more accurate than humans. E.g. each car produced is exactly the same.	Robots can break down.
They handle repetitive tasks which frees people to do more skilled work	

2.5 Design and Control Issues, Manipulation and Control, Resolution of Robots

Basic terms/parameters in robotics:

❖ **Links and Joints** - Connections are the robot's solid structural components and joints are the moveable connections between them [8].

❖ **Degree of Freedom (DoF)** - Every joint on the robot gives a degree of freedom. Each DF may be a slider, rotary actuator or other actuator type. Robots are usually free by 5 or 6 degrees. 3 degrees of freedom permit positioning in 3D space, while the remaining 2 or 3 are utilized for end effector orientation. 6 degrees of freedom are sufficient to enable the robot to reach all 3D spatial locations and directions. 5 DoF needs 2D space restrictions, or else it restricts guidance. For handling equipment like arc welders, 5 DoF robots are frequently utilized.

❖ **Orientation Axes** - Basically, while the tool is kept at a given location, the direction in which it may be directed is determined. The common orientation axes utilized are roll, pitch and yaw. Looking at the following picture, it becomes evident that the tool may be positioned in space in any direction. (Imagine sitting at a plane. If the aircraft flips, you'll go upside down. The pitch will vary to take-off and land and the aircraft will yaw in a crosswind.)

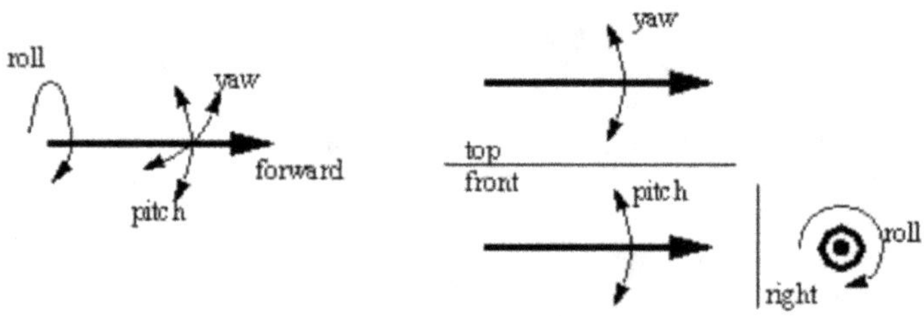

Figure 2.9: Basic terms/parameters in robotics.

❖ **Position Axes** - The tool, independent of orientation, may be moved to a variety of locations in space. Various robot geometries are adapted to various work geometries. (more later).

❖ **Tool Centre Point (TCP)** - The tool centre point is placed either on the robot, or the tool. Typically, the TCP is used when referring to the robot's location, as well as the focus point of the tool. (e.g., the TCP may be at the tip of a welding torch) The TCP can be given in cartesian, cylindrical, spherical, etc. coordinates depending on the robot. As tools are replaced, we will frequently reprogram the robot for the TCP.

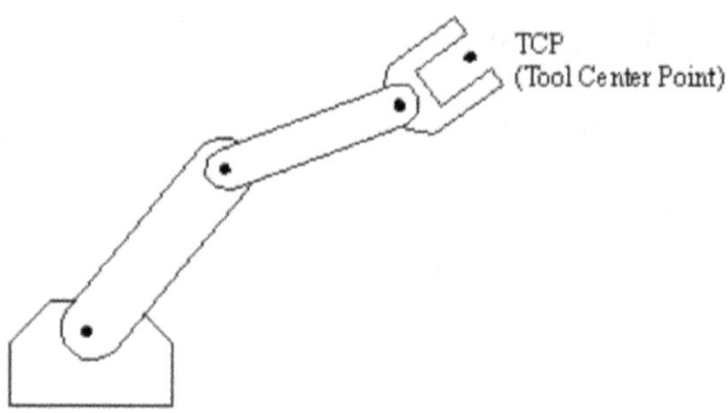

Figure 2.10 Tool centre point (TCP).

❖ **Work envelope/Workspace** - The robot tends to have a fixed, and restricted geometry. The work envelope is the limit of locations in space that the robot can reach. For a cartesian robot (like an overhead crane) the workspace could be a square, for more advanced robots the workspace might be a form that appears like a 'clump of intersecting bubbles'.

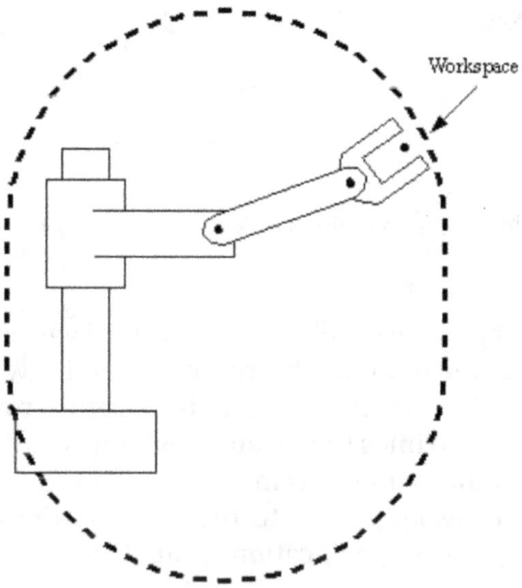

Figure 2.11 Work envelope of arm robot.

❖ **Speed** – Speed denotes either to the greatest velocity that is attainable by the TCP, or by particular joints. This value is not precise in most robots and will fluctuate across the workspace as the geometry of the robot changes (and therefore the dynamic effects) (and hence the dynamic effects). The number will typically represent the highest safety speed achievable. Some robots enable the maximum rated speed (100 percent) to be passed, however it should be done with extreme care.

❖ **Payload/Carrying capacity**- The payload shows the greatest mass the robot can lift either before failure of the robotics, or significant loss of precision. It is possible to surpass the maximum payload, and still have the robot function, although this is not recommended. When the robot is accelerating rapidly, the payload should be smaller than the maximum mass. This is affected by the ability to firmly grip the part, and also the robot structure, and the actuators. The end of arm tooling should be considered to be part of the payload.

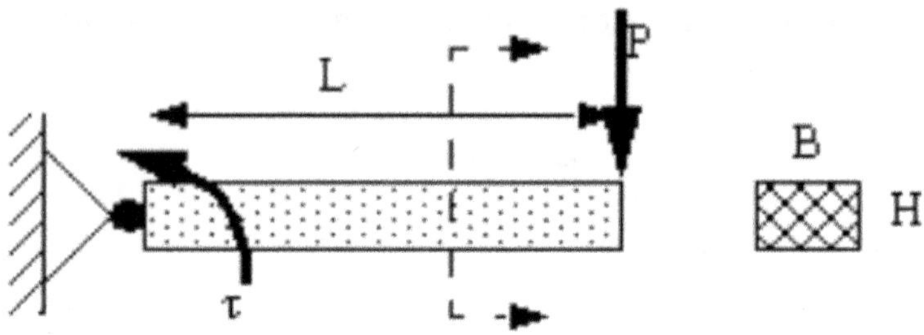

Figure 2.12 Payload of arm robot's link.

❖ **Repeatability** - There will be some natural variance in the robot mechanism. This means that the robot will not always stop at the same position if it is instructed repeatedly to return to the same point. Repeatability is +/- 3 times the default position deviation or 99.5 percent of all repeatability measurements fall. This figure varies across the workspace, especially close to the limits of the workspace, but manufacturers give the specifications a single value.

❖ **Accuracy** - This is defined by the workspace resolution. If the robot is ordered to travel to an area, it is often off by a certain amount, the accuracy of the maximum distance shall be considered. This is an effect of a not necessarily continuous control system.

❖ **Settling Time** - While the robot moves quickly, the robot reaches the final position but slows down and approaches slowly. The adjustment time is the necessary time for the robot to really be within a certain distance.

❖ **Control Resolution** - This is the smallest change that can be detected by or by the bigger feedback sensors. If a rotary joint has an encoder that measures every 0.01 degree of rotation and a direct drive servo motor is used to operate the joint with a 0.5-degree precision, then the control resolution is approximately 0.5 degrees.

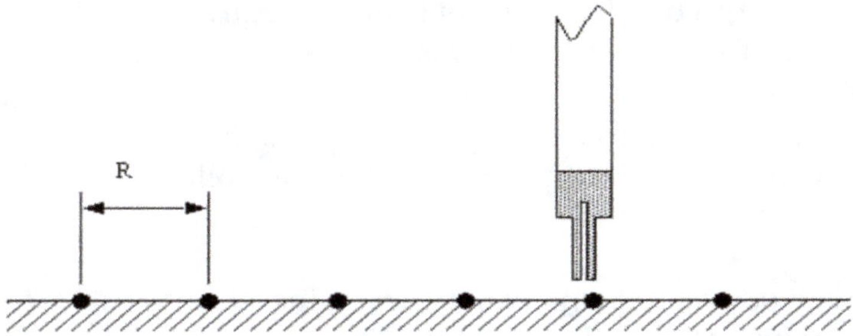

Figure 2.13 Control resolution.

❖ **Coordinates** - The robot may move; thus, positioning must be defined. Please note that the co-ordinates represent both the origin location and the axis orientation.

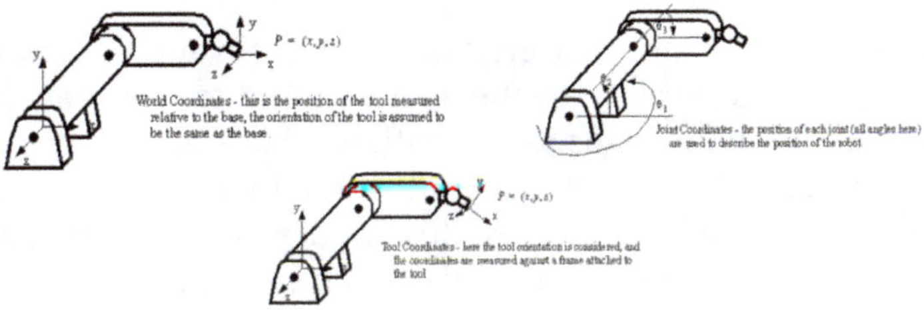

Figure 2.14 Coordinates of location of end effector.

❖ **Number of axes** – To reach any point in a plane, two axes are needed, three axes are required to reach any space point. Three more axes (yaw, pitch, roll) are needed to completely control the orientation of the end of the arm (i.e., the wrist). Some designs (e.g., robot SCARA) have constraints with cost, speed and accuracy in motion.

❖ **Kinematics** - the real setup of stiff robot parts and joints, which defines the potential movement of the robot. Robot film classes include Articulated, Cartesian, Parallel and SCARA.

❖ **Precision** - Precision is a random error description, a statistical variability metric. The accuracy of a measuring system linked to reproducibility and repeatability is that repeated measurements are shown to have the same results under unchanging circumstances.

❖ **Motion control** – The robot only has to return repeatedly to a restricted set of pre-taught locations for certain applications such as basic pick-and-place assembly. Motion has to be constantly regulated for complex applications, such as welding and finishing (spray painting) to follow the track across space with controlled orientation and speed.

❖ **Power source** – Some robots are using electric motors, while others are using hydraulic actuators. The former are quicker, the latter are more powerful and useful for applications such as spray painting where a spark may start an explosion, but the arm can prevent inflammatory vapors and other pollutants from entry by the internal air pressurization. It is quite improbable that any hydraulic robots are currently on the market. Additional screens, brushless electric motors and spark-proof shielding make it easier to build devices that operate in an explosive atmosphere in the surroundings.

❖ **Drive** – Some robots link electric motors through gears to the joints while others directly connect the engine to the joint (direct drive). The use of gears leads to quantifiable 'backlash' - free circulation in an axis. Smaller robot arms often use high-speed DC motors that typically need high gear ratios; this has a backlash drawback. The harmonic drive is frequently utilized in such situations.

❖ **Compliance** - This is a measure of how much the robot axis moves in angle or distance when it is applied with a force. Because of the compliance with the maximum payload position of a robot it is somewhat lower than when no payload is carried. Compliance may also be accountable for over-compliance while hauling heavy weights, which would need a reduction in acceleration [1, 7].

2.6 Robot Programming

Setup or scheduling of the movements and sequences of an industrial robot is usually learned by connecting the robot controller to a laptop, desktop or (internal or Internet) network.

❑ ***Software:*** The computer is equipped with the appropriate software interface. The usage of a computer reduces the programming process considerably. Depending on the architecture of the system, specialized robot software runs on the robot controller or computer.

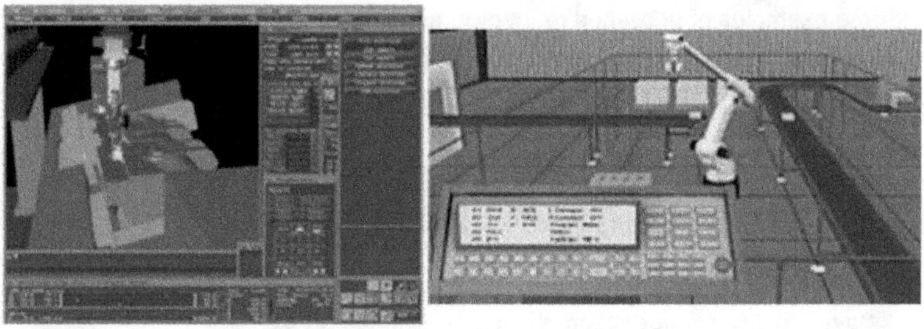

Figure 2.15 Robotic software.

Robotic Languages and Software

The robot software is a set of encoded commands or instructions which tell a robot what tasks to perform on a mechanical device and an electronic system. Robot software is used for independent tasks. Many software systems and frameworks were proposed to facilitate the programming of robots. Some robot software is designed to develop smart mechanical devices. Feedback loops, control, pathfinding, filtering of data, location and sharing of data are common tasks.

Examples include:

Some examples of published robot programming languages are shown below.

Task in plain English:

```
Move to P1 (a general safe position)
Move to P2 (an approach to P3)
Move to P3 (a position to pick the object)
Close gripper
Move to P4 (an approach to P5)
Move to P5 (a position to place the object)
Open gripper
Move to P1 and finish
```

```
PROGRAM PICKPLACE
   1. MOVE P1
   2. MOVE P2
   3. MOVE P3
   4. CLOSEI 0.00
   5. MOVE P4
   6. MOVE P5
   7. OPENI 0.00
   8. MOVE P1
.END
```

Example of Stäubli VAL3 program:

```
begin
   movej(p1,tGripper,mNomSpeed)
   movej(appro(p3,trAppro),tGripper,mNomSpeed)
   movel(p3,tGripper,mNomSpeed)
   close(tGripper)
   movej(appro(p5,trAppro),tGripper,mNomSpeed)
   movel(p5,tGripper,mNomSpeed)
   open(tGripper)
   movej(p1,tGripper,mNomSpeed)
end
```

In Other Platform

Epson RC+ (example for a vacuum pickup)

```
Function PickPlace
        Jump P1
        Jump P2
        Jump P3
        On vacuum
        Wait .1
        Jump P4
        Jump P5
        Off vacuum
        Wait .1
        Jump P1
Fend
```

ROBOFORTH (a language based on FORTH).

```
: PICKPLACE
P1
P3 GRIP WITHDRAW
P5 UNGRIP WITHDRAW
P1
;
```

Programming Languages in Industrial Robotics

Table 2.2 Various robotic languages with manufacturing company

Robot brand	Language name
ABB	RAPID
Comau	PDL2
Fanuc	Karel
Kawasaki	AS
Kuka	KRL
Stäubli	VAL3
Yaskawa	Inform

2.7 Control Systems, Spatial Resolution, Accuracy, and Repeatability

Control systems:

The motion control system used to control the movement of the end-effector or tool with regard to robotics.

1. Limited robots sequence (Non-Servo)
2. Point-to-point playback robots (Servo)
3. Robots with continuous path control play back,
4. Smart robots. Intelligent.

1. Limited sequence robots (Non-servo)

Limited robots do not provide a servo with the relative inclination positions of the joints regulated, but instead are controlled by limits and mechanical stops. Usually, no feedback is connected with a restricted sequence robot to show that the intended position, generally thin robots like pick-and-place activities, requires easy movement.

2. Point to point motion

These robots are able to regulate speed and motion route from the beginning to the conclusion of the journey. It utilizes sophisticated control programs, PLC computers (programmable logic controller) to manage the movements. Point-to-point movement robots can execute a motion cycle consisting of a sequence of desired point position. The robot is hard and recorded.

3. Continuous path motion

In these robots, the route followed by the robot is regulated by a motion cycle. The robot passes through a succession of narrow spaces describing the intended route, e.g., spray, arc welding and complicated assembling.

4. Intelligent robots

This kind of robots not only interact with a programmed motion cycle but also in an intelligent manner with the surroundings for years. It makes rational choices based on the operating sensor data. Robots are typically programmed using the symbolic language English that is not like a computer programming language.

2.8 Precision of Movement (or) Parameters of Robot

The previous discussion on reaction speed and stability deals with robot's dynamic performance. Another performance measurement is the precise movement of the robot. We define accuracy according to three characteristics:

1. Space resolution
2. Precision
3. Repetitiveness

Some assumptions are:

1. Definitions must apply with no hand connected to the wrist at the end of the robot's wrist.
2. The terms must apply in the worst case to the circumstances under which the robot's accuracy is at its disposal. This usually implies that the robot's arm in the case of a joint arm or polar configurable is completely extended.
3. Third, in connection with a point-to-point robot, our definitions will be refined [3].

1. Spatial resolution

The robot's spatial resolution is the smallest increase in movement into which the robot can split its working volume. Spatial resolution relies on two factors: control resolution of the system and mechanical imperfections of the robot. These variables may be easily conceptualized in terms of a robot with 1 degree of freedom.

The no of increments $= 2^n$

Where n = the number of bits in the control memory.

$$The\ control\ resolution = \frac{Total\ moments\ range}{The\ number\ of\ increments}$$

2. Accuracy

Accuracy relates to the capacity of a robot to place its wrist end inside the volume of work at a specified target location. The accuracy of a robot may be determined with regard to spatial resolution, since how the robot can precisely specify control increments for each of its joint movements relies on the capacity to reach a particular target point.

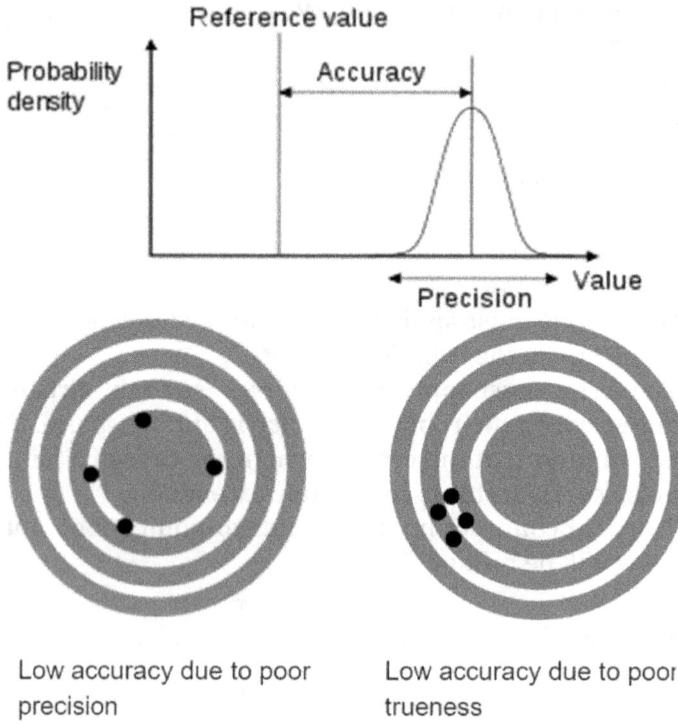

Figure 2.16 Robotic accuracy and precision.

3. Repeatability

Repeatability is the robot's ability to place its wrist or to connect an end effector to a location in space called repeatability. Repeatability and precision refer to two distinct robot precision characteristics. Accuracy concerns the ability of the robot to be trained to accomplish a particular goal. The actual programmed position is likely to be different from the goal, since control resolution constraints Repeatability relate to the ability of the robot to return to the programmed location when ordered.

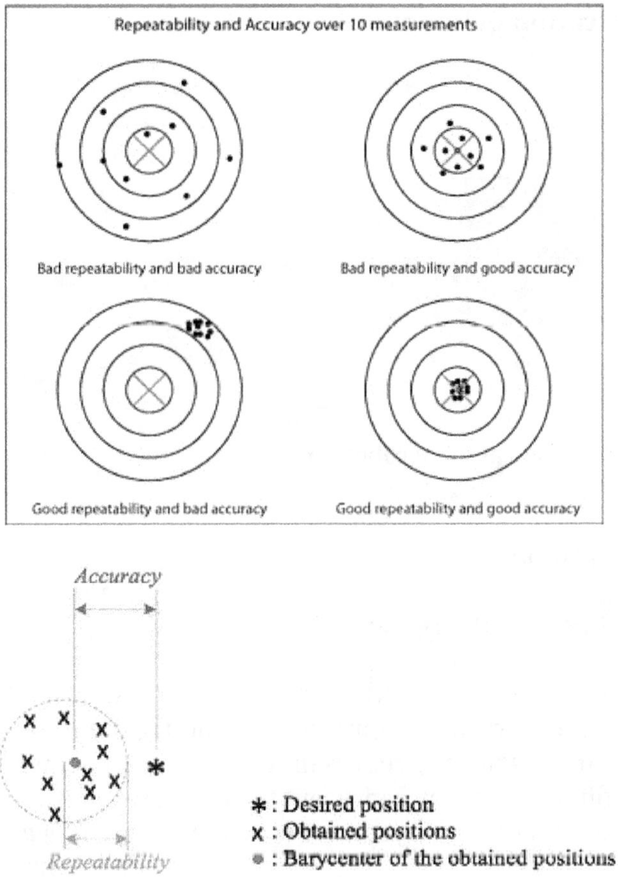

Figure 2.17 Robotic repeatability.

Difference between accuracy and repeatability:

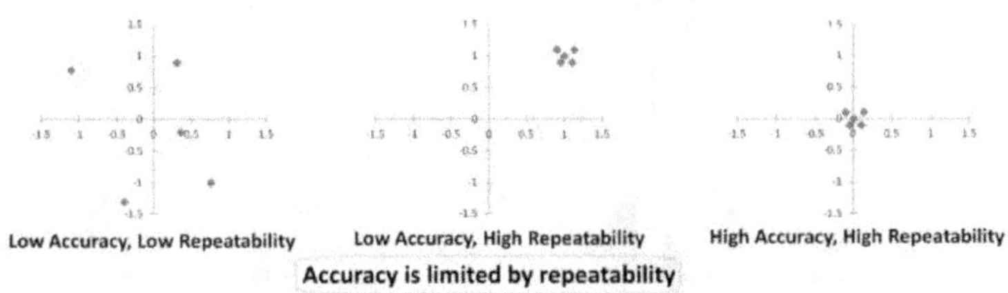

Figure 2.18 Comparison of robotic accuracy and repeatability.

2.9 Classification of Robots

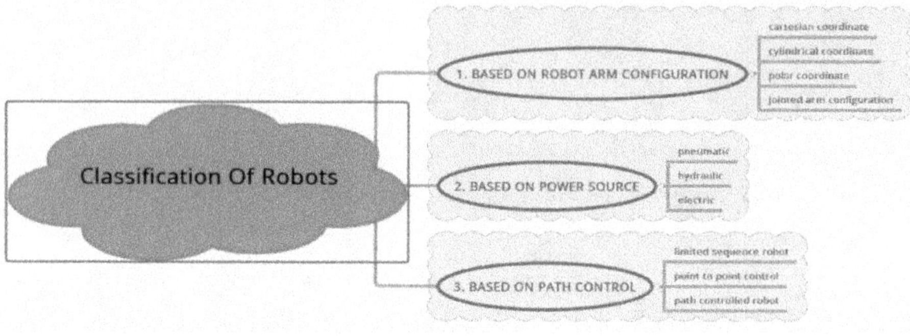

Figure 2.19 Classification of robots.

Arm Configuration

Cartesian Coordinate System

- In this there are three orthogonal directions X, Y and Z.
- X-coordinate axis may represent left and right motion.
- Y-coordinate axis may represent forward and backward motion.
- Z-coordinate axis may represent up and down motions.
- Example of Cartesian System is *Overhead Crane Movement* [4].

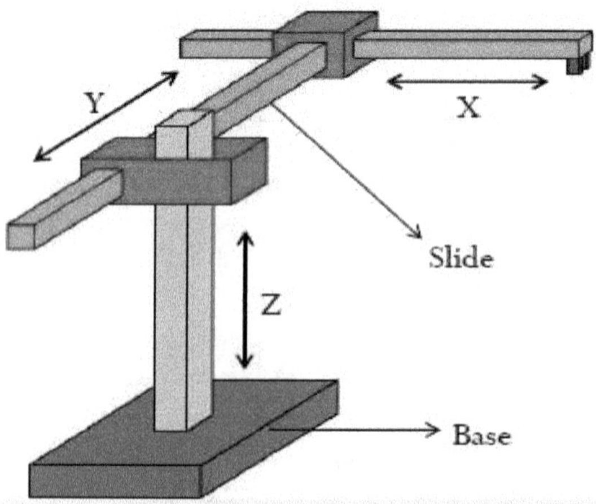

Figure 2.20 Cartesian coordinate system.

Working Envelope of Cartesian Configuration Robot

The Cartesian configuration work envelope is a rectangular prism. The robot may manipulate its maximal payload across the whole volume.

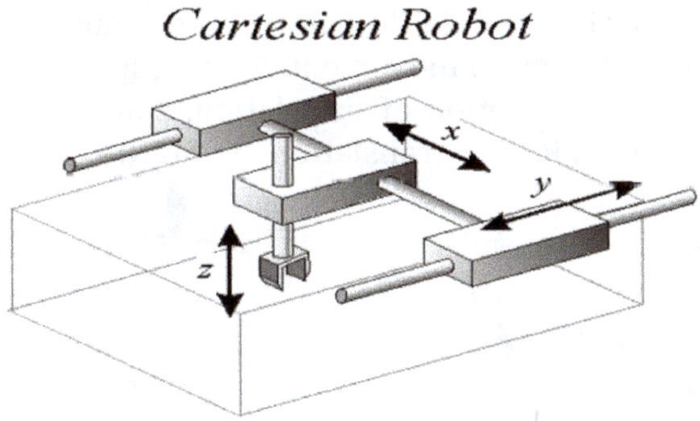

Figure 2.21 Cartesian configuration robot.

Advantages:

- Travel along the x axis may expand the work envelope.
- Linear movement and thus easier control.
- High degree of precision and repetitiveness owing to its structure.
- Can bear higher weights since load capacity does not vary from work envelope at the various positions.

Disadvantages:

- Movement is restricted to just one direction at a time.

Applications:

- Pick and put operation
- Adhesive applications
- Assembly and sub assembly
- Nuclear material handling
- Welding

Cylindrical Configuration Robot

It uses a vertical column and a diaphragm that you can move up and down the column. The robot arm is attached to the slide to move it radially to the column. The robot is able to achieve a working space that approximates the cylinder by rotating the column. There are two linear motions and a rotary movement. Angular motion along vertical axis; angular motion; Motion of translation, z along the z-direction that is up and down; Radial, translation in or out.

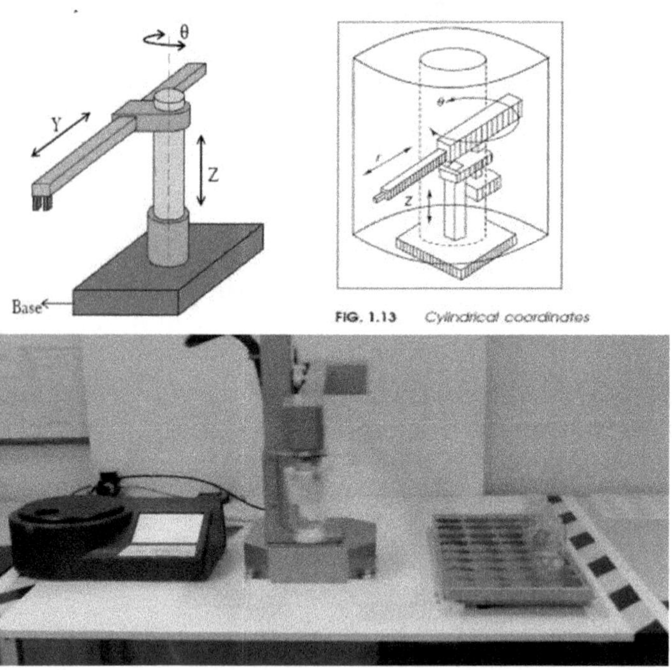

Figure 2.21 Cylindrical configuration robot.

Working Envelope of Cylindrical Configuration Robot

As the name suggests, the working envelope of this configuration is a cylinder.

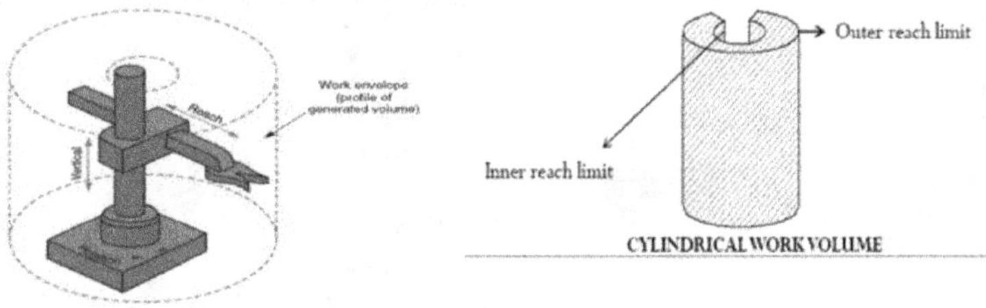

Figure 2.22 Working envelope of cylindrical configuration robot.

Advantages

- The results are larger than the rectangular manipulator.
- The vertical structure retains space on the ground.
- Can carry large payloads.

Disadvantages

- Repetitiveness and precision in the direction of rotating motion are lower.
- Requires an advanced control system.

Applications

- Assembly
- Application for coating
- Die cast
- Application foundation and forging
- Loading and unloading of the machine [3,4]

Polar Configuration Robot

The arm is lifted or lowered by a horizontal pivot. The pivot is attached to a spinning base. The several joints enable the robot to move its arm in a spherical space; thus, it is also referred to as "sphere coordinate robot." It has one linear movement and two rotating movements. The UNIMATE 2000 series is a spherical robot example.

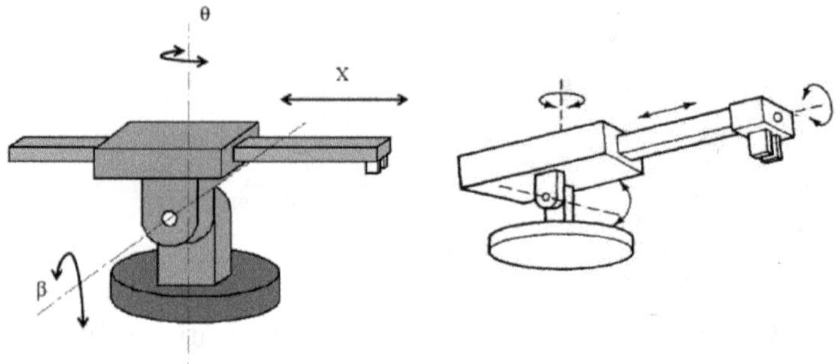

Figure 2.23 Polar configuration robot.

Work Volume of a Polar Configuration Robot

The volume of work of a polar configuration robot is a sphere. It consists of one linear movement and two corner movements. The linear motion r is a radial translation in or out. The angular movement corresponds to a base rotation, — that is, around a vertical axis. Another angular move, the β, along an axis perpendicular to the vertical axis through the foundation and often called the elbow rotation.

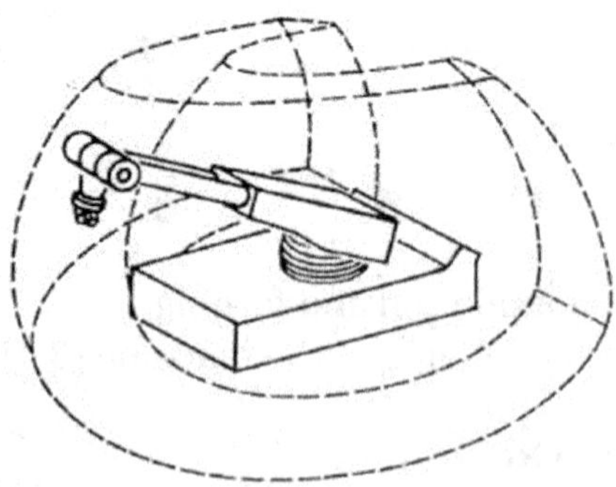

Figure 2.24 Polar configuration robot with work envelope.

Advantages

- Larger envelope of work than the straight or cylindrical form.
- Structure vertical retains less space.

Disadvantages

- Repeatability and precision in the direction of rotational movement are likewise reduced.
- More advanced control system is required.

Applications

- Die cast
- Forging
- Handling of glass
- Molding for injection
- Stacking and dislodging [2]

Jointed Arm Configuration

It has a similar configuration to that of a human arm. These components are linked by two shoulder- and elbow-related rotational joints. At the end of the forearm a wrist is connected, which provides many more joints. A commercially available Cincinnati Milacron T3 (Model 776) robot.

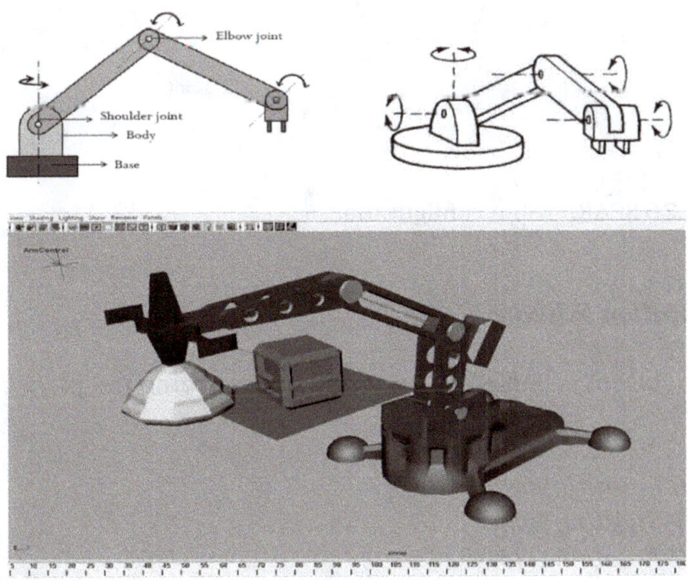

Figure 2.25 Jointed arm configuration.

Work Envelope of Jointed-Arm Configuration

- It is comparable to the configuration of a human arm.
- It comprises of a vertical column that swivels (rotation) around the base utilizing a T-joint.
- Shoulder joint (R-joint) is situated at the top of the column.
- The output connection is an elbow joint (another R joint) (another R joint).

SCARA Robot

SCARA is a specific kind of array of joints. It means a selective compliance automatic robot arm (or) articulated robot arm for selective compliance. It is similar to the jointed-arm, except that the vertical axis is used in vertical insertion procedures for shoulder and elbow joints.

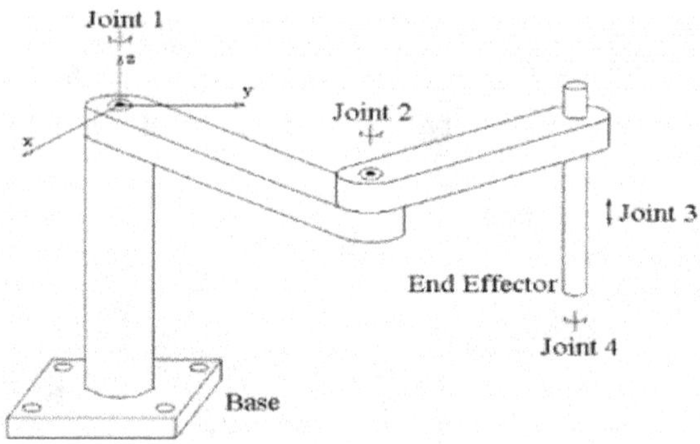

Figure 2.26 SCARA robot configuration.

Joint Notation Scheme

A joint notation scheme can be used for describing the physical configuration.

- *L-Linear Joint*
- *T-Twisting Joint*
- *R-Rotational Joint*
- *V-Revolving Joint*

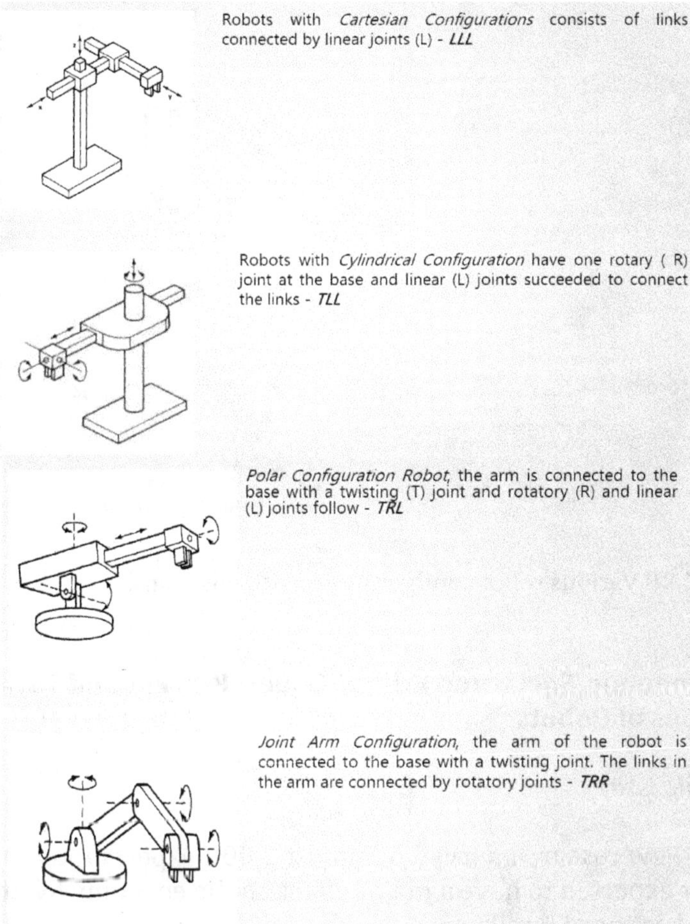

Figure 1.27 Joint notation schemes.

Table 2.3 Robot configuration with notations [2]

Robot Configuration (Arm and Body)	Symbol
Polar Configuration	TRL
Cylindrical Configuration	TLL, LTL, LVL
Cartesian Configuration	LLL
Joint Arm Configuration	TRR, VVR

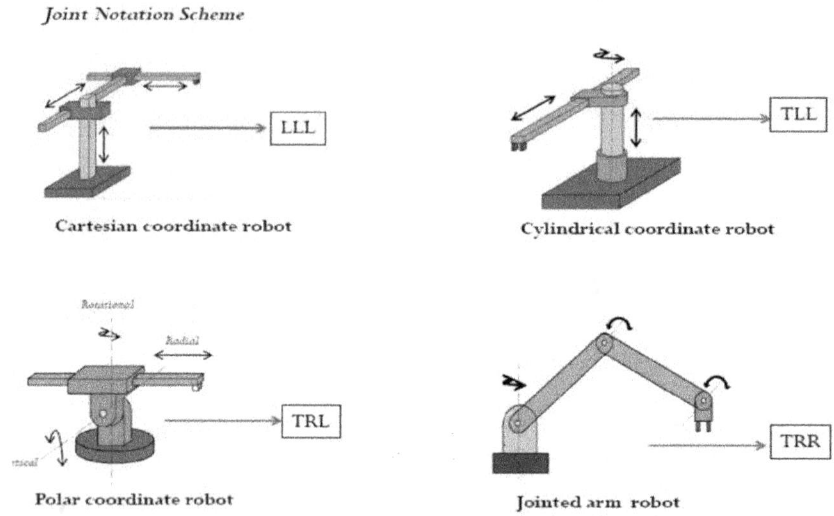

Figure 2.28 Various robot configurations with notations.

2.11 Economic, Social and Ethical Issues; Present and Future Applications of Robots

Economic Issues

A 2014 Pew research survey of almost 2,000 experts found that 48 per cent expected to have a major effect on the economy by 2025 robots and artificial intelligence.

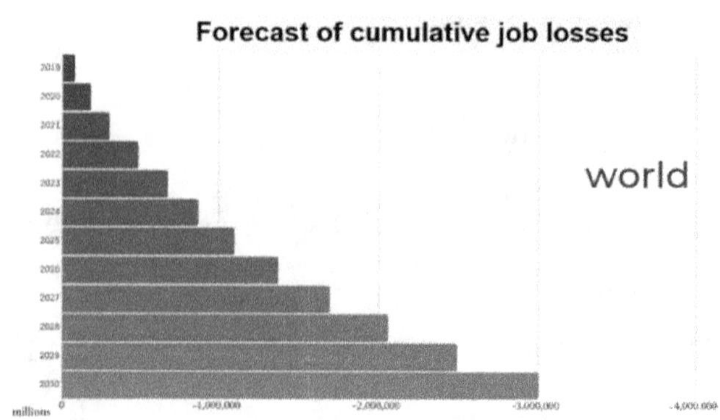

Figure 2.29 New industrial robot installations (thousands).

Positive Impacts

Robots take your jobs. But for decades they have been infringing on production activities and now literally enter into jobs like driving, transportation and inventory management. While some labor sectors may be negatively affected, robotics and automation can improve productivity, reduce manufacturing costs and generate new employment in the technology industry. Some things to note about the beneficial effects on humans are mentioned like the increase in productivity of machines and growth of the gross domestic product.

Ethical Issues

Ethics

Good and evil science. In the field of ethics it is necessary to identify what is good or evil. The aim of ethics is to explain how human beings should be compared, act and be with and around others.

Legal

Under law, not in breach of law or anything (Dictionary of law), the "proposal for a green paper" is the document that provides most of the definitions of law problems.

Since the ideas of robots and robotics have been born, it is obvious that a machine capable of carrying out activities in close touch with people and involving them directly has to be limited by a specific set of ethical, legal and social norms. One of the most notable efforts to determine what the robot needs to follow is the creation of the famous three robotic laws of Asimov.

Steps to Overcome the Issues

A strong legal framework, clearly outlining the duties is essential required (e.g., conduct and priority in unexpected circumstances, what occurs with damage inflicted by the robot to human) (e.g., behaviour and priorities in unforeseen situations, what happens regarding injuries caused by the robot to human). It essential to establish the appropriate design principles based on a shared and responsible vision of what the robotic society should be for humans. The ethical problem surrounding the teaching of young people to utilize robots. Personal Robots & Robot Companions: "Insincere" emotional connection; Replacement of human social interaction by robots. Personal Robots: Replacement of nursing assistance, professions. Similarly, need to portray the robot as being complimentary to the human help. Robot Companions: Fake impression of reality of human interaction causing loneliness. Autonomous cars: Safety; Certification of vehicles; Insurance; Black box to recorder data. Security: Can a robot injure / harm a human (even a criminal)? Security Behaviour and priorities in unexpected circumstance. Security: Data protection for surveillance data; Alert or intervention choices; Dual usage of technology in systems with offensive capacity

2.12 Present and Future Applications of Robots

In the near future, many of these will be industrial robots. However, robots and autonomous systems are gradually expected to have widespread exploitation in society in the future including self-driving vehicles and service robots at work and at home. The tricky question to answer is how soon we will see a shift [6].

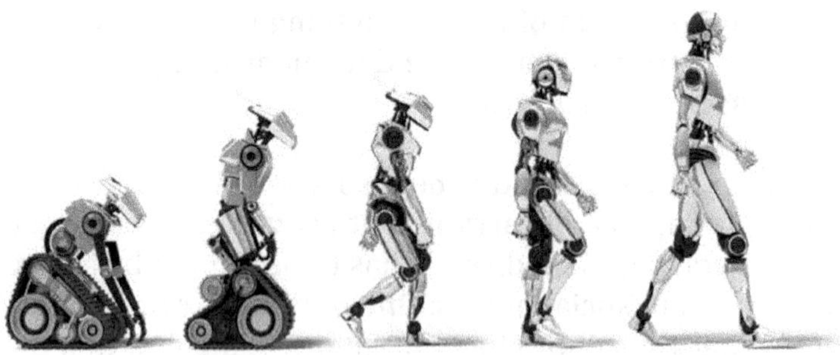

Figure 2.30 Evolution of robots.

2.13 Applications of Robots

The technologies around us take many forms and have different levels of development and impact on our lives. A gross categorization might be as follows:

Industrial robots: these have been around for many years and have had an enormous impact on production. It is preprogrammed mainly by a human instructor and consists of a robot arm with a certain degree of freedom.

Service robots: a robot that functions semi- or totally autonomously for useful tasks for human beings or equipment but excludes industrial automation applications. They are currently used in selected environments such as hospital in-house transport, lawn mowing and vacuum cleaning.

Artificial intelligence: software capable of adapting technology through learning so that systems are able to understand, understand and act in the best way possible. In recent years, the deployment of artificial intelligence has increased significantly in a number of business areas, including customer service and decision support.

Robot movement between the past and the future: The technological transition from industrial robots to service robots represents an evolution into more personalized and increasingly autonomous systems. This means flexible robots which are capable of performing tasks in an uncontrolled.

2.14 Basic Mathematics used in Robotics

Linear Algebra (History):

The origin of matrices and determinants dates back to the second century BC, although traces date from the fourth century BC. However, the concepts did not reach popular math until the end of the 16th century. About 300 BC the Babylonians explored issues leading to simultaneous linear equations. The Chinese got considerably closer to the matrices than the Babylonians, between 200 BC and 100 BC. The earliest known example of matrix techniques is in fact given in the Nine Chapters on Mathematical Art published during the Han dynasty. Cardano studied two-by-two determinants in Europe towards the end of the 16th century, whereas Leibniz considered bigger determinants and Seki investigated them in Japan about 100 years later.

What is a Matrix?

A matrix is a group of elements arranged in columns and rows. In other words, a matrix is an ordered rectangular array of numbers, arranged in rows and columns.

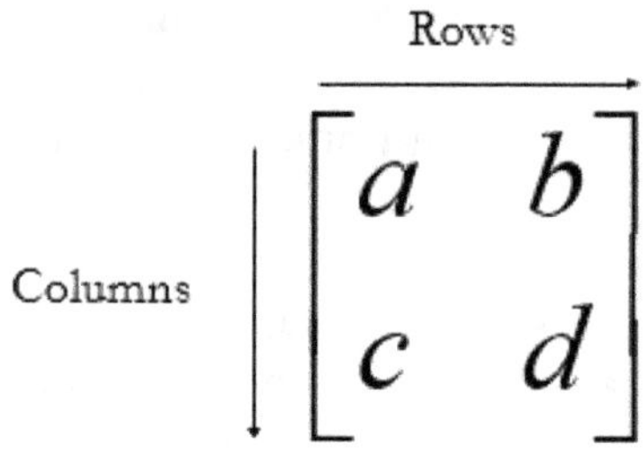

The diagonal elements are a and d. The off-diagonal elements are b and c. Matrices are in many respects like simple numbers: they may be added, subtracted, multiplied and reversed in certain instances (divided) [11].

General Form of Matrix

$$\boldsymbol{A} = \begin{bmatrix} a_{11} & \cdots & a_{1n} \\ a_{21} & \cdots & a_{2n} \\ \cdots\cdots\cdots & \cdots\cdots\cdots & \cdots\cdots\cdots \\ a_{m1} & \cdots & a_{mn} \end{bmatrix} = (a_{ij})_{mn}$$

Order of a Matrix

The matrix's size or order is specified by its number of rows and column counts. If a matrix, A, contains m rows and n columns, A is a m*n matrix. The integers are termed their elements in a matrix. The element in the ith row and j^{th} column of a matrix is generally denoted by a^{ij}.

A matrix with m rows and n columns is written $(a_{ij})_{mn}$ or (a_{ij}).

Row Matrix

- A matrix with just one row is called a row matrix (or row vector).

$$A = [a_1 a_2 .., a_n] = a_j \; (1 \times n)$$

Column Matrix

- A matrix with just one column is called a column matrix.

$$A = \begin{bmatrix} a_1 \\ a_2 \\ \cdots \\ a_m \end{bmatrix} = a_i \; (m \times 1)$$

Matrices of the Same Order

Two matrices which have the same number of rows and columns are said to be matrices of the same order.

Equal Matrices

Two matrices $A = (a_{ij})$ and $B = (b_{ij})$ are said to be equal if, and only if, each element a_{ij} of A is equal to the corresponding element b_{ij} of B.

In symbolic form, this reads:

$$A = B \leftrightarrow a_{ij} = b_{ij} \text{ for all i andj}$$

From this it follows that equal matrices are of the same order, but matrices of the same order are not necessarily equal [9].

Null matrix

Any matrix, all of whose elements are zero, is called a null or zero matrix and is denoted by O.

$$[0], \begin{bmatrix} 0 & 0 \\ 0 & 0 \end{bmatrix}, \begin{bmatrix} 0 & 0 & 0 \\ 0 & 0 & 0 \\ 0 & 0 & 0 \end{bmatrix}, \begin{bmatrix} 0 & 0 & 0 & 0 \\ 0 & 0 & 0 & 0 \\ 0 & 0 & 0 & 0 \\ 0 & 0 & 0 & 0 \end{bmatrix}$$

Matrix Addition

A new matrix C may be defined as the additive combination of matrices A and B,

where: C=A+B is defined by:

$c_{ij}=a_{ij}+b_{ij}$

Note: all three matrices are of the same dimensions.

Addition:

If $\quad A = \begin{bmatrix} a_{11} & a_{12} \\ a_{21} & a_{22} \end{bmatrix}$

and $\quad B = \begin{bmatrix} b_{11} & b_{12} \\ b_{21} & b_{22} \end{bmatrix}$

then $\quad C = \begin{bmatrix} a_{11}+b_{11} & a_{12}+b_{12} \\ a_{21}+b_{21} & a_{22}+b_{22} \end{bmatrix}$

Matrix Addition Example

$$A + B = \begin{bmatrix} 3 & 4 \\ 5 & 6 \end{bmatrix} + \begin{bmatrix} 1 & 2 \\ 3 & 4 \end{bmatrix} = \begin{bmatrix} 4 & 6 \\ 8 & 10 \end{bmatrix} = C$$

Multiplication by a Scalar

- If A is a given matrix and λ a scalar then λA is the matrix each of whose elements is λ times the corresponding element of A.

Identity Matrix

- Square matrix with ones on the diagonal and zeros elsewhere.

$$I = \begin{bmatrix} 1 & 0 & 0 & 0 \\ 0 & 1 & 0 & 0 \\ 0 & 0 & 1 & 0 \\ 0 & 0 & 0 & 1 \end{bmatrix}$$

Transpose Matrix

- Rows become columns and columns become rows

$$A' = \begin{bmatrix} a_{11} & a_{21} & ,\ldots, & a_{m1} \\ a_{12} & a_{22} & ,\ldots, & a_{m2} \\ \ldots & \ldots & \ldots & \ldots \\ a_{1n} & a_{2n} & ,\ldots, & a_{mn} \end{bmatrix}$$

Square Matrix

- Same number of rows and columns

$$B = \begin{bmatrix} 5 & 4 & 7 \\ 3 & 6 & 1 \\ 2 & 1 & 3 \end{bmatrix}$$

Matrix Subtraction

C = A - B

is defined by

$$\{C_{ij}\} = \{A_{ij}\} - \{B_{ij}\}$$

Matrix Multiplication

- Let A and B be two matrices. If the number of columns in A is equal to the number of rows in B we say that A and B are conformable for the matrix product AB.
- If A is order $m \times n$ and B is of order $n \times p$, then the product AB is defined and is a matrix of order $m \times p$.

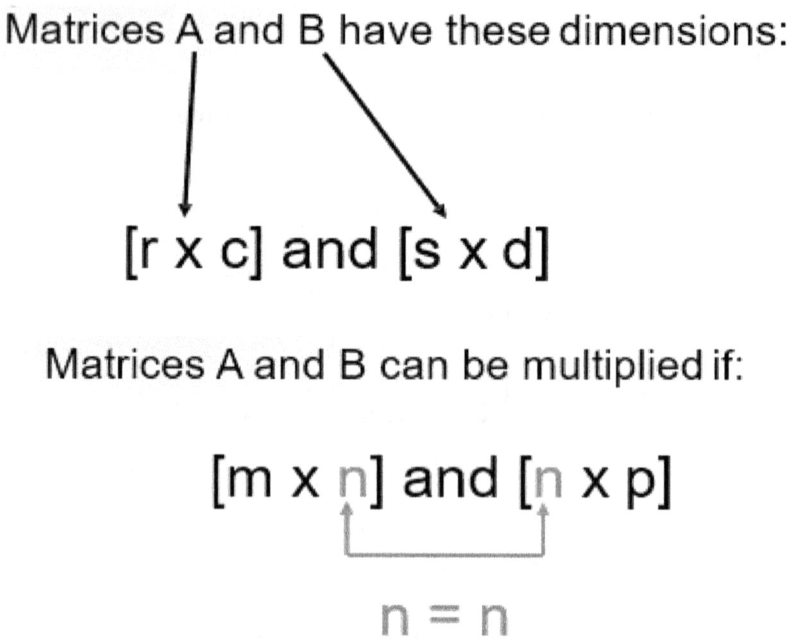

The resulting matrix will have the dimensions:

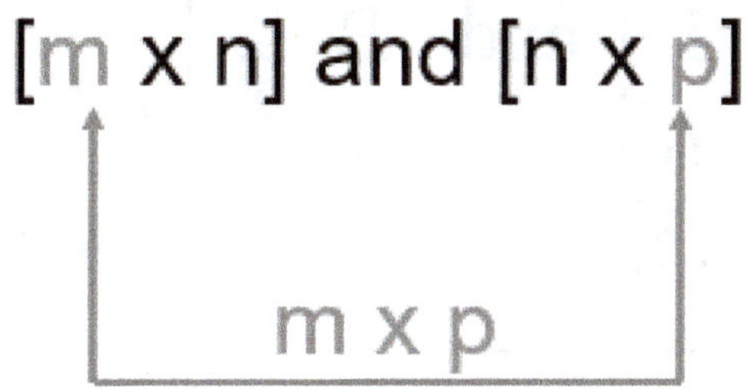

Computation: A x B = C

$$A = \begin{bmatrix} a_{11} & a_{12} \\ a_{21} & a_{22} \end{bmatrix}_{[2 \times 2]}$$

$$B = \begin{bmatrix} b_{11} & b_{12} & b_{13} \\ b_{21} & b_{22} & b_{23} \end{bmatrix}_{[2 \times 3]}$$

$$C = \begin{bmatrix} a_{11}b_{11} + a_{12}b_{21} & a_{11}b_{12} + a_{12}b_{22} & a_{11}b_{13} + a_{12}b_{23} \\ a_{21}b_{11} + a_{22}b_{21} & a_{21}b_{12} + a_{22}b_{22} & a_{21}b_{13} + a_{22}b_{23} \end{bmatrix}_{[2 \times 3]}$$

$$A = \begin{bmatrix} 2 & 3 \\ 1 & 1 \\ 1 & 0 \end{bmatrix}_{[3 \times 2]} \text{ and } B = \begin{bmatrix} 1 & 1 & 1 \\ 1 & 0 & 2 \end{bmatrix}_{[2 \times 3]}$$

A and B can be multiplied

$$C = \begin{bmatrix} 2*1+3*1=5 & 2*1+3*0=2 & 2*1+3*2=8 \\ 1*1+1*1=2 & 1*1+1*0=1 & 1*1+1*2=3 \\ 1*1+0*1=1 & 1*1+0*0=1 & 1*1+0*2=1 \end{bmatrix} = \begin{bmatrix} 5 & 2 & 8 \\ 2 & 1 & 3 \\ 1 & 1 & 1 \end{bmatrix}$$

[3 x 3]

$$A = \begin{bmatrix} 2 & 3 \\ 1 & 1 \\ 1 & 0 \end{bmatrix} \text{ and } B = \begin{bmatrix} 1 & 1 & 1 \\ 1 & 0 & 2 \end{bmatrix}$$

[3 x 2] [2 x 3]

Result is 3 x 3

$$C = \begin{bmatrix} 2*1+3*1=5 & 2*1+3*0=2 & 2*1+3*2=8 \\ 1*1+1*1=2 & 1*1+1*0=1 & 1*1+1*2=3 \\ 1*1+0*1=1 & 1*1+0*0=1 & 1*1+0*2=1 \end{bmatrix} = \begin{bmatrix} 528 \\ 213 \\ 111 \end{bmatrix}$$

[3 x 3]

Note: If A is an m × n and B is n × p matrix, then AB is an m × p matrix. Hence, we see that BA is defined only when p=m [10].

The Inverse of a Matrix

Definition:

Let A be a square matrix. A matrix B such that AB=I=BA is called the inverse matrix of A and is denoted by A^{-1}.

So, if A^{-1} exists, we have $AA^{-1}=I=A^{-1}A$ and the matrix is said to be invertible.

If a matrix has no inverse, then it is said to be non-invertible.

$$A^{-1}A = AA^{-1} = I$$

Like a reciprocal in scalar math

Like the number one in scalar math

2.15 What is Mapping?

Mapping is altering the description of a space point (or vector) from frame to frame. The second frame has three options for the first frame, i.e.,

a) The second frame is rotated w.r.t. The first frame is rotated, its origin is identical.
b) The second frame is shifted from the first, respectively the axes of both frames stay parallel.
c) The first and moved from the second frame is rotated w.r.t. i.e., the second frame is translated and oriented as well.

The next sections mimic these scenarios. It should be noted that Mapping affects the point and not the point itself description.

Mappings Involving Translated Frames

✓ We have a position defined by the vector ^{B}P (a point P described in coordinate frame {B}).
✓ We have to express this point in space in terms of coordinate frame {A}, when {A} has the same orientation as {B}.
✓ In this case, {B} differs from {A} only by a translation which is given by $^{A}P_{BORG}$ (a vector which locates the origin of {B} relative to {A}) [3].
✓ The description of point P, ^{B}P, relative to coordinate frame {A}, ^{A}P, is determined by simple vector addition, i.e.,

- $^{A}P = {}^{B}P + {}^{A}P_{BORG}$
- Alternatively, the description of point P, ^{B}P, relative to coordinate frame {A}, ^{A}P, can also be determined by using a simple transformation matrix, i.e.,
  $^{A}P = T.{}^{B}P$
  where, T is the Translation Matrix which is given by

$$T = \begin{matrix} 1 & 0 & 0 & tx \\ 0 & 1 & 0 & ty \\ 0 & 0 & 1 & tz \\ 0 & 0 & 0 & 1 \end{matrix}$$

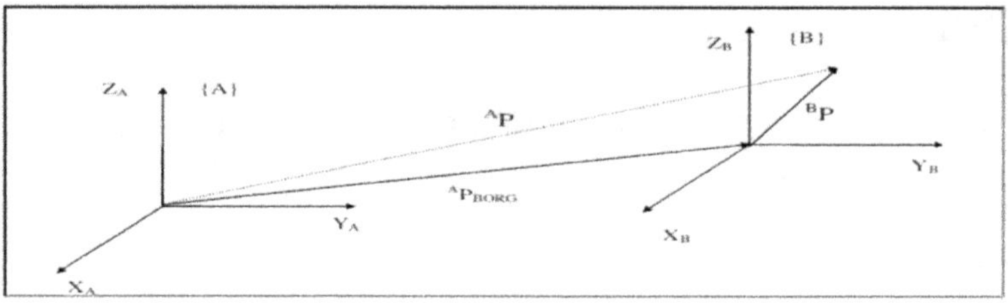

Figure 2.31 Mappings involving translated frames [3].

Mappings Involving Rotated Frames

✓ The origins of the two coordinate frames {A} and {B} are coincident but they have a different orientation (rotated by a certain angle).
✓ The description of a point P in coordinate frame {B}, ^{B}P, relative to coordinate frame {A}, AP, can be determined by using a simple transformation matrix, i.e.,

- $^{A}P = R.^{B}P$
  where, R is the Rotation Matrix which is given by

$$R = \begin{bmatrix} -sin\gamma & 0 & cos\gamma & 0 \\ 0 & 1 & 0 & 0 \\ cos\gamma & 0 & sin\gamma & 0 \\ 0 & 0 & 0 & 1 \end{bmatrix} \begin{bmatrix} 1 & 0 & 0 & 0 \\ 0 & cos\beta & -sin\beta & 0 \\ 0 & sin\beta & cos\beta & 0 \\ 0 & 0 & 0 & 1 \end{bmatrix} \begin{bmatrix} cos\alpha & -sin\alpha & 0 & 0 \\ -sin\alpha & cos\alpha & 0 & 0 \\ 0 & 0 & 1 & 0 \\ 0 & 0 & 0 & 1 \end{bmatrix}$$

Mappings Involving Rotated Frames

$$\text{Where, } R(\boldsymbol{\alpha}) = \begin{bmatrix} cos\alpha & -sin\alpha & 0 & 0 \\ -sin\alpha & cos\alpha & 0 & 0 \\ 0 & 0 & 1 & 0 \\ 0 & 0 & 0 & 1 \end{bmatrix}$$

$$R(\boldsymbol{\beta}) = \begin{bmatrix} 1 & 0 & 0 & 0 \\ 0 & cos\beta & -sin\beta & 0 \\ 0 & sin\beta & cos\beta & 0 \\ 0 & 0 & 0 & 1 \end{bmatrix}$$

$$\text{Thus, } R = R(\boldsymbol{\gamma}).R(\boldsymbol{\beta}).R(\boldsymbol{\alpha})$$

$$R(\boldsymbol{\gamma}) = \begin{bmatrix} -sin\gamma & 0 & cos\gamma & 0 \\ 0 & 1 & 0 & 0 \\ cos\gamma & 0 & sin\gamma & 0 \\ 0 & 0 & 0 & 1 \end{bmatrix}$$

Mappings Involving Rotated Frames

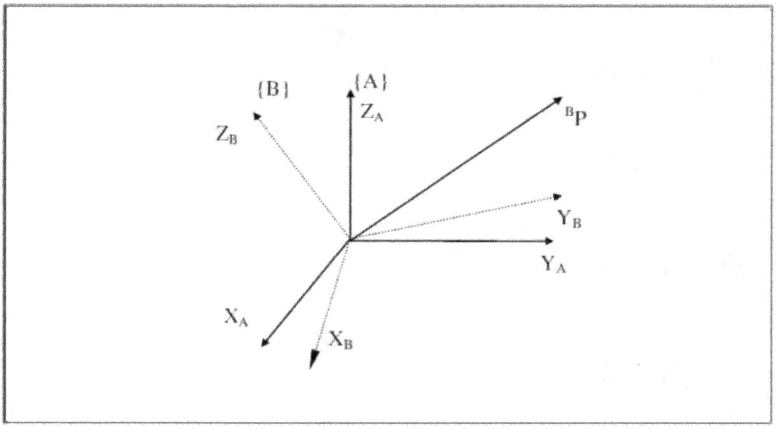

Figure 2.32 Mappings involving rotated frames.

Mappings Involving General Frames

✓ We will explain the mapping involving general frames by taking the origin of frame {B}, located at a distance of vector called ${}^{A}P_{BORG}$ from frame {A}; and {A} is rotated with respect to {B} as described by R [5].
✓ The following equation describes the mapping of a vector from its description in one frame to a description in a second frame, which is given by
${}^{A}P = R.{}^{B}P + {}^{A}P_{BORG}$
✓ The above equation can also be implemented in a matrix form by using a composite transformation matrix, i.e.,
${}^{A}P = C.{}^{B}P$
where, C is the Composite Transformation Matrix which is given by
$C = R.T = R(\boldsymbol{\gamma}).R(\boldsymbol{\beta}).R(\boldsymbol{\alpha}).T$

So,

So,

$$C = \begin{bmatrix} -sin\gamma & 0 & cos\gamma & 0 \\ 0 & 1 & 0 & 0 \\ cos\gamma & 0 & sin\gamma & 0 \\ 0 & 0 & 0 & 1 \end{bmatrix} \begin{bmatrix} 1 & 0 & 0 & 0 \\ 0 & cos\beta & -sin\beta & 0 \\ 0 & sin\beta & cos\beta & 0 \\ 0 & 0 & 0 & 1 \end{bmatrix} \begin{bmatrix} cos\alpha & -sin\alpha & 0 & 0 \\ -sin\alpha & cos\alpha & 0 & 0 \\ 0 & 0 & 1 & 0 \\ 0 & 0 & 0 & 1 \end{bmatrix} \begin{bmatrix} 1 & 0 & 0 & tx \\ 0 & 1 & 0 & ty \\ 0 & 0 & 1 & tz \\ 0 & 0 & 0 & 1 \end{bmatrix}$$

Thus,

$${}^{A}P = R.T.{}^{B}P = R(\boldsymbol{\gamma}).R(\boldsymbol{\beta}).R(\boldsymbol{\alpha}).T.{}^{B}P$$

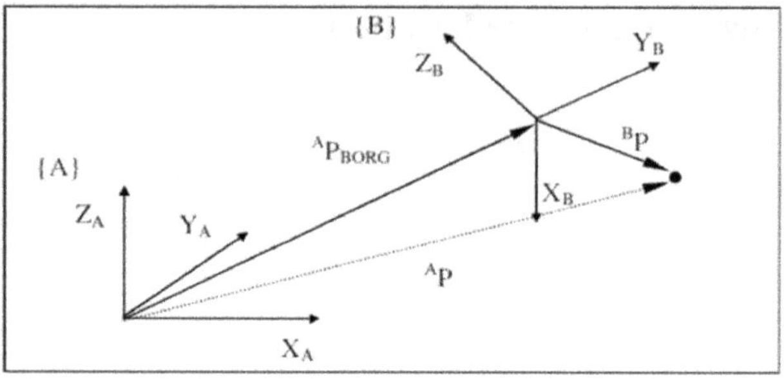

Figure 2.33 Mappings involving general frames.

Homogeneous Transformation Matrix

- The homogeneous transformation matrix is used to define both the position and the position of the space coordinate frames.
- The Homogeneous Transformation matrix is a four x four matrix, which carries an object specified in a homogenous coordinative system and may be seen as two sub-matrices, that is, a translation matrix and a rotation matrix.
- A homogeneous matrix for transformation may be seen as a composite matrix in a homogenous coordinate system [5].

References

1. Maheu, V., Archambault, P. S., Frappier, J., & Routhier, F. (2011, June). Evaluation of the JACO robotic arm: Clinico-economic study for powered wheelchair users with upper-extremity disabilities. In 2011 IEEE International Conference on Rehabilitation Robotics (pp. 1-5). IEEE.
2. Kimura, T. (2017). Robotics and AI in the sociology of religion: A human in imago roboticae. Social Compass, 64(1), 6-22.
3. Savsani, P., Jhala, R. L., & Savsani, V. J. (2014). Comparative study of different metaheuristics for the trajectory planning of a robotic arm. IEEE Systems Journal, 10(2), 697-708.
4. Liarokapis, M. V., Artemiadis, P. K., Katsiaris, P. T., Kyriakopoulos, K. J., & Manolakos, E. S. (2012, May). Learning human reach-to-grasp strategies: Towards EMG-based control of robotic arm-hand systems. In 2012 IEEE International Conference on Robotics and Automation (pp. 2287-2292). IEEE.

5. Roshanianfard, A., Noguchi, N., & Kamata, T. (2019). Design and performance of a robotic arm for farm use. International Journal of Agricultural and Biological Engineering, 12(1), 146-158.
6. Kostic, D., De Jager, B., Steinbuch, M., & Hensen, R. (2004). Modeling and identification for high-performance robot control: An RRR-robotic arm case study. IEEE Transactions on Control Systems Technology, 12(6), 904-919.
7. Awang, M. S., & Abdullah, M. Z. (2011). Robotic neurosurgery: a preliminary study using an active vision-guided robotic arm for bone drilling and endoscopic manoeuvres. The Malaysian Journal of Medical Sciences: MJMS, 18(2), 53.
8. Ligutan, D. D., Abad, A. C., & Dadios, E. P. (2018). Adaptive robotic arm control using artificial neural network. In 2018 IEEE 10th International Conference on Humanoid, Nanotechnology, Information Technology, Communication and Control, Environment and Management (HNICEM) (pp. 1-6). IEEE.
9. Bhupender, B., & Rahul, R. (2016). Study and Analysis of Design Optimization and Synthesis of Robotic ARM. International Journal of Advanced Engineering, Management and Science, 2(5), 239459.
10. Li, X., Lian, Q., Li, D., Xin, H., & Jia, S. (2017). Development of a robotic arm-based hydrogel additive manufacturing system for in-situ printing. Applied Sciences, 7(1), 73.
11. Moro, C., Lin, S., Nejat, G., & Mihailidis, A. (2019). Social robots and seniors: a comparative study on the influence of dynamic social features on human–robot interaction. International Journal of Social Robotics, 11(1), 5-24.
12. Yakub, F., Khudzari, A. Z. M., & Mori, Y. (2014). Recent trends for practical rehabilitation robotics, current challenges and the future. International Journal of Rehabilitation Research, 37(1), 9-21.
13. Danaher, J., & McArthur, N. (Eds.). (2017). Robot sex: Social and ethical implications. MIT Press.
14. Lin, P., Abney, K., & Bekey, G. (2011). Robot ethics: Mapping the issues for a mechanized world. Artificial Intelligence, 175(5-6), 942-949.
15. Yang, G. Z., Bellingham, J., Dupont, P. E., Fischer, P., Floridi, L., Full, R., & Wood, R. (2018). The grand challenges of Science Robotics. Science robotics, 3(14), eaar7650.
16. Lin, P., Abney, K., & Bekey, G. A. (Eds.). (2012). Robot ethics: the ethical and social implications of robotics. Intelligent Robotics and Autonomous Agents series.
17. Veruggio, G., Operto, F., & Bekey, G. (2016). Roboethics: Social and ethical implications. In Springer handbook of robotics (pp. 2135-2160). Springer, Cham.

3

Kinematic Modelling of Robotic Manipulator using Different Methods

Sasmita Nayak[1], Dipti Acharjya[2], Biswajit Parida[3], Abhishek Barua[4], and Deepak Kumar Bagal[5]

[1]*Department of Mechanical Engineering, Government College of Engineering, Kalahandi, Bhawanipatna, Odisha, India*
[2]*Department of Mechanical Engineering, College of Engineering and Technology, Bhubaneswar, Odisha, India*
[3]*Department of Mechanical Engineering, Centre for Advanced Post Graduate Studies, BPUT, Rourkela, Odisha, India*
[4]*Department of Mechanical Engineering, Centre for Advanced Post Graduate Studies, BPUT, Rourkela, Odisha, India*
[5]*Department of Electrical Engineering, VSSUT, Burla, Odisha, India*

3.1 Mechanical Structure and Notations

Manipulator

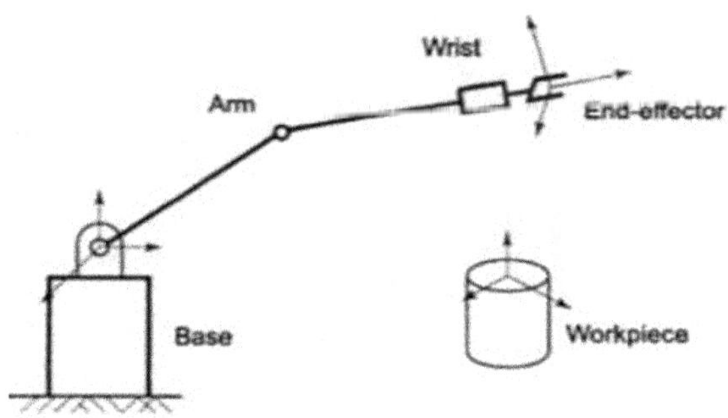

Figure 3.1 A manipulator.

A manipulator's mechanical structure is comprised of the base, arm, wrist, and end-effector components. Its mechanical structure is made up of rigid bodies (links) that are linked together by joints. It is divided into three parts: an arm that provides movement and reachability, a wrist that provides orientation, and an end-effector that carries out the necessary job [1].

Links and Joints

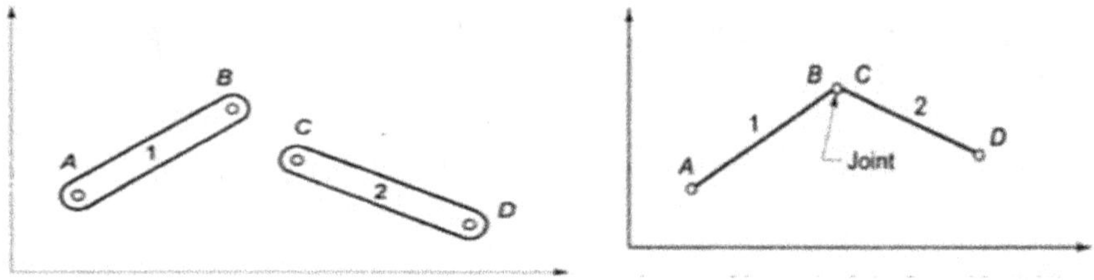

Figure 3.2 Difference between link and joint.

Figure 3.3 Types of motion of link.

Degrees of Freedom

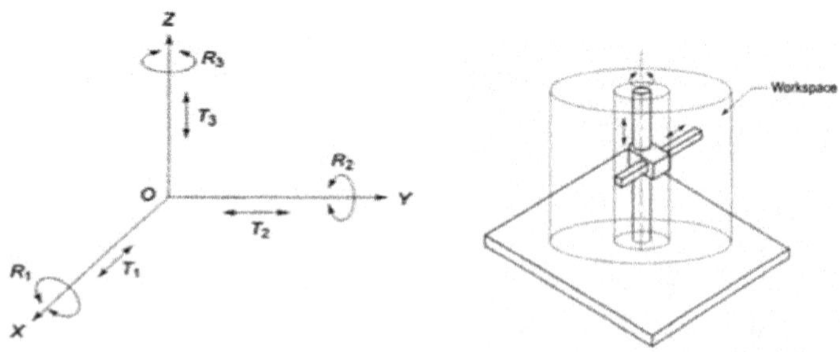

Figure 3.4 DoF of Robotic system.

The maximum number of separate motions that an item may make in three-dimensional space. A rigid body free in space has six degrees of freedom - three translations (T1, T2, T3), which indicate linear movements along three perpendicular axes, define the location of the body in space, and three rotations (T1, T2, T3), which represent rotations around three perpendicular axes. The orientation of the body in space is specified by three rotations (R1, R2, and R3), which indicate angular movements about three axes and represent angular motions about three axes [2-20].

IMPORTANT: The degree of freedom of a kinematic chain is equal to the number of joints in the chain, unless otherwise stated [19].

Frames

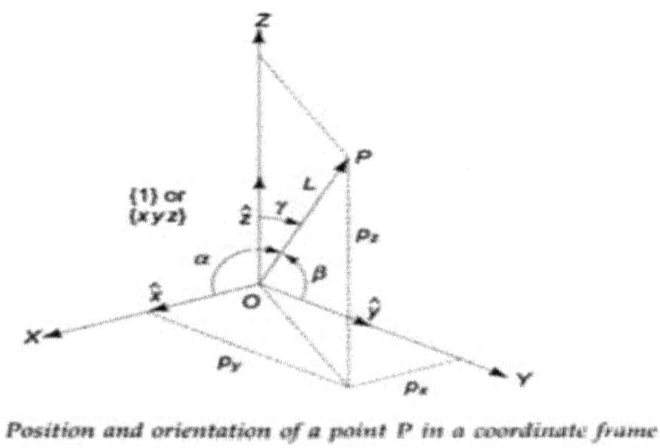

Figure 3.5 Position vector of point P.

The following is the vector representing the location of point P in the coordinate system [16]:

$$\vec{P} = \overrightarrow{OP} = p_x\hat{x} + p_y\hat{y} + p_z\hat{z}$$

3.2 Mapping

It is the process of transferring the description of a point in space from one frame to another frame from one frame to another frame. It is possible to choose between three options:

1. Mapping between translated frames
2. Mapping between rotated frames
3. Mapping between rotated and translated frames

Mapping between Translated Frames

O1P =O1O2 + O2P

=>1P = 2P+ 1D2

$$^{1}\boldsymbol{P} = \begin{bmatrix} 1 & 0 & 0 & d_x \\ 0 & 1 & 0 & d_y \\ 0 & 0 & 1 & d_z \\ 0 & 0 & 0 & 1 \end{bmatrix} \begin{bmatrix} ^{2}p_u \\ ^{2}p_v \\ ^{2}p_w \\ 1 \end{bmatrix}$$

$$^{1}\boldsymbol{P} = {}^{1}\boldsymbol{T}_2 \, {}^{2}\boldsymbol{P}$$

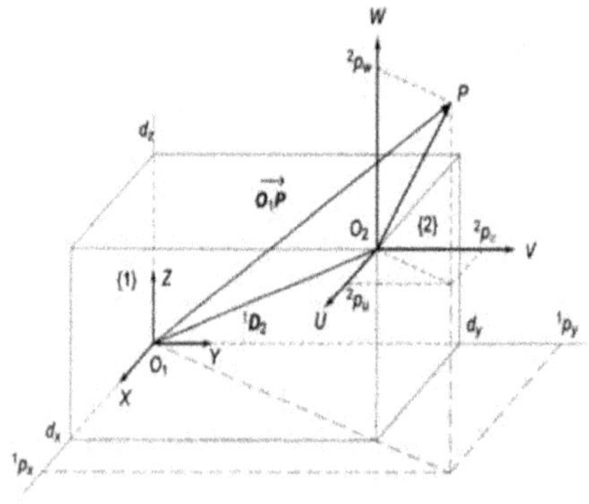

Figure 3.6 Mapping between translated frames.

Mapping between Rotated Frames

$$^{1}p_x = \hat{x} \cdot {}^{2}\boldsymbol{P} = \hat{x} \cdot {}^{2}p_u \hat{u} + \hat{x} \cdot {}^{2}p_v \hat{v} + \hat{x} \cdot {}^{2}p_w \hat{w}$$
$$^{1}p_y = \hat{y} \cdot {}^{2}\boldsymbol{P} = \hat{y} \cdot {}^{2}p_u \hat{u} + \hat{y} \cdot {}^{2}p_v \hat{v} + \hat{y} \cdot {}^{2}p_w \hat{w}$$
$$^{1}p_z = \hat{z} \cdot {}^{2}\boldsymbol{P} = \hat{z} \cdot {}^{2}p_u \hat{u} + \hat{z} \cdot {}^{2}p_v \hat{v} + \hat{z} \cdot {}^{2}p_w \hat{w}$$

This can be written in matrix form as

$$\begin{bmatrix} ^{1}p_x \\ ^{1}p_y \\ ^{1}p_z \end{bmatrix} = \begin{bmatrix} \hat{x}\cdot\hat{u} & \hat{x}\cdot\hat{v} & \hat{x}\cdot\hat{w} \\ \hat{y}\cdot\hat{u} & \hat{y}\cdot\hat{v} & \hat{y}\cdot\hat{w} \\ \hat{z}\cdot\hat{u} & \hat{z}\cdot\hat{v} & \hat{z}\cdot\hat{w} \end{bmatrix} \begin{bmatrix} ^{2}p_x \\ ^{2}p_y \\ ^{2}p_z \end{bmatrix}$$

In compressed vector-matrix notation Eq. (2.8) is written as

$$^{1}\boldsymbol{P} = {}^{1}\boldsymbol{R}_2 \, {}^{2}\boldsymbol{P}$$

where

$$^{1}\boldsymbol{R}_2 = \begin{bmatrix} \hat{x}\cdot\hat{u} & \hat{x}\cdot\hat{v} & \hat{x}\cdot\hat{w} \\ \hat{y}\cdot\hat{u} & \hat{y}\cdot\hat{v} & \hat{y}\cdot\hat{w} \\ \hat{z}\cdot\hat{u} & \hat{z}\cdot\hat{v} & \hat{z}\cdot\hat{w} \end{bmatrix}$$

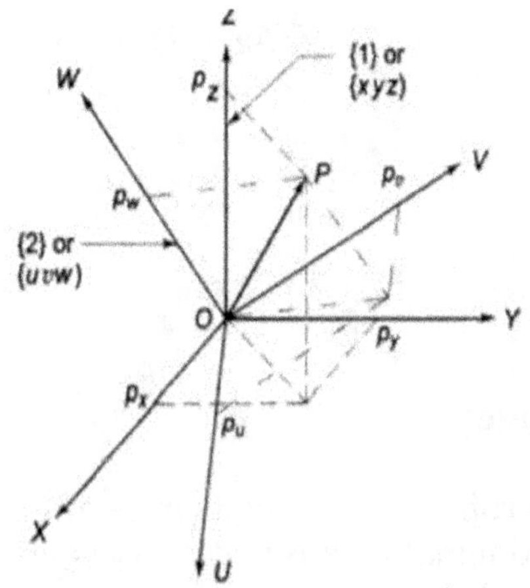

Figure 3.7 Mapping between rotated frames.

Mapping between Rotated and Translated Frames

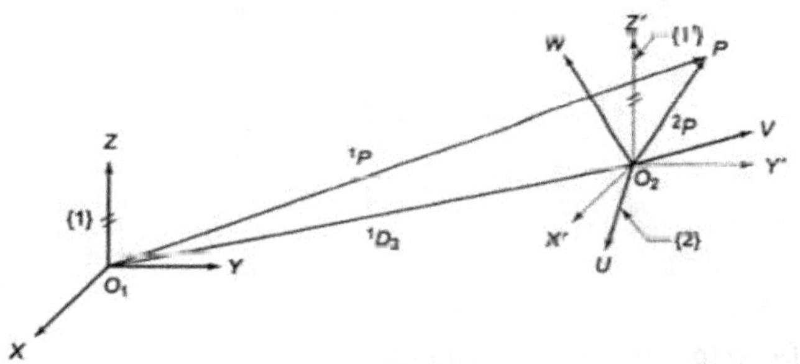

Figure 3.8 Mapping between rotated and translated frames.

$$T = \left[\begin{array}{c|c} \text{Rotation matrix} & \text{Translation vector} \\ (3 \times 3) & (3 \times 1) \\ \hline \text{Perspective} & \text{Scale factor} \\ \text{transformation matrix} & (1 \times 1) \\ (1 \times 3) & \end{array}\right]$$

$$
{}^{1}T_{2} = \begin{bmatrix} \hat{x}.\hat{u} & \hat{x}.\hat{v} & \hat{x}.\hat{w} & d_x \\ \hat{y}.\hat{u} & \hat{y}.\hat{v} & \hat{y}.\hat{w} & d_y \\ \hat{z}.\hat{u} & \hat{z}.\hat{v} & \hat{z}.\hat{w} & d_z \\ 0 & 0 & 0 & 1 \end{bmatrix}
$$

${}^{1}R_{2}$ ${}^{1}D_{2}$

Scale factor σ

$$
{}^{1}P = {}^{1}T_{2}\,{}^{2}P
$$

3.3 Kinematic Model

It is the study of a robot's movements in relation to a reference system that is called kinematics. It is necessary to create a mathematical model of the manipulator in order to programme the tool motion and joint-link motions. The kinematic model is defined as the relationship between the joint-variables and the position and orientation of the end-effector (or joint). The Homogeneous Transformation Matrix (HTM) is used to represent the kinematic model [15].

Joint-Link Parameters

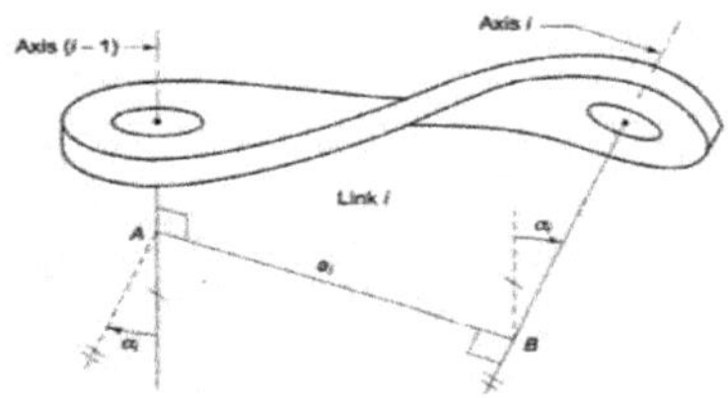

Figure 3.9 Description of link parameters.

Link Parameters: Link Length & Link Twist

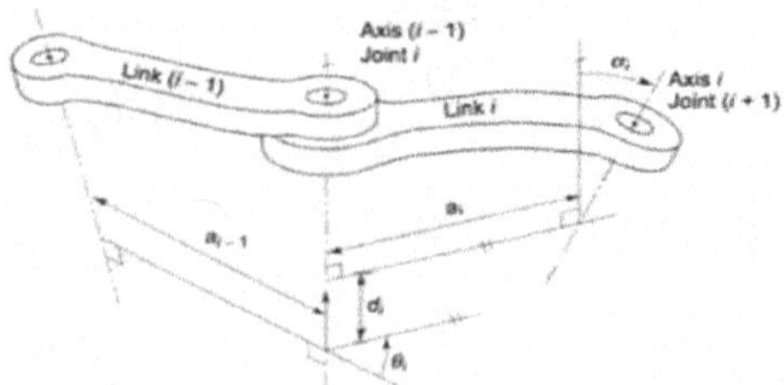

Figure 3.10 Joint-link parameters for joint i and link i [16].

3.4 Direct and Inverse Kinematics

The problem of kinematics modelling can be divided into two parts:

1. Kinematics in direct motion: This model calculates the positions and orientations of the end-effector as a function of the joint-link parameters and is used to model direct motion.
2. Inverse Kinematics: An example of inverse kinematics is when an end-effector is placed in a specific position and orientation, and it is necessary to find a set of joint-variables that will move the end-effector to the specified position and orientation [6, 11, 19].
3.

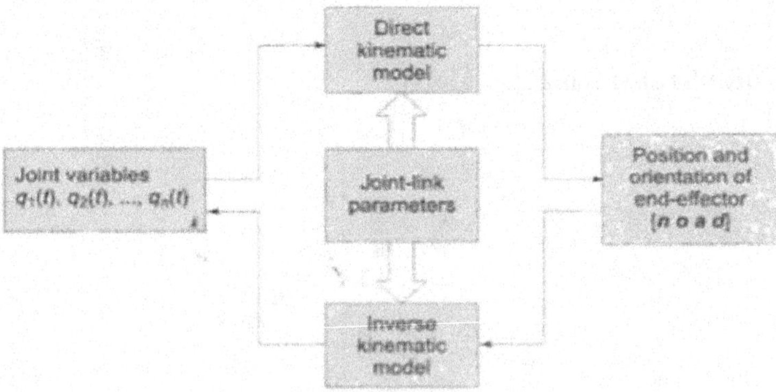

Figure 3.11 Direct and inverse kinematics model.

Denavit-Hartenberg Notation (assigning frames)

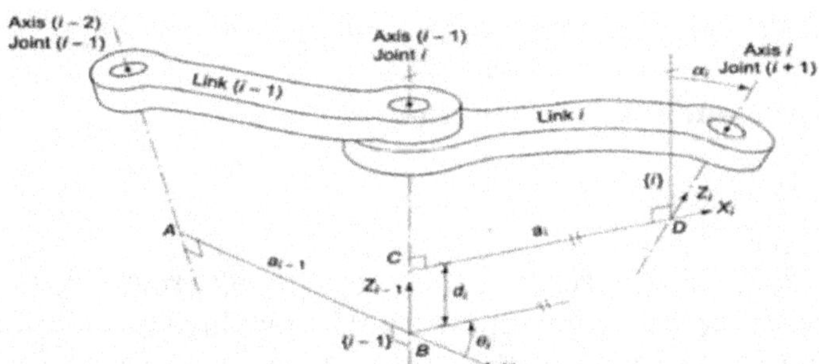

Figure 3.12 DH conversion for assigning frames to links.

Kinematic Relationship b/w Adjacent Links

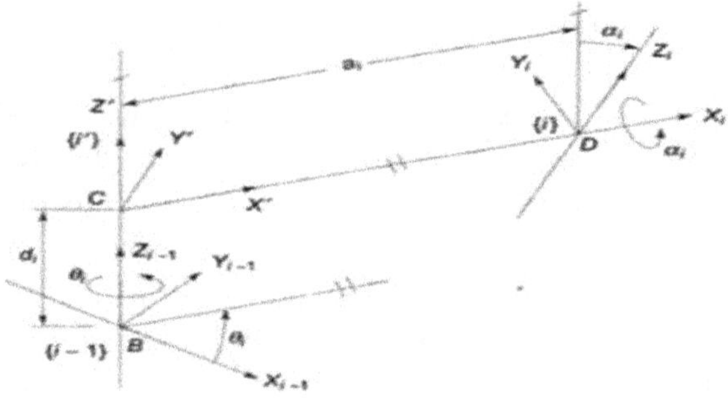

Figure 3.13 Geometric relationship between corresponding links.

Manipulator Transformation Matrix

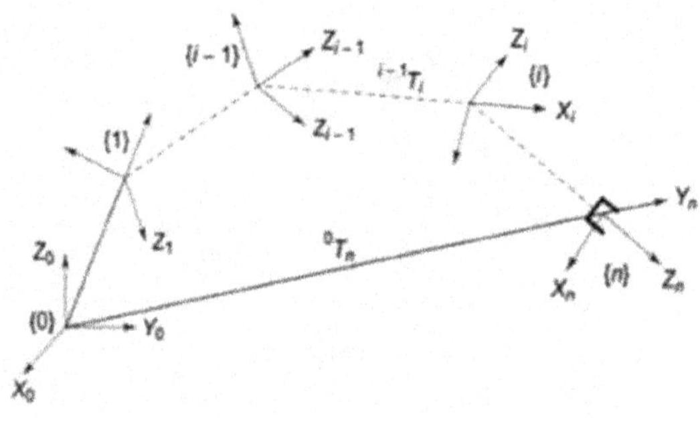

Figure 3.14 Location of end-effector frame relative to base frame.

The above equation is a function of the joint-link parameters. This equation is called the Kinematic model.

3.5 Forward and Inverse Kinematics

What is "Kinematics"?

While studying kinematics, it is important to note that it does not take into account the forces or moments that produce the motion of the bodies. Robot kinematics is the analytical study of the motion of a robot manipulator, also known as a robot manipulator [18].

Types of Kinematics in ROBOTICS

Figure 3.15 Types of kinematics model of robotics.

Kinematics Tasks in Robotics

In manipulator robotics, there are two kinematic tasks that must be completed:

- Direct (or forward) kinematics – The joint relations (rotations and translations) for the robot arm are specified. What is the direction and location of the end effector in relation to the task?
- Inverse (or backward) kinematics – The intended end effector location and orientation are specified. What are the joint rotations and orientations required to accomplish this task.

3.6 Coordinate Frames

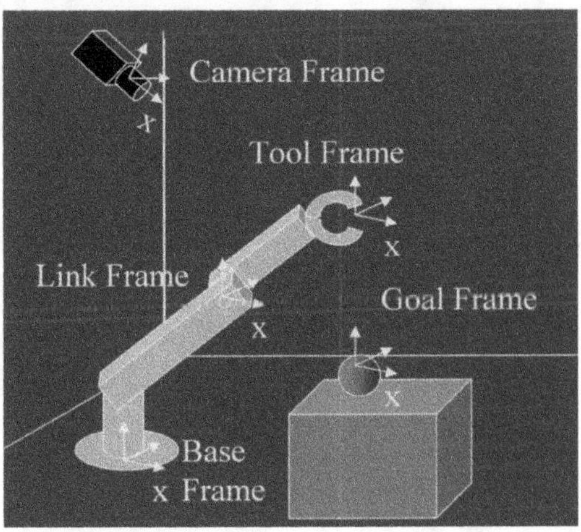

Figure 3.16 Coordinate frame of robot.

Two Basic Joints

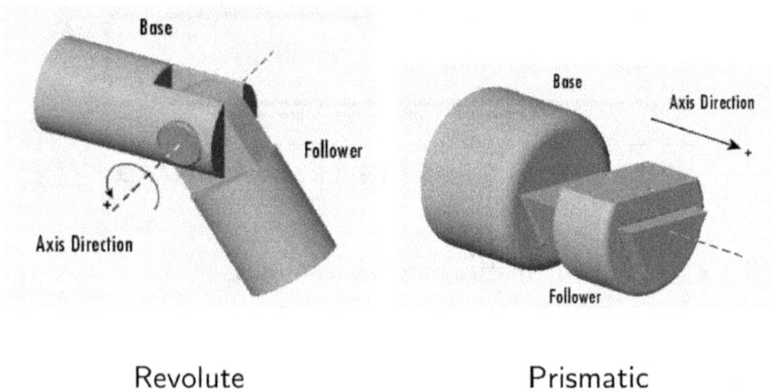

Figure 3.17 Basic revolute and prismatic joints.

Direct Kinematics

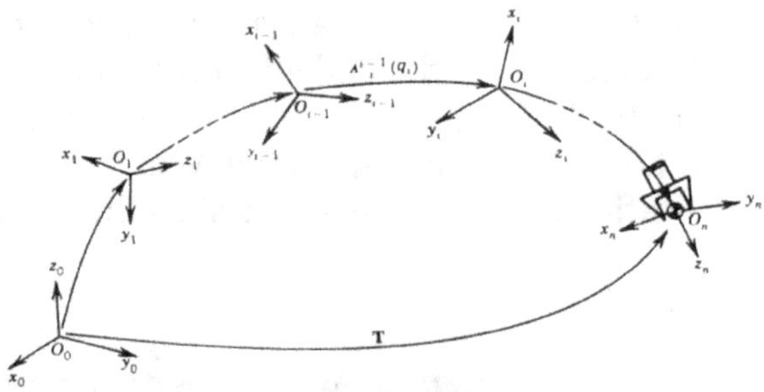

Figure 3.18 Direct kinematic model.

- One joint: $\mathbf{x}_i = A\mathbf{x}_{i-1}$.
- Chain of joints: $\mathbf{x}_{n-1} = A_{n-1}\, A_{n-2}\, \ldots\, A_1\, A_0\, \mathbf{x}_0$.
- Easy to compute (matrix multiplication).
- Unique solution.

Inverse Kinematics

- For a kinematic mechanism, the inverse kinematic problem is difficult to solve.
- The robot controller must solve a set of non-linear simultaneous algebraic equations.
- Source of problems:
    - Non-linear equations ($\sin$, $\cos$ in rotation matrices).
    - The existence of multiple solutions.
    - The possible non-existence of a solution.
    - Singularities.

Example of Kinematics in 2D

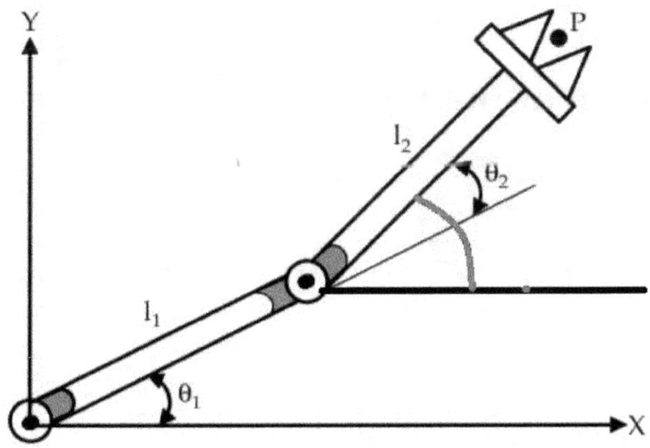

Figure 3.19 Kinematic model of 2D type.

3.7 Difference Between Forward and Backward Kinematics in 2-Dimension

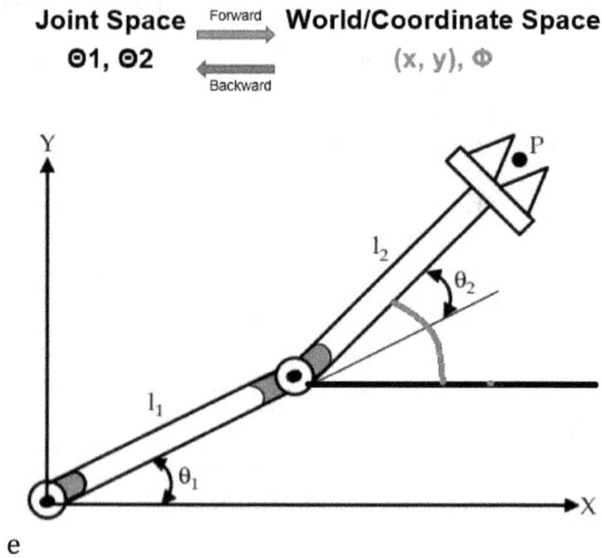

Figure 3.20 Kinematic model of 2D type with notations.

Two Frames Kinematic Relationship

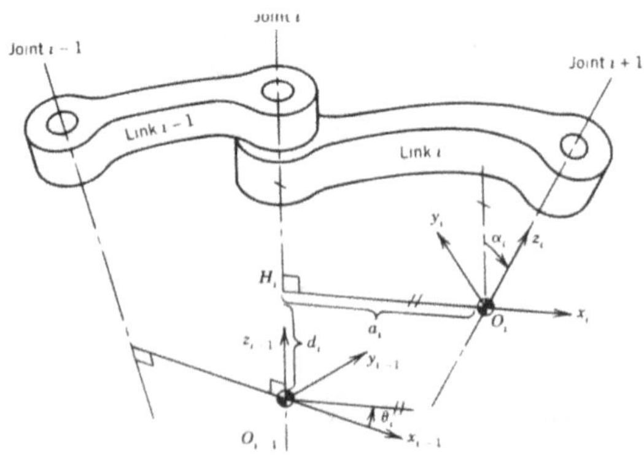

Figure 3.21 Two frames kinematic model relationship.

- There is a kinematic relationship between two frames, basically a translation and a rotation.
- This relationship is represented by a 4 × 4 homogeneous transformation matrix.

3.8 Homogeneous Transformation

	r_1	r_2	r_3	Δx	
3x3 rotation matrix	r_4	r_5	r_6	Δy	3x1 translation
	r_7	r_8	r_9	Δz	
1x3 perspective	0	0	0	1	global scale

Homogeneous Transformation Matrix

$$\left[\begin{array}{ccc|c} \cos\theta_n & -\sin\theta_n\cos\alpha_n & \sin\theta_n\sin\alpha_n & a_n\cos\theta_n \\ \sin\theta_n & \cos\theta_n\cos\alpha_n & -\cos\theta_n\sin\alpha_n & a_n\sin\theta_n \\ 0 & \sin\alpha_n & \cos\alpha_n & d_n \\ \hline 0 & 0 & 0 & 1 \end{array}\right] = \left[\begin{array}{ccc|c} & R & & T \\ \hline 0 & 0 & 0 & 1 \end{array}\right]$$

Virtual robot in Roboanalyzor

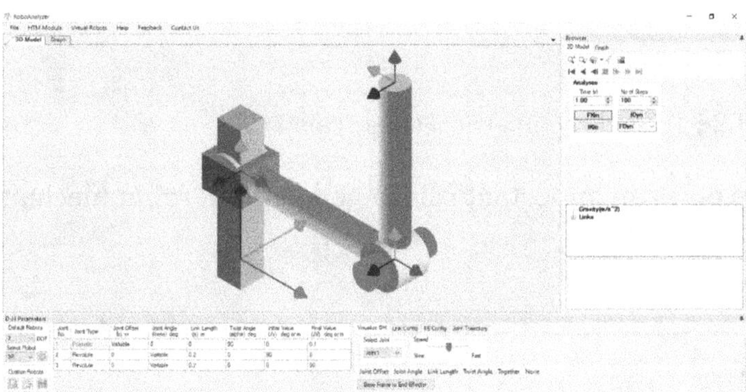

Figure 3.22 Virtual robot in Roboanalyzor software.

3.8 Inverse Geometry and Workspace of Robot Mechanisms

Triangle

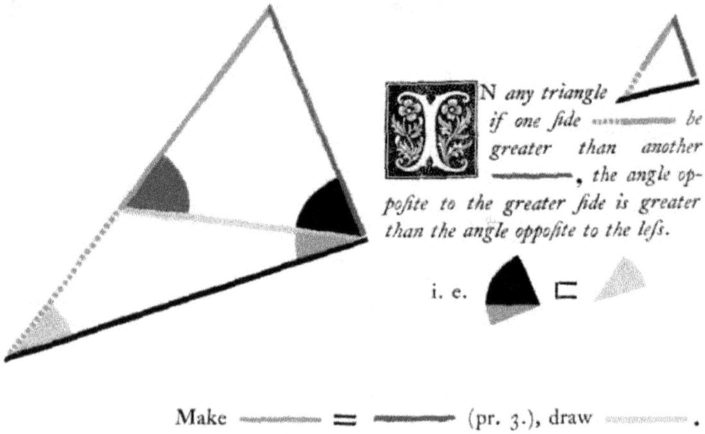

Figure 3.23 Triangle geometry.

Triangle plays an important role in Euclidean geometry.

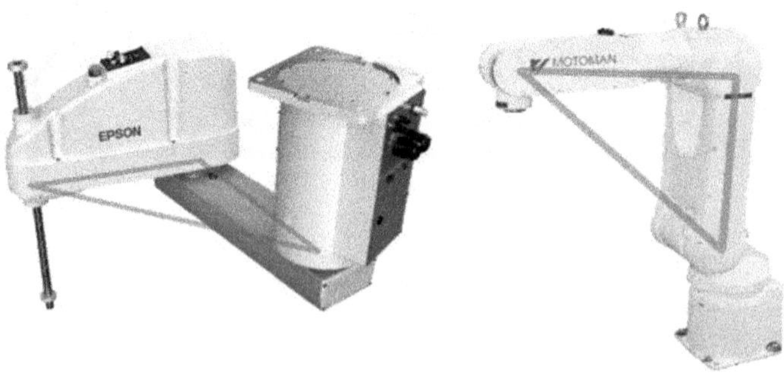

Figure 3.24 Triangle geometrical shape robot [5].

Triangle plays an important role in geometry of robot mechanisms.

Two-segment Planar Robot

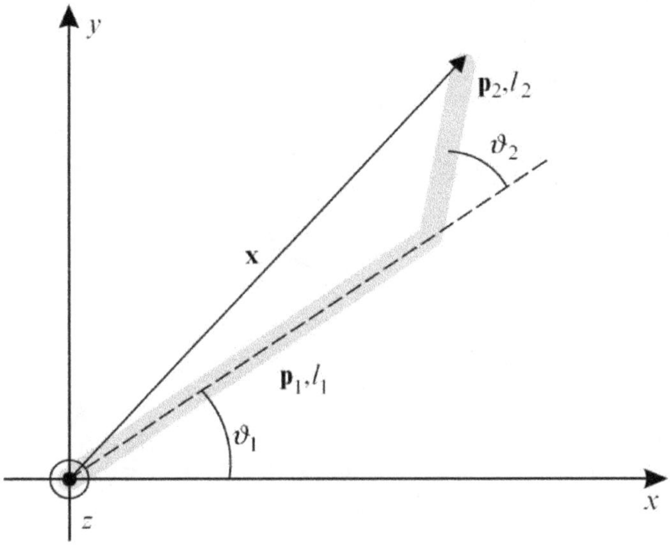

$$\begin{bmatrix} x \\ y \end{bmatrix} = \begin{bmatrix} l_1 \cos \vartheta_1 + l_2 \cos(\vartheta_1 + \vartheta_2) \\ l_1 \sin \vartheta_1 + l_2 \sin(\vartheta_1 + \vartheta_2) \end{bmatrix}$$

When solving the inverse geometry of robot, we calculate the joint angles ϑ_1 and ϑ_2 from the known position of the robot end-point x, y.

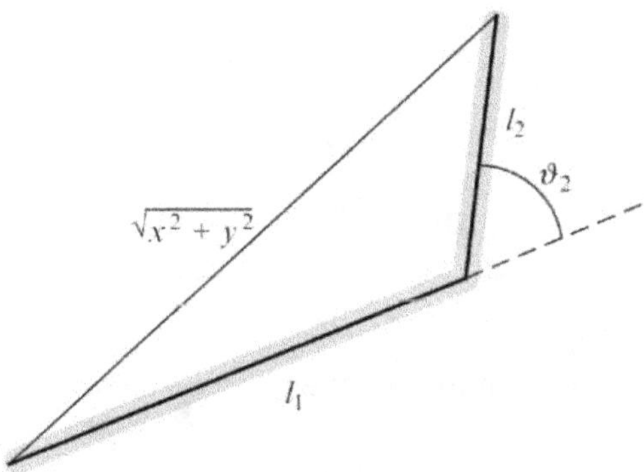

$$\vartheta_2 = \arccos \frac{x^2 + y^2 - l_1^2 - l_2^2}{2 l_1 l_2}$$

The angle in the second joint of the two-segment robot is calculated by the use of the law of cosines.

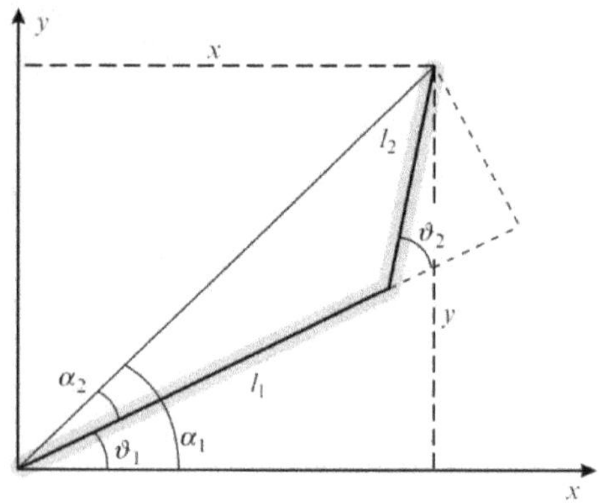

$$\vartheta_1 = \arctan\left(\frac{y}{x}\right) - \arctan\left(\frac{l_2 \sin \vartheta_2}{l_1 + l_2 \cos \vartheta_2}\right)$$

The angle in the first joint is calculated as the difference of the angles ϑ_1 and ϑ_2 [5].

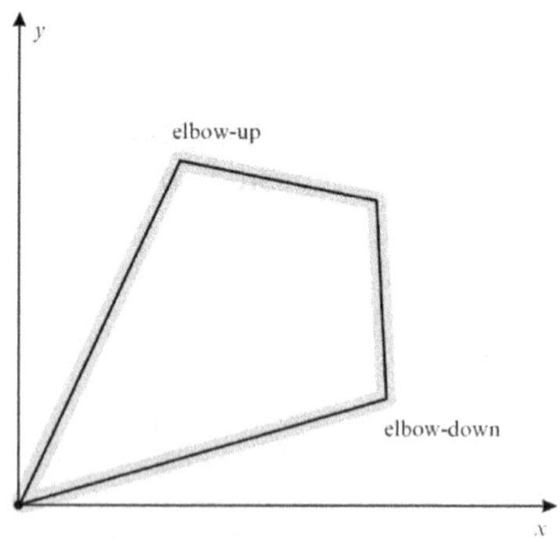

When calculating the joint angles we have two configurations, "lbow-up" and "elbow-down".

Three-segment Planar Robot

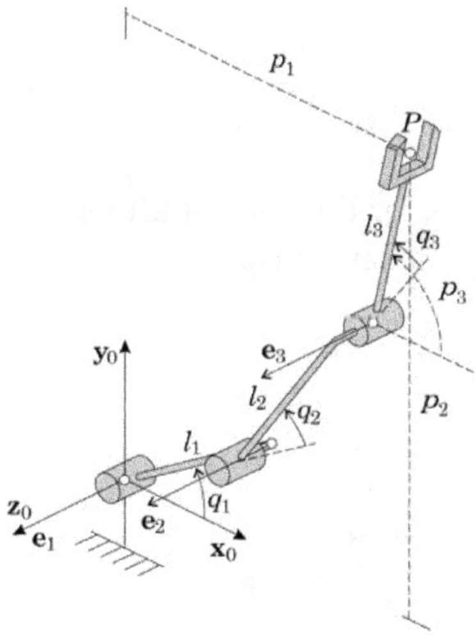

$$p_1 = l_1 c_1 + l_2 c_{12} + l_3 c_{123}$$

$$p_2 = l_1 s_1 + l_2 s_{12} + l_3 s_{123}$$

$$p_3 = q_1 + q_2 + q_3$$

$$q_3 = p_3 - q_1 - q_2$$

When solving the inverse geometry of robot, we calculate the internal coordinates q_1, q_2, q_3 from the known position p_1, p_2 and orientation p_3 of the end-effector.

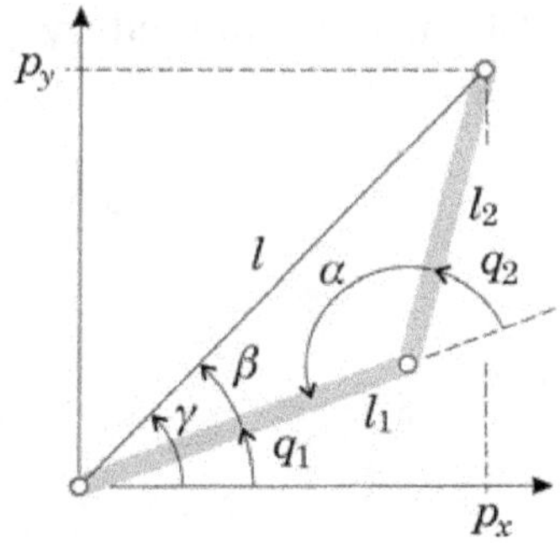

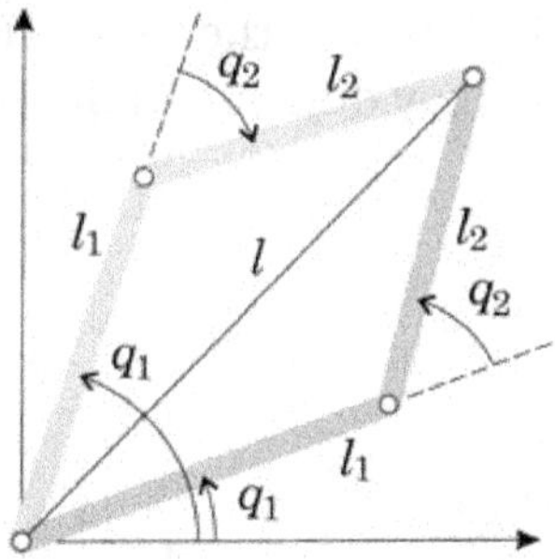

$$\alpha = \arccos\left(\frac{l_1^2 + l_2^2 - l^2}{2l_1 l_2}\right)$$

$$q_2 = \pm(\pi - \alpha)$$

While defining $l^2 = p_x^2 + p_y^2$ the two solutions for the second joint angle q_2 are obtained by the law of cosines [5, 7].

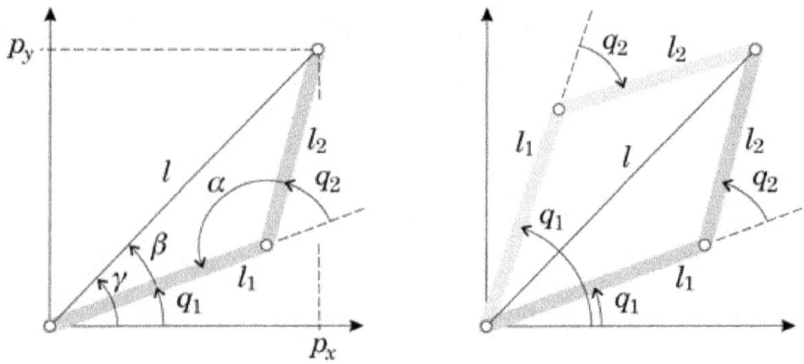

$$\beta = \arccos\left(\frac{l_1^2 + l^2 - l_2^2}{2l_1 l}\right)$$

$$q_2 > 0 \quad \Longrightarrow \quad q_1 = \gamma - \beta$$

$$q_2 < 0 \quad \Longrightarrow \quad q_1 = \gamma + \beta$$

$$\gamma = \arctan_2 \frac{p_y}{p_x}$$

The solutions for the angle in the first joint are obtained by law of cosines. They depend on the selected solution for the second joint angle q_2.

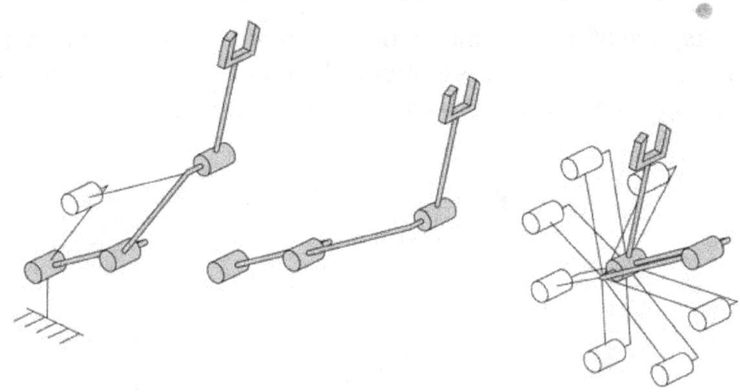

Usually there exist two configurations. When the second joint is extended ($q_2 = 0$), only single solution exists. When $l_1 = l_2$ and $q_2 = \pm\pi$, there is infinite number of configurations [6].

Two-dimensional Robot Workspace

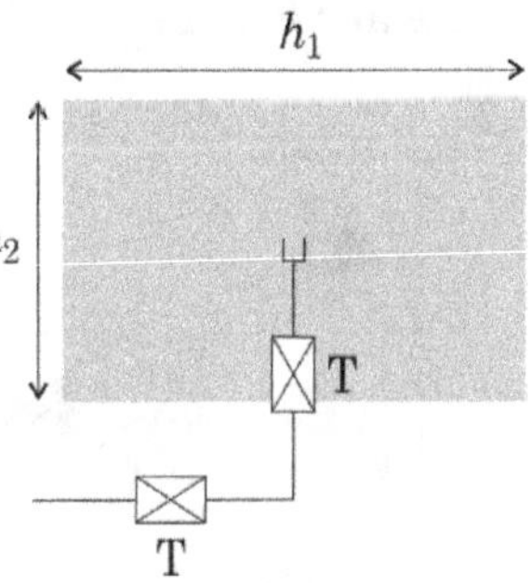

The workspace of a robot mechanism is the spatial volume which is reachable by its end-point.

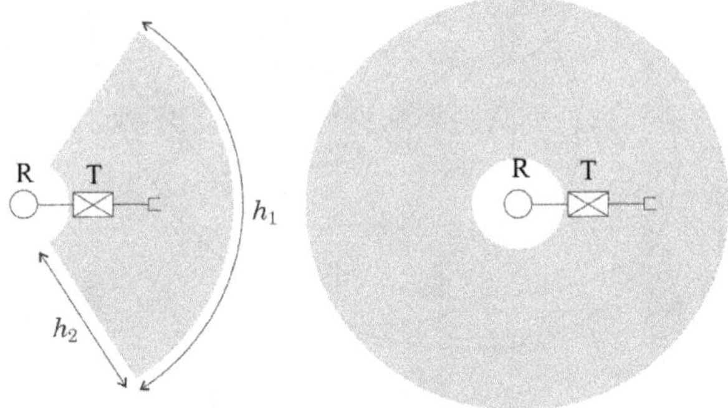

The workspace of a robot mechanism depends on the number of degrees of freedom, their arrangement, the lengths of the segments and constraints in the motion of particular joint coordinates.

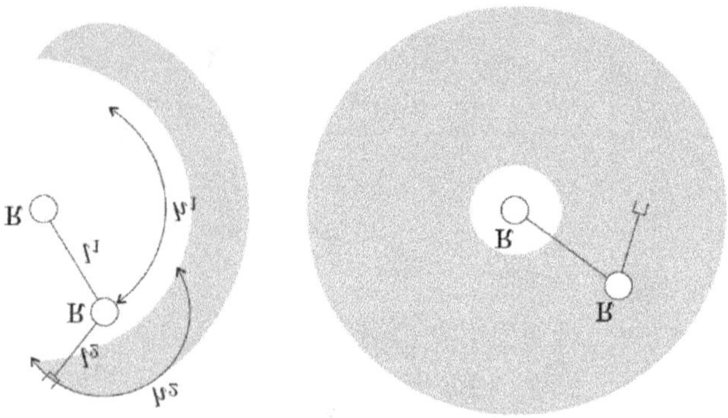

The reachable workspace of a planar mechanism with two rotational joints (2R) is determined with arc h_2 which is expanded around the first rotational axis along the arc h_1.

Workspace of 2R Robot Mechanism

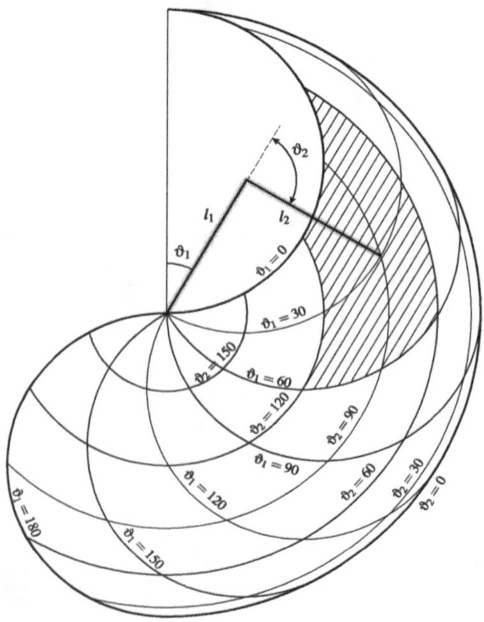

The work space can be described by a mesh of two types of circles. The circles depending on the angle ϑ_1 have their radii of equal length while their centers travel around the origin of the coordinate frame. The circles depending on ϑ_2 angle have all their centers in the origin of the frame, while their radii depend on the lengths of both segments and the angle between them.

The shape of workspace is presented for

$$l_1 = l_2$$
$$0° \leq \vartheta_1 \leq 180°$$
$$0° \leq \vartheta_2 \leq 180°$$

and

$$0° \leq \vartheta_1 \leq 60°$$
$$60° \leq \vartheta_2 \leq 120°$$

The area of the workspace can be replaced by the area of a corresponding sector of a ring.

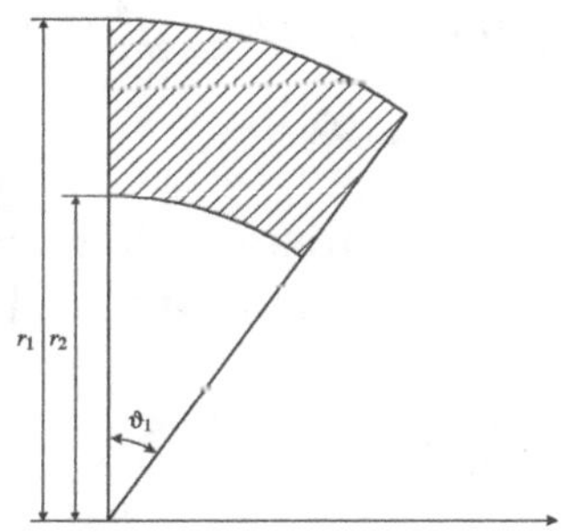

$$A = \frac{\vartheta_1 \pi}{360}\left(r_1^2 - r_2^2\right)$$
$$\vartheta_1 = \vartheta_{1\max} - \vartheta_{1\min}$$
$$r_1^2 = l_1^2 + l_2^2 + 2l_1 l_2 \cos \vartheta_{2\min}$$
$$r_2^2 = l_1^2 + l_2^2 + 2l_1 l_2 \cos \vartheta_{2\max}$$

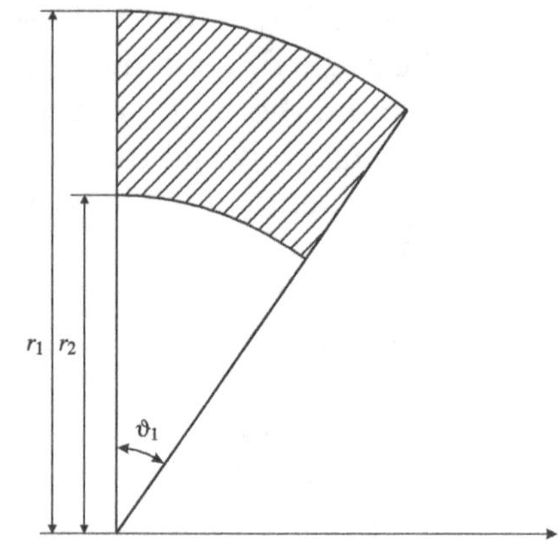

$$0 \le \vartheta_2 \le 30 \qquad A = 0,07$$
$$30 \le \vartheta_2 \le 60 \qquad A = 0,19$$
$$60 \le \vartheta_2 \le 90 \qquad A = 0,26$$
$$90 \le \vartheta_2 \le 120 \qquad A = 0,26$$
$$120 \le \vartheta_2 \le 150 \qquad A = 0,19$$
$$150 \le \vartheta_2 \le 180 \qquad A = 0,07$$

Different values of the working areas are obtained for equal ranges of the angle ϑ_2, $0° \le \vartheta_1 \le 30°$, and for $l_1 = l_2 = 1$.

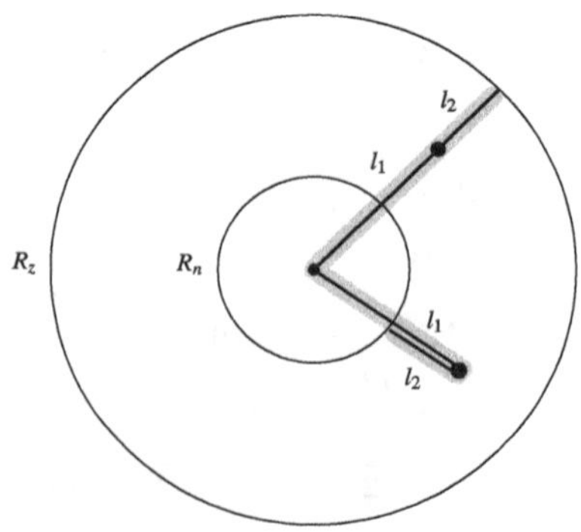

$$0^\circ \le \vartheta_1 \le 360^\circ$$
$$0^\circ \le \vartheta_2 \le 360^\circ$$
$$A = \pi R_z^2 - \pi R_n^2$$
$$R_n = l_1 - l_2 = 2l_1 - R_z$$
$$A = \pi R_z^2 - \pi\left(2l_1 - R_z\right)^2$$
$$\frac{\partial A}{\partial l_1} = 2\pi\left(2l_1 - R_z\right) = 0$$
$$l_1 = l_2$$

The largest working area of the 2R mechanism occurs for equal lengths of both segments.

Workspace of 3R Planar Robot Mechanism

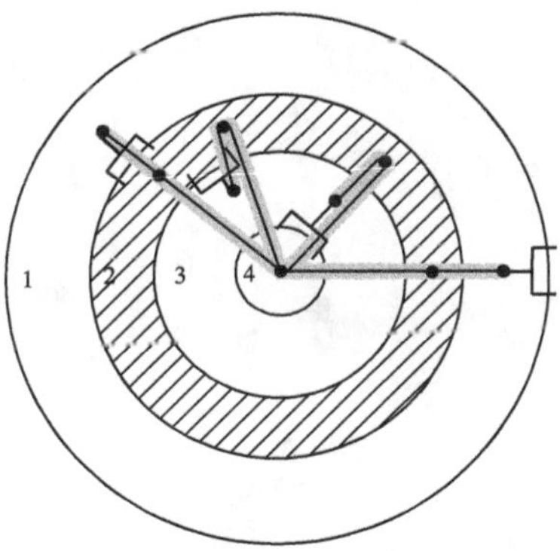

The reachable robot workspace represents all the points that can be reached by the robot end-point. The dexterous workspace comprises all the points that can be reached with an arbitrary orientation of the robot end-effector.

Three-dimensional Robot Workspace

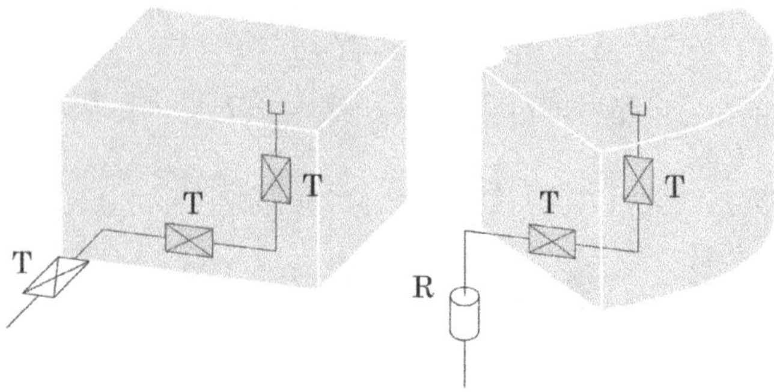

When adding translation to 2T mechanism, the Cartesian mechanism is obtained. When adding rotation to 2T mechanism, the cylindrical mechanism is obtained.

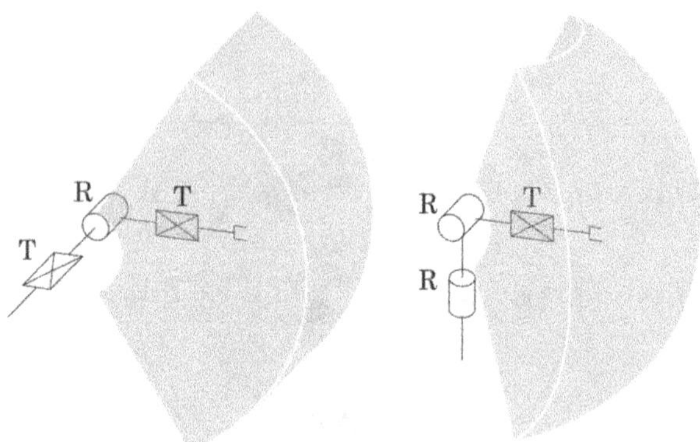

When adding translation to RT mechanism, the cylindrical mechanism is obtained. When adding rotation to RT mechanism, the spherical mechanism is obtained.

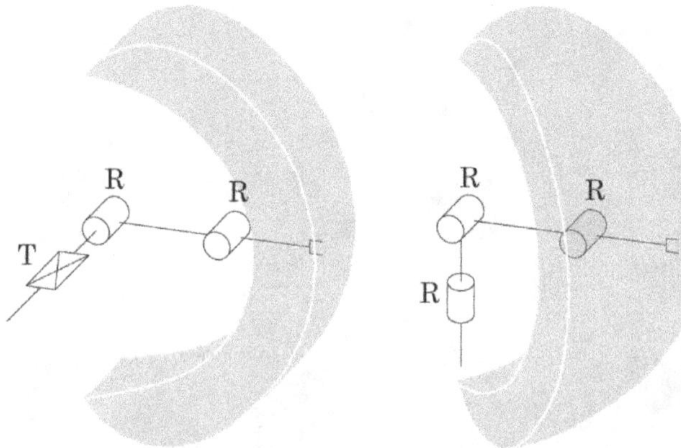

When adding translation to RR mechanism, the so called SCARA mechanism is obtained. When adding rotation to RR mechanism, the anthropomorphic mechanism is obtained [5, 7].

Robot Workspace

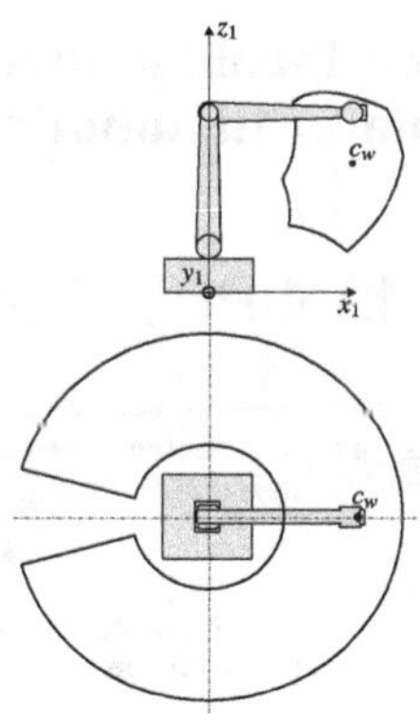

The robot manufacturer is required to clearly show the maximal reachable workspace of an industrial robot in at least two planes.

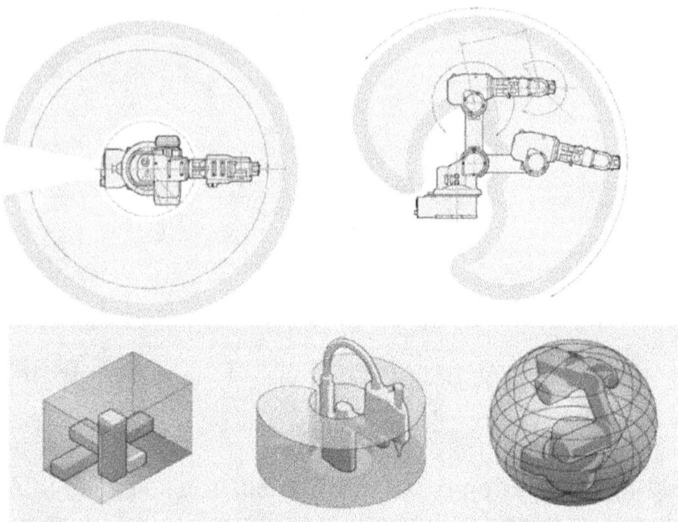

Robot workspace plays an important role when selecting an industrial robot for an anticipated task [5, 7, 14].

3.10 Manipulator Jacobian, Jacobian inverse and Applications of Jacobian Inverse Manipulator in Robotics

WHAT IS A MANIPULATOR?

- A manipulator is a series of links, which connects the hand to the base, with each link connected to the next by an actuated joint. If a coordinate frame is attached to each link, the relationship between two links can be described with a homogeneous transformation matrix.
- In robotics a manipulator is a device used to manipulate materials without direct contact. The applications were originally for dealing with radioactive or biohazardous materials, using robotic arms, or they were used in inaccessible places.
- It is an arm-like mechanism that consists of a series of segments, usually sliding or jointed called cross-slides which grasp and move objects with a number of degrees of freedom.
- Additionally, manipulator tooling gives the lift assist the ability to pitch, roll, or spin the part for appropriate placement. An example would be removing a part from a press in the horizontal and then pitching it up for vertical placement in a rack or rolling a part over for exposing the back of the part.

What is a Manipulator?

•A robot manipulator is constructed using rigid links connected by joints with one fixed end and one free end to perform a given task (e.g., to move a box from one location to the next).

•The joints to this robotic manipulator are the movable components, which enables relative motion between the adjoining links.

Jacobian

•Jacobian is used in change of variables in multiple integrals.

•Jacobian is basically a determinant.

•Jacobian is used to transform variables in one coordinate frame to variables in another coordinate frame.

The Jacobian matrix in Robotics

•We use the Jacobian Matrix to find the velocity of an end effector.

•We use a notation ξ (chi) to represent the velocity of the end effector in the base frame.

•The velocity of the end effector at the base frame has 2 components:

•Linear velocity of the end effector(V_n^0).
•Angular velocity of the end effector(ω_n^0).

•The linear velocity has 3 components x,y,z.
•The angular velocity has 3 components rotation about x,rotation about y,rotation about z.

Two Degree of Freedom Planar Manipulator

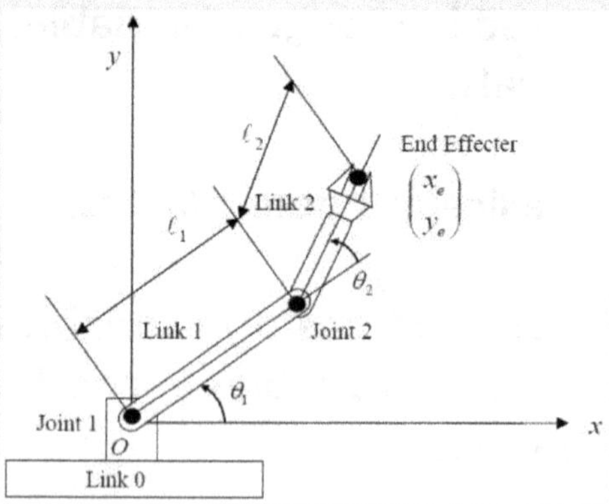

$$\xi = \begin{bmatrix} V_n^0 \\ \omega_n^0 \end{bmatrix}$$

We consider a parameter q as a generalisation of joint variation.
For each revolute joint q is 'ϴ' for the joint.
For a prismatic joint q is displacement of the joint 'd'.

q' is the time derivative of each joint variable.
q' is a vector with one column and number of rows equal to the number of joints.

For an articulated manipulator, we have 3 revolute joints, so we have

$$q' = \begin{bmatrix} \theta_1' \\ \theta_2' \\ \theta_3' \end{bmatrix}$$

For a prismatic joint as 3rd joint of a manipulator we have

$$q' = \begin{bmatrix} \theta_1' \\ \theta_2' \\ d \end{bmatrix}$$

We seek the expressions of the form:

$$\begin{aligned} v_n^0 &= J_v \dot{q} \\ \omega_n^0 &= J_\omega \dot{q} \end{aligned}$$

$$\begin{bmatrix} v_n^0 \\ \omega_n^0 \end{bmatrix} = J_n^0 \dot{q}$$

$$J_n^0 = \left[\frac{J_v}{J_\omega} \right].$$

•The above matrix is called the Manipulator Jacobian or Jacobian for short.

• It is a 6 × n matrix where n is the number of links.

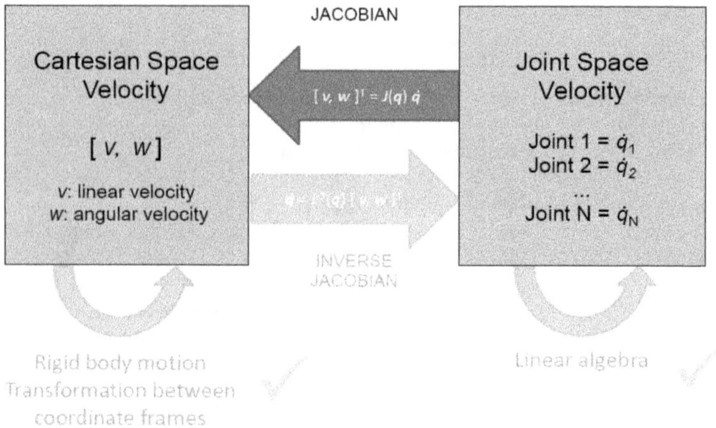

Difference between Jacobian and Inverse Jacobian

- The configuration of a robot in terms of its joint angles is "Joint Space" and
- The configuration of a robot with its end effector in space as it's "Cartesian Space" configuration.
- We can use the concepts and mathematics to figure out the position of the end effector in space, given that we know the joint angles, using a tool called JACOBIAN.

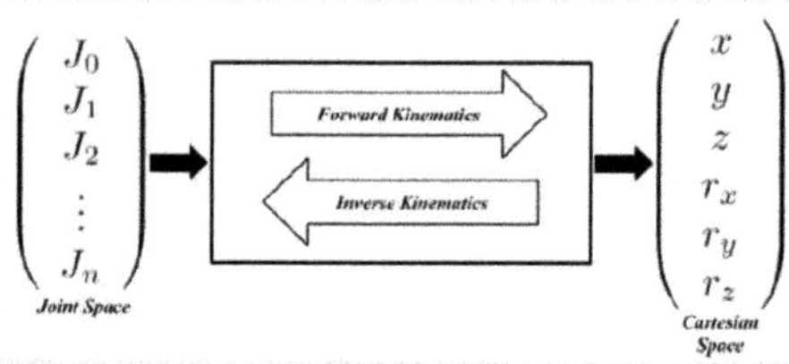

Applications of Jacobian

The robotics has been instrumental in the various domains such as –

- Industries – Robots are used for handling material, cutting, welding, color coating, drilling, polishing, etc.
- Military – Autonomous robots can reach inaccessible and hazardous zones during war. A robot named Daksh, developed by Defense Research and Development Organization (DRDO), is in function to destroy life-threatening objects safely.
- Medicine – The robots are capable of carrying out hundreds of clinical tests simultaneously, rehabilitating permanently disabled people, and performing complex surgeries such as brain tumors.
- Exploration – The robot rock climbers used for space exploration, underwater drones used for ocean exploration are to name a few.
- Entertainment – Disney's engineers have created hundreds of robots for movie making.

Jacobian Inverse, Jacobian Singularity, Static Analysis

Kinematic Relations

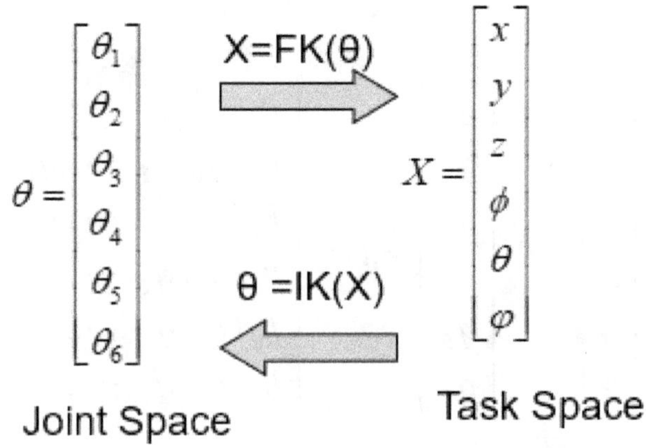

Location of the tool can be specified using a joint space or a cartesian space description.

Velocity Relations

- Relation between joint velocity and Cartesian velocity.
- JACOBIAN matrix J(θ) [2].

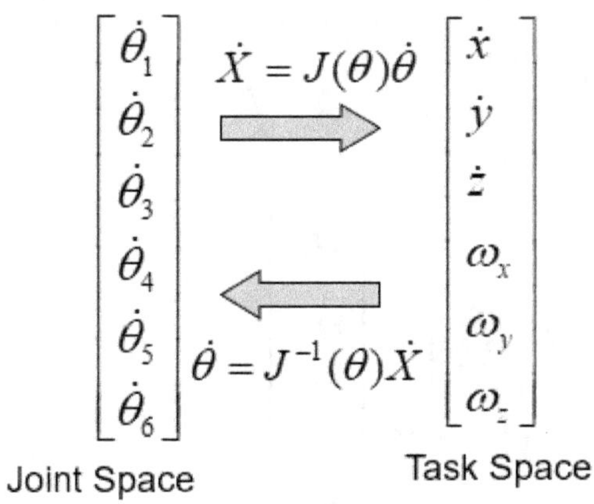

Jacobian

- Suppose a position and orientation vector of a manipulator is a function of 6 joint variables: (from forward kinematics)

$X = h(q)$

$$X = \begin{bmatrix} x \\ y \\ z \\ \phi \\ \theta \\ \varphi \end{bmatrix} = h\left(\begin{bmatrix} q_1 \\ q_2 \\ q_3 \\ q_4 \\ q_5 \\ q_6 \end{bmatrix}\right)_{6\times 1} = \begin{bmatrix} h_1(q_1, q_2, \cdots, q_6) \\ h_2(q_1, q_2, \cdots, q_6) \\ h_3(q_1, q_2, \cdots, q_6) \\ h_4(q_1, q_2, \cdots, q_6) \\ h_5(q_1, q_2, \cdots, q_6) \\ h_6(q_1, q_2, \cdots, q_6) \end{bmatrix}_{6\times 1}$$

Jacobian Matrix

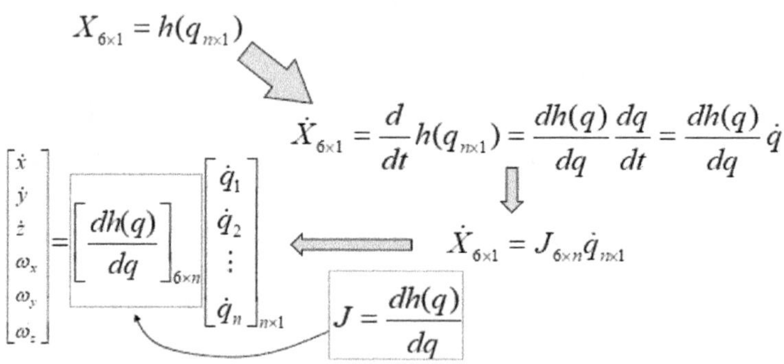

$$\begin{bmatrix} \dot{x} \\ \dot{y} \\ \dot{z} \\ \omega_x \\ \omega_y \\ \omega_z \end{bmatrix} = \begin{bmatrix} \dfrac{dh(q)}{dq} \end{bmatrix}_{6\times n} \begin{bmatrix} \dot{q}_1 \\ \dot{q}_2 \\ \vdots \\ \dot{q}_n \end{bmatrix}_{n\times 1}$$

$$J = \left(\frac{dh(q)}{dq} \right)_{6\times n} = \begin{bmatrix} \frac{\partial h_1}{\partial q_1} & \frac{\partial h_1}{\partial q_2} & \cdots & \frac{\partial h_1}{\partial q_n} \\ \frac{\partial h_2}{\partial q_1} & \frac{\partial h_2}{\partial q_2} & \cdots & \frac{\partial h_2}{\partial q_n} \\ \vdots & \vdots & \vdots & \vdots \\ \frac{\partial h_6}{\partial q_1} & \frac{\partial h_6}{\partial q_2} & \cdots & \frac{\partial h_6}{\partial q_n} \end{bmatrix}_{6\times n}$$

Jacobian is a function of q, it is not a constant [2, 6].

Linear velocity Angular velocity

$$\dot{X} = \begin{bmatrix} \dot{x} \\ \dot{y} \\ \dot{z} \\ \omega_x \\ \omega_y \\ \omega_z \end{bmatrix} = \begin{bmatrix} V \\ \Omega \end{bmatrix} \qquad V = \begin{bmatrix} \dot{x} \\ \dot{y} \\ \dot{z} \end{bmatrix} \qquad \Omega = \begin{bmatrix} \omega_x = \dot{\phi} \\ \omega_y = \dot{\theta} \\ \omega_z = \dot{\psi} \end{bmatrix}$$

$$\dot{q}_{n\times 1} = \begin{bmatrix} \dot{q}_1 \\ \dot{q}_2 \\ \vdots \\ \dot{q}_n \end{bmatrix}_{n\times 1}$$

The Jacobian Equation

$$\dot{X} = J_{6\times n}\dot{q}_{n\times 1}$$

History of Jacobian Concept

- A **Jacobian** is the matrix equivalent of the derivative
  - the derivative of a vector-valued function of a vector with respect to a vector.

$$\boldsymbol{y} = F(\boldsymbol{x}) \quad \boldsymbol{x} \in \mathbb{R}^n, \ \boldsymbol{y} \in \mathbb{R}^m$$

- The Jacobian is an m x n matrix

$$\mathbf{J} = \frac{\partial \boldsymbol{F}}{\partial \boldsymbol{x}} = \begin{pmatrix} \frac{\partial y_1}{\partial x_1} & \cdots & \frac{\partial y_1}{\partial x_n} \\ \vdots & \ddots & \vdots \\ \frac{\partial y_m}{\partial x_1} & \cdots & \frac{\partial y_m}{\partial x_n} \end{pmatrix}$$

$$y = f(x)$$
$$\frac{df}{dx} = \frac{dy}{dx}$$

$$\xi = \mathcal{K}(\boldsymbol{q}) \quad \xi \in \mathbb{R}^2, \boldsymbol{q} \in \mathbb{R}^2$$

Example: Jacobian Matrix

- 2-DOF planar robot arm
- Given $\boldsymbol{l_1}, \boldsymbol{l_2}$, ***Find: Jacobian***

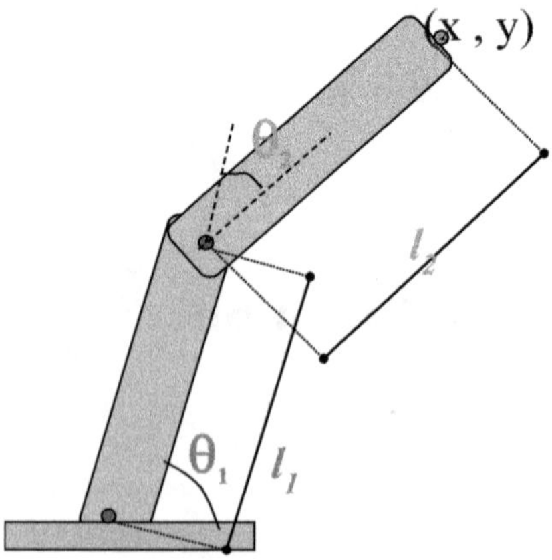

$$\begin{bmatrix} x \\ y \end{bmatrix} = \begin{bmatrix} l_1 \cos\theta_1 + l_2 \cos(\theta_1+\theta_2) \\ l_1 \sin\theta_1 + l_2 \sin(\theta_1+\theta_2) \end{bmatrix} = \begin{bmatrix} h_1(\theta_1,\theta_2) \\ h_2(\theta_1,\theta_2) \end{bmatrix}$$

$$\dot{Y} = \begin{bmatrix} \dot{x} \\ \dot{y} \end{bmatrix} = J \begin{bmatrix} \dot{\theta}_1 \\ \dot{\theta}_2 \end{bmatrix}$$

$$J = \begin{bmatrix} \frac{\partial h_1}{\partial \theta_1} & \frac{\partial h_1}{\partial \theta_2} \\ \frac{\partial h_2}{\partial \theta_1} & \frac{\partial h_2}{\partial \theta_2} \end{bmatrix} = \begin{bmatrix} -l_1 \sin\theta_1 - l_2 \sin(\theta_1+\theta_2) & -l_2 \sin(\theta_1+\theta_2) \\ l_1 \cos\theta_1 + l_2 \cos(\theta_1+\theta_2) & l_2 \cos(\theta_1+\theta_2) \end{bmatrix}$$

Singularities:

- The inverse of the jacobian matrix cannot be calculated when det [J(θ)] = 0
- Singular points are such values of θ that cause the determinant of the Jacobian to be zero [3]

3.11 Singularity Configuration

- Find the singularity configuration of the 2-DOF planar robot arm

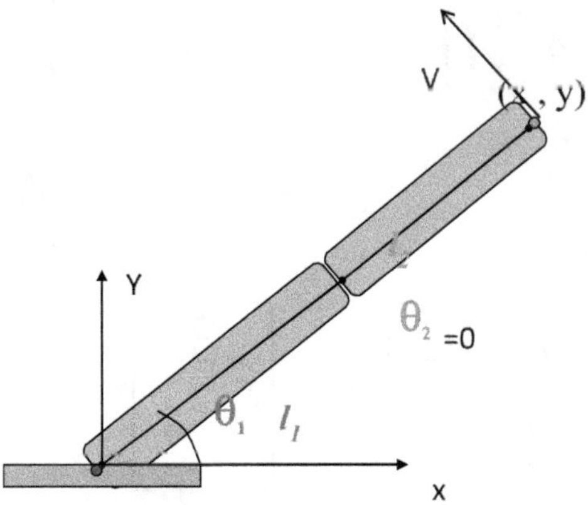

Determinant (J) =0 $\Longrightarrow$ Not full rank

$$\dot{X} = \begin{bmatrix} \dot{x} \\ \dot{y} \end{bmatrix} = J \begin{bmatrix} \dot{\theta}_1 \\ \dot{\theta}_2 \end{bmatrix}$$

$$J = \begin{bmatrix} -l_1 \sin\theta_1 - l_2 \sin(\theta_1 + \theta_2) & -l_2 \sin(\theta_1 + \theta_2) \\ l_1 \cos\theta_1 + l_2 \cos(\theta_1 + \theta_2) & l_2 \cos(\theta_1 + \theta_2) \end{bmatrix}$$

Det(J)=0

$$\theta_2 = 0$$

Pseudo-inverse

Let A be a mxn matrix, and let A^+ be the pseudoinverse of A. If A is of full rank, then A^+ can be computed as:

$$A^+ = \begin{cases} A^T[AA^T]^{-1} & m \le n \\ A^{-1} & m = n \\ [A^T A]^{-1} A^T & m \ge n \end{cases}$$

Jacobian Matrix

- Example: Find X s.t.

$$\begin{bmatrix} 1 & 0 & 2 \\ 1 & -1 & 0 \end{bmatrix} x = \begin{bmatrix} 3 \\ -2 \end{bmatrix}$$

$$A^+ = A^T[AA^T]^{-1} = \begin{bmatrix} 1 & 1 \\ 0 & -1 \\ 2 & 0 \end{bmatrix} \begin{bmatrix} 5 & 1 \\ 1 & 2 \end{bmatrix}^{-1} = \frac{1}{9} \begin{bmatrix} 1 & 4 \\ 1 & -5 \\ 4 & -2 \end{bmatrix}$$

$$x = A^{+}b = \frac{1}{9}\begin{bmatrix} -5 \\ 13 \\ 16 \end{bmatrix}$$

Inverse Jacobian

$$\dot{X} = J\dot{q} = \begin{bmatrix} J_{11} & J_{12} & \cdots & J_{16} \\ J_{21} & J_{22} & \cdots & J_{26} \\ \vdots & \vdots & \vdots & \vdots \\ J_{61} & J_{62} & \cdots & J_{66} \end{bmatrix} \begin{bmatrix} \dot{q}_1 \\ \dot{q}_2 \\ \dot{q}_3 \\ \dot{q}_4 \\ \dot{q}_5 \\ \dot{q}_6 \end{bmatrix}$$

$$\dot{q} = J^{-1}\dot{X}$$

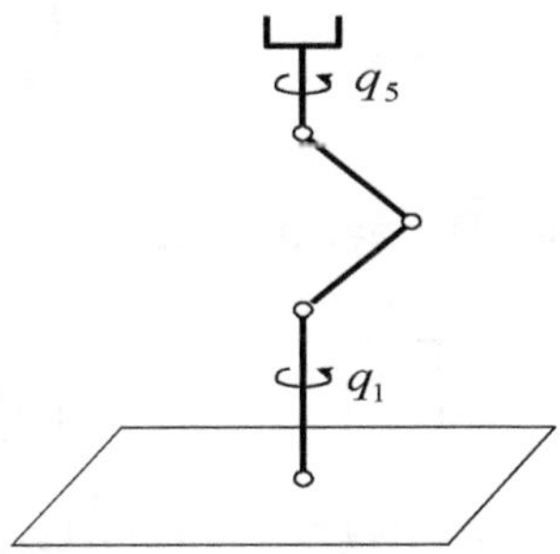

- Singularity
  - rank(J)<n: Jacobian Matrix is less than full rank
  - Jacobian is non-invertable
  - Boundary Singularities: occur when the tool tip is on the surface of the work envelop.
  - Interior Singularities: occur inside the work envelope when two or more of the axes of the robot form a straight line, i.e., collinear [3, 4].

- At Singularities
  - The manipulator end effector can't move in certain directions.
  - Bounded End-Effector velocities may correspond to unbounded joint velocities.
  - Bounded joint torques may correspond to unbounded End-Effector forces and torques [9].

- If $A=\begin{bmatrix} a_x \\ a_y \\ a_z \end{bmatrix}, B=\begin{bmatrix} b_x \\ b_y \\ b_z \end{bmatrix}$
- Then the cross product

$$A\times B=\begin{vmatrix} i & j & k \\ a_x & a_y & a_z \\ b_x & b_y & b_z \end{vmatrix}=\begin{bmatrix} a_y b_z - a_z b_y \\ -(a_x b_z - a_z b_x) \\ a_x b_y - a_y b_x \end{bmatrix}$$

Remember DH parameter

$$A_i=\begin{bmatrix} c\theta_i & -c\alpha_i s\theta_i & s\alpha_i s\theta_i & a_i c\theta_i \\ s\theta_i & c\theta_i c\alpha_i & -s\alpha_i c\theta_i & a_i s\theta_i \\ 0 & s\alpha_i & c\alpha_i & d_i \\ 0 & 0 & 0 & 1 \end{bmatrix}$$

The transformation matrix T

$$T_0^i = A_1 A_2 \ldots\ldots A_i$$

$$J=\begin{bmatrix} J_1 & J_2 & \ldots & J_n \end{bmatrix}$$

where if joint *(i)* is revolute

$$J_i=\begin{bmatrix} Z_{i-1}\times(O_n-O_{i-1}) \\ Z_{i-1} \end{bmatrix}$$

and if joint *(i)* is prismatic

$$J_i=\begin{bmatrix} Z_{i-1} \\ 0 \end{bmatrix}$$

3.12 2-DOF Planar Robot Arm

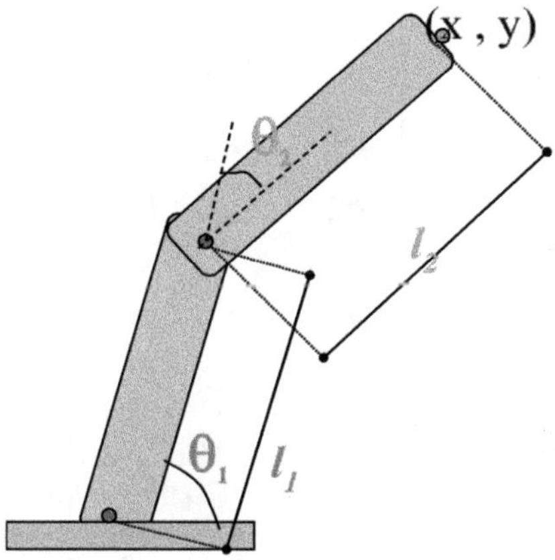

Given *l1, l2, Find: Jacobian*

- *Here, n=2,*

Link	a_i	α_i	d_i	θ_i
1	a_1	0	0	θ_1^*
2	a_2	0	0	θ_2^*

* variable

$$A_i = \begin{bmatrix} c\theta_i & -c\alpha_i s\theta_i & s\alpha_i s\theta_i & a_i c\theta_i \\ s\theta_i & c\theta_i c\alpha_i & -s\alpha_i c\theta_i & a_i s\theta_i \\ 0 & s\alpha_i & c\alpha_i & d_i \\ 0 & 0 & 0 & 1 \end{bmatrix}$$

$$A_1 = \begin{bmatrix} c_1 & -s_1 & 0 & a_1c_1 \\ s_1 & c_1 & 0 & a_1s_1 \\ 0 & 0 & 1 & 0 \\ 0 & 0 & 0 & 1 \end{bmatrix}.$$

$$A_2 = \begin{bmatrix} c_2 & -s_2 & 0 & a_2c_2 \\ s_2 & c_2 & 0 & a_2s_2 \\ 0 & 0 & 1 & 0 \\ 0 & 0 & 0 & 1 \end{bmatrix}$$

Where $(\theta_1 + \theta_2)$ denoted by θ_{12} and $\cos(\theta_1 + \theta_2)$ by c_{12}

$$T_1^0 = A_1.$$

$$T_2^0 = A_1A_2 = \begin{bmatrix} c_{12} & -s_{12} & 0 & a_1c_1 + a_2c_{12} \\ s_{12} & c_{12} & 0 & a_1s_1 + a_2s_{12} \\ 0 & 0 & 1 & 0 \\ 0 & 0 & 0 & 1 \end{bmatrix}$$

$$O_0 = \begin{bmatrix} 0 \\ 0 \\ 0 \end{bmatrix}, O_1 = \begin{bmatrix} a_1 \cos\theta_1 \\ a_1 \sin\theta_1 \\ 0 \end{bmatrix}, O_2 = \begin{bmatrix} a_1 \cos\theta_1 + a_2 \cos(\theta_1 + \theta_2) \\ a_1 \sin\theta_1 + a_2 \sin(\theta_1 + \theta_2) \\ 0 \end{bmatrix}$$

$$Z_0 = Z_1 = \begin{bmatrix} 0 \\ 0 \\ 1 \end{bmatrix}$$

Jacobian Manipulator Types

$$J_1 = \begin{bmatrix} -a_1\sin\theta_1 - a_2\sin(\theta_1+\theta_2) \\ a_1\cos\theta_1 + a_2\cos(\theta_1+\theta_2) \\ 0 \\ 0 \\ 0 \\ 1 \end{bmatrix} \qquad J_2 = \begin{bmatrix} -a_2\sin(\theta_1+\theta_2) \\ a_2\cos(\theta_1+\theta_2) \\ 0 \\ 0 \\ 0 \\ 1 \end{bmatrix}$$

The required Jacobian matrix **J**

$$J = [J_1 \quad J_2]$$

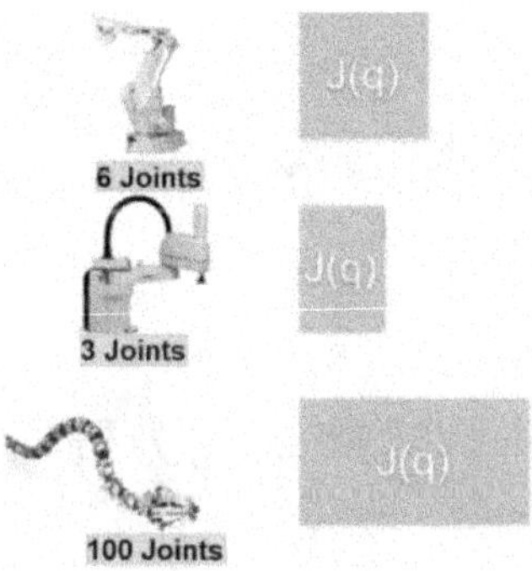

- Fully actuated (6x6)
  - full access to SE(3)
  - Jacobian is square
- Under actuated (6xN), N < 6
  - limited access to SE(3)
  - remove some spatial degrees of freedom
- Over actuated (6xN), N > 6
  - full access to SE(3)
  - use pseudo inverse
  - has **spare** joints
  - can do null-space motion

Jacobian Velocity

Tool tip velocity

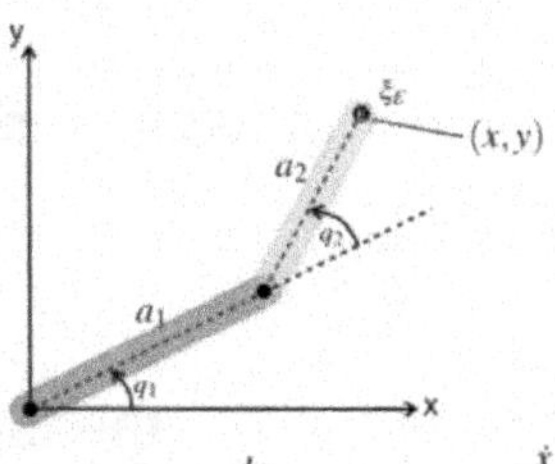

$$\dot{q}_j = \frac{dq_j}{dt}$$

$$\begin{pmatrix} x \\ y \end{pmatrix} = \begin{pmatrix} a_2\cos(q_1+q_2) + a_1\cos q_1 \\ a_2\sin(q_1+q_2) + a_1\sin q_1 \end{pmatrix}$$

- Joint angles are functions of time

$$q_1 = q_1(t) \qquad q_2 = q_2(t)$$

- Compute the derivative (chain rule)

$$\dot{x} = -a_1\dot{q}_1\sin q_1 - a_2(\dot{q}_1+\dot{q}_2)\sin(q_1+q_2)$$
$$\dot{y} = a_1\dot{q}_1\cos q_1 + a_2(\dot{q}_1+\dot{q}_2)\cos(q_1+q_2)$$

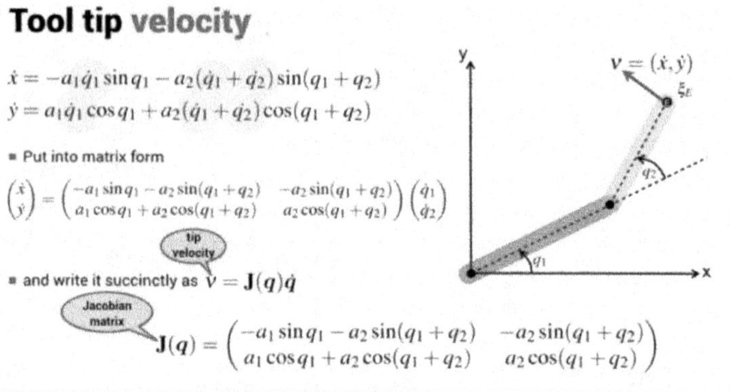

Jacobian Inverse Velocity

- The relation between the joint and end-effector velocities:

$$\dot{X} = J(q)\dot{q}$$

where j (m×n). If J is a square matrix (m=n), the joint velocities:

$$\dot{q} = J^{-1}(q)\dot{X}$$

- If $m<n$, let pseudoinverse J^{+} where $= J^{T}[JJ^{T}]^{-1}$

$$\dot{q} = J^{+}(q)\dot{X}$$

Jacobian Acceleration

The relation between the joint and end-effector velocities:

$$\dot{X} = J(q)\dot{q}$$

Differentiating this equation yields an expression for the acceleration:

$$\ddot{X} = J(q)\ddot{q} + [\frac{d}{dt}J(q)]\dot{q}$$

Given $\ddot{x}$ of the end-effector acceleration, the joint $\dot{q}$ acceleration

$$J(q)\ddot{q} = \ddot{X} - [\frac{d}{dt}J(q)]\dot{q}$$

$$\ddot{q} = J^{-1}(q)[\ddot{X} - \frac{d}{dt}J(q)]\dot{q}]$$

3.13 Analysis of Kinematic Model

Definition:

The kinematic model studies the motion of a robot mechanism regardless of forces and torque that cause it. It allows to compute the position and orientation of robot manipulator's end-effector relative to the base of the manipulator as a function of the joint variables [1-20].

Forward Kinematic Model:

The forward kinematic problem asks: once the joint position, velocity, acceleration are known, compute the corresponding variables of the end- effector in a given reference frame (e.g. A Cartesian frame). The forward kinematic model can be defined as a function "f" is defined between the joint space R^n and the workspace R^m:

$$x=f(q);\ x \in R^m,\ q \in R^n$$

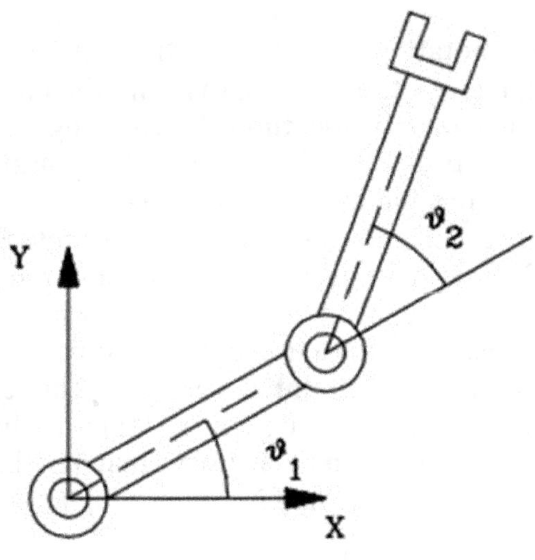

Inverse Kinematic Model

Inverse Kinematic Problem: computation of the relevant variables (positions, velocities, accelerations) from the workspace to the joint space.

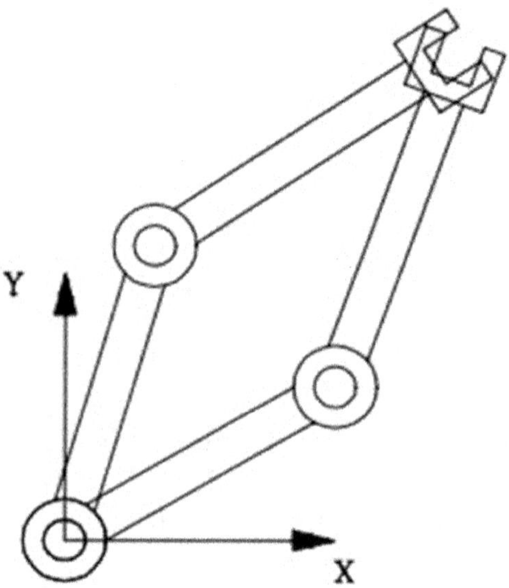

Denavit-Hartenberg Notations

In DH convention notation system, each link can be represented by two parameters, namely the link length $\boldsymbol{a_i}$ and the link twist angle $\boldsymbol{\alpha_i}$. The link twist angle $\boldsymbol{\alpha_i}$ indicates the axis twist angle of two adjacent joints $\boldsymbol{i}$ and $\boldsymbol{i}$ - **1**. Joints are also described by two parameters, namely the link offset $\boldsymbol{d_i}$, which indicates the distance from a link to next link along the axis of joint $\boldsymbol{i}$, and the joint revolute angle $\boldsymbol{\Theta}_i$, which is the rotation of one link with respect to the next about the joint axis. Usually, three of these four parameters are fixed while one variable is called joint variable. For a revolute joint, the joint variable is parameter $\boldsymbol{\Theta}_i$, while for a prismatic joint, it will be $\boldsymbol{d_i}$. The four parameters of each link can be specified using DH notation method. With these parameters, link homogeneous transform matrix which transforms link coordinate frame $\boldsymbol{i}$-**1** to frame $\boldsymbol{i}$.

$$
{}^{i-1}A_i(\theta_i, d_i, a_i, \alpha_i) = R_z(\theta_i)T_z(d_i)T_x(a_i)R_x(\alpha_i)
$$

$$
= \begin{bmatrix} c\theta_i & -s\theta_i c\alpha_i & s\theta_i s\alpha_i & a_i c\theta_i \\ c\theta_i & c\theta_i c\alpha_i & -c\theta_i s\alpha_i & a_i s\theta_i \\ 0 & s\alpha_i & c\alpha_i & d_i \\ 0 & 0 & 0 & 1 \end{bmatrix}
$$

Here **sin(Θ)** is abbreviated as ***s*Θ** and **cos(Θ)** as ***c*Θ**, R_j denotes rotation about axis ***j***, and T_j denotes translation along axis ***j***.

Uses

The robotic kinematics is essential for describing an end effector's position, orientation as well as motion of all the joints. In order to deal with the complex geometry of a robot manipulator, the properly chosen coordinate frames are fixed to various parts of the mechanism and then we can formulate the relationships between these frames. The manipulator kinematics mainly studies how the locations of these frames change as the robot joints move [1, 3].

3.14 Analysis of Dynamic Model

Robot dynamics is the study of the relation between the applied forces/torques and the resulting motion of an industrial manipulator. The dynamic model of a robot studies the relation between the joint actuator torques and the resulting motion.

Direct Dynamic Model

Problem: Once the forces/torques applied to the joints, as well as the joint positions and velocities are known, compute the joint accelerations.

$$
\ddot{\mathbf{q}} = f(\mathbf{q}, \dot{\mathbf{q}}, \tau)
$$

and then

$$
\dot{\mathbf{q}} = \int \ddot{\mathbf{q}}\, dt, \qquad \mathbf{q} = \int \dot{\mathbf{q}}\, dt
$$

Inverse Dynamic Model

Problem: Once the joint accelerations, velocities and positions are known, compute the corresponding forces/torques.

$$\tau = f^{-1}(\ddot{\mathbf{q}}, \dot{\mathbf{q}}, \mathbf{q}) = g(\ddot{\mathbf{q}}, \dot{\mathbf{q}}, \mathbf{q})$$

3.15 Formulation Methods

There are two commonly used methods for formulating the dynamics, based on the specific geometric and inertial parameters of the robot:

- the Lagrange–Euler (L–E) formulation
- the Newton–Euler (N–E) method

Both are equivalent, as both describe the dynamic behaviour of the robot motion but are specifically useful for different purposes.

Lagrange-Euler Formulation

The L–E method is based on simple and systematic methods to calculate the kinetic and potential energies of a rigid body system. The equations of dynamic motion for a robot manipulator are highly nonlinear, consisting of inertial and gravity terms, and are dependent on the link physical parameters and configuration (i.e., position, angular velocity and acceleration). This provides the closed form of the robot dynamics and is therefore applicable to the analytical computation of robot dynamics, and therefore can be used to design joint-space (or task-space, using transformation via the Jacobian) control strategies. The L–E formulation may also be used for forward and inverse dynamic calculation, but this requires the calculation of a large number of coefficients in $M(q)$ and $C(q, \dot{q})$, which may take a long time. This makes this method somewhat unsuitable for online dynamic calculations.

Newton-Euler Method

The N-E formulation is based on a balance of all the forces acting on the generic link of the manipulator; this forms a set of equations with a recursive solution and was developed. A forward recursion propagates link velocities and accelerations, then a backward recursion propagates the forces and torques along the manipulator chain. This is developed as a more efficient method than L–E and is based on the principle of the manipulator being a serial chain; when a force is applied to one link, it may also produce motion in connected links. Due to this effect, there may be considerable duplication of calculation, which can be avoided if expressed in a recursive form. This reduction in computational load greatly reduces calculation time, allowing the forward and inverse dynamics calculations to be performed in real-time; therefore, it can enable real-time torque control methods of robot manipulators.

Uses

Dynamics modelling is crucial for analysing and synthesizing the dynamic behaviour of robot. An accurate dynamics model of a robot manipulator is useful in many ways: for the design of motion control systems, analysis of mechanical design, simulation of manipulator motion, etc.

3.16 Kinematic Model vs Dynamic Model

• KINEMATIC MODEL	• DYNAMIC MODEL
• Studies the motion of a robot mechanism regardless of forces and torque that cause it.	• Studies the relation between the joint actuator torques and the resulting motion.
• DH notations are used.	• LE and NE methods are used.
• Essential for describing an end-effector's position, orientation as well as motion of all the joints.	• Crucial for analyzing and synthesizing the dynamic behaviour of robot.

D-H transformation matrix for adjacent coordinate frames, *i* and *i-1*. The position and orientation of the *i*-th frame coordinate can be expressed in the *(i-1)*th frame by the following 4 successive elementary transformations:

$$T_{i-1}^{i} = T(z_{i-1}, d_i)R(z_{i-1}, \theta_i)T(x_i, a_i)R(x_i, \alpha_i)$$

$$= \begin{bmatrix} C\theta_i & -C\alpha_i S\theta_i & S\alpha_i S\theta_i & a_i C\theta_i \\ S\theta_i & C\alpha_i C\theta_i & -S\alpha_i C\theta_i & a_i S\theta_i \\ 0 & S\alpha_i & C\alpha_i & d_i \\ 0 & 0 & 0 & 1 \end{bmatrix}$$

Kinematics Equations

chain product of successive coordinate transformation matrices of T_{i-1}^{i}

T_0^n specifies the location of the n-th coordinate frame w.r.t. the base coordinate system

$$T_0^n = T_0^1 T_1^2 \dots T_{n-1}^n$$

Orientation matrix

$$= \begin{bmatrix} R_0^n & P_0^n \\ 0 & 1 \end{bmatrix} = \begin{bmatrix} n & s & a & P_0^n \\ 0 & 0 & 0 & 1 \end{bmatrix}$$

Position vector

3.17 Dynamic Analysis and Forces

The dynamics, related with accelerations, loads, masses and inertias.

$$\sum \overline{F} = m \cdot \overline{a} \qquad \sum \overline{T} = I \cdot \overline{\alpha}$$

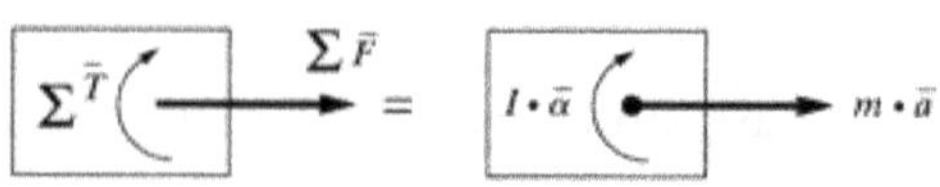

The actuator can accelerate a robot's links for exerting enough forces and torques at a desired acceleration and velocity. By the dynamic relationships that govern the motions of the robot, considering the external loads, the designer can calculate the necessary forces and torques. Mathematical equations describing the dynamic behavior of the manipulator

- For computer simulation
- Design of suitable controller
- Evaluation of robot structure
- Joint torques
- Robot motion, i.e., acceleration, velocity, position [2, 3].

Lagrange-Euler Formulation

$$\frac{d}{dt}\left(\frac{\partial L}{\partial \dot{q}_i}\right) - \frac{\partial L}{\partial q_i} = \tau_i$$

Lagrange function is defined

$$L = K - P$$

- *K*: Total kinetic energy of robot
- *P*: Total potential energy of robot

q_i : Joint variable of i-th joint

$\dot{q}_i$: first time derivative of q_i

τ_i : Generalized force (torque) at i-th joint

Kinetic energy

Single particle: $k = \frac{1}{2} m v^2$

Rigid body in 3-D space with linear velocity (V) , and angular velocity (ω) about the center of mass

$$k = \frac{1}{2} m V^T V + \frac{1}{2} I \omega^T \omega$$

$$I \equiv \begin{bmatrix} \int (y^2 + z^2) dm & -\int xy dm & -\int xz dm \\ -\int xy dm & \int (x^2 + z^2) dm & -\int yz dm \\ -\int xz dm & -\int yz dm & \int (x^2 + y^2) dm \end{bmatrix}$$

I : Inertia Tensor:

Diagonal terms: moments of inertia $\quad I_{xx} = \int (y^2 + z^2) dm$

Off-diagonal terms: products of inertia $\quad I_{xy} = \int (xy) dm$

Lagrange-Euler Formulation

Lagrange Dynamics:

The Lagrangian, “L”, of any system is defined as:

$$L = K - P \tag{1}$$

where:

$K = Kinetic\ Energy\ of\ the\ System,$

$P = Potential\ Energy\ of\ the\ System.$

Lagrangian Dynamics for 2-Link Manipulator

The Lagrangian, “L”, of any system is defined as:

$$L = K - P$$

$K = Kinetic\ Energy\ of\ the\ System,$

$P = Potential\ Energy\ of\ the\ System.$

Furthermore, we have [2]:

$$F_i = \frac{d}{dt}\left(\frac{\partial L}{\partial \dot{q}_i}\right) - \frac{\partial L}{\partial q_i}$$

Lagrange Equation

$$F_i = \frac{d}{dt}\left(\frac{\partial L}{\partial \dot{q}_i}\right) - \frac{\partial L}{\partial q_i}$$

$F_i = generalized\ force\ coordinate\ (force\ or\ torque\ of\ the\ i^{th}\ link),$

$q_i = generalized\ position\ coordinate (linear\ \&\ angular)\ of\ the\ i^{th}\ link,$

$\dot{q}_i = generalized\ velocity\ vector.$

3.18 Two-link Manipulator

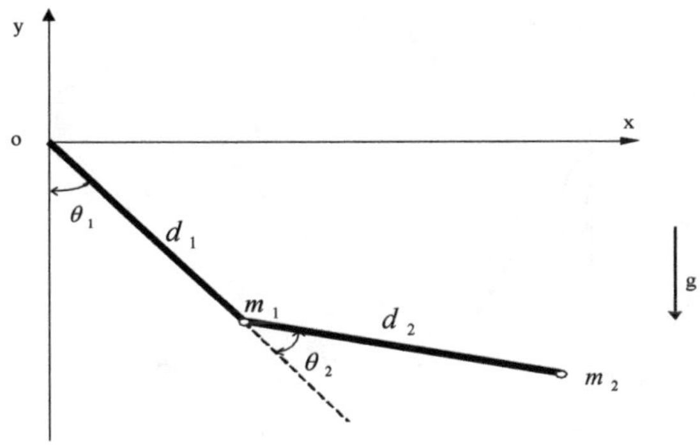

Kinetic energy of n-1 link

$$K_1 = \frac{1}{2} m_1 v_1^2$$

$$v_1 = d_1 \omega_1 = d_1 \dot{\theta}_1$$

$$K_1 = \frac{1}{2} m_1 d_1^2 \dot{\theta}_1^2$$

Potential energy of link-1

- potential energy of link-1 can be presented as:

$$P_1 = m_1 g h_1$$

$$h_1 = -d_1 \cos(\theta_1)$$

$$P_1 = -m_1 g d_1 \cos(\theta_1)$$

- kinetic energy of link-2

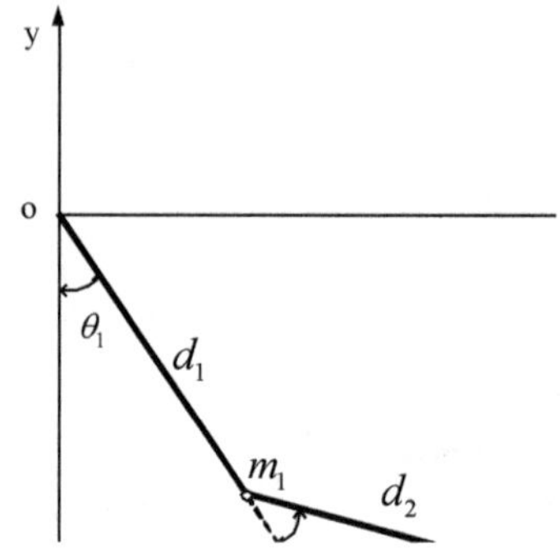

$$x_2 = d_1 \sin(\theta_1) + d_2 \sin(\theta_1 + \theta_2)$$
$$y_2 = -d_1 \cos(\theta_1) - d_2 \cos(\theta_1 + \theta_2)$$

$$\dot{x}_2 = \frac{d}{dt}(x_2) = d_1 \cos(\theta_1)\dot{\theta}_1 + d_2 \cos(\theta_1 + \theta_2)(\dot{\theta}_1 + \dot{\theta}_2)$$
$$\dot{y}_2 = \frac{d}{dt}(y_2) = d_1 \sin(\theta_1)\dot{\theta}_1 + d_2 \sin(\theta_1 + \theta_2)(\dot{\theta}_1 + \dot{\theta}_2)$$

$$(v_2)^2 = (v_{2x})^2 + (v_{2y})^2 = \dot{x}_2^2 + \dot{y}_2^2$$
$$= d_1^2\dot{\theta}_1^2 + d_2^2(\dot{\theta}_1 + \dot{\theta}_2)^2 + 2d_1d_2\dot{\theta}_1(\dot{\theta}_1 + \dot{\theta}_2)[\cos(\theta_1)\cos(\theta_1 + \theta_2) + \sin(\theta_1)\sin(\theta_1 + \theta_2)]$$

$$\cos(A)\cos(B) + \sin(A)\sin(B) = \cos(A - B) = \cos(B - A)$$

$$(v_2)^2 = d_1^2\dot{\theta}_1^2 + d_2^2(\dot{\theta}_1 + \dot{\theta}_2)^2 + 2d_1d_2\dot{\theta}_1(\dot{\theta}_1 + \dot{\theta}_2)\cos(\theta_2)$$

Lagrangian Dynamics of 2-Link Mmanipulator

Kinetic energy of link – 2 is:

$$K_2 = \frac{1}{2} m_2 v_2^2$$
$$= \frac{1}{2} m_2 d_1^2 \dot{\theta}_1^2 + \frac{1}{2} m_2 d_2^2 (\dot{\theta}_1 + \dot{\theta}_2)^2 + m_2 d_1 d_2 \cos(\theta_2) \dot{\theta}_1 (\dot{\theta}_1 + \dot{\theta}_2)$$

Potential energy of link 2 is:

P_2 = - $m_2 g d_1 C_1$ – $m_2 g d_2 C_{12}$

where $C_1 = \cos(\theta_1)$
$C_{12} = \cos(\theta_1 + \theta_2)$

Total kinetic and potential energy

$$K = K_1 + K_2$$
$$= \frac{1}{2} m_1 d_1^2 \dot{\theta}_1^2 + \frac{1}{2} m_2 d_1^2 \dot{\theta}_1^2 + \frac{1}{2} m_2 d_2^2 (\dot{\theta}_1 + \dot{\theta}_2)^2 + m_2 d_1 d_2 \cos(\theta_2) \dot{\theta}_1 (\dot{\theta}_1 + \dot{\theta}_2)$$

Total potential energy

Potential energy: $P_1 + P_2$

$$P = P_1 + P_2$$
$$= -m_1 g d_1 \cos(\theta_1) - m_2 g d_1 \cos(\theta_1) - m_2 g d_2 \cos(\theta_1 + \theta_2)$$

3.19 Lagrangian of the system

$$L = K - P$$
$$= \frac{1}{2}(m_1 + m_2) d_1^2 \dot{\theta}_1^2 + \frac{1}{2} m_2 d_2^2 (\dot{\theta}_1 + \dot{\theta}_2)^2 + m_2 d_1 d_2 \cos(\theta_2) \dot{\theta}_1 (\dot{\theta}_1 + \dot{\theta}_2)$$
$$+ (m_1 + m_2) g d_1 \cos(\theta_1) + m_2 g d_2 \cos(\theta_1 + \theta_2)$$

$$F_i = \frac{d}{dt}\left(\frac{\partial L}{\partial \dot{q}_i}\right) - \frac{\partial L}{\partial q_i}$$

$$\frac{\partial L}{\partial q_1} = \frac{\partial L}{\partial \theta_1} = -(m_1 + m_2) g d_1 \sin(\theta_1) - m_2 g d_2 \sin(\theta_1 + \theta_2)$$

$$\frac{\partial L}{\partial \dot{q}_1} = \frac{\partial L}{\partial \dot{\theta}_1} = (m_1 + m_2) d_1^2 \dot{\theta}_1 + m_2 d_2^2 (\dot{\theta}_1 + \dot{\theta}_2) + m_2 d_1 d_2 \cos(\theta_2)[2\dot{\theta}_1 + \dot{\theta}_2]$$

$$\begin{aligned} \frac{d}{dt}\left(\frac{\partial L}{\partial \dot{q}_1}\right) &= (m_1 + m_2) d_1^2 \ddot{\theta}_1 + m_2 d_2^2 (\ddot{\theta}_1 + \ddot{\theta}_2) + m_2 d_1 d_2 \cos(\theta_2)[2\ddot{\theta}_1 + \ddot{\theta}_2] \\ &\quad - m_2 d_1 d_2 \sin(\theta_2)[2\ddot{\theta}_1 + \ddot{\theta}_2] \\ &= [(m_1 + m_2) d_1^2 + m_2 d_2^2 + 2 m_2 d_1 d_2 \cos(\theta_2)]\ddot{\theta}_1 + [m_2 d_2^2 + m_2 d_1 d_2 \cos(\theta_2)]\ddot{\theta}_2 \\ &\quad - 2 m_2 d_1 d_2 \sin(\theta_1)\dot{\theta}_1\dot{\theta}_2 - m_2 d_1 d_2 \sin(\theta_2)\dot{\theta}_2^2 \end{aligned}$$

Torque T_1

$$F_i = \frac{d}{dt}\left(\frac{\partial L}{\partial \dot{q}_i}\right) - \frac{\partial L}{\partial q_i}$$

$$T_1 = \frac{d}{dt}\frac{\partial L}{\partial \dot{\theta}_1} - \frac{\partial L}{\partial \theta_1}$$

$$\begin{aligned} T_1 &= [(m_1 + m_2) d_1^2 + m_2 d_2^2 + 2 m_2 d_1 d_2 \cos(\theta_2)]\ddot{\theta}_1 \\ &\quad + [m_2 d_2^2 + m_2 d_1 d_2 \cos(\theta_2)]\ddot{\theta}_2 - 2 m_2 d_1 d_2 \sin(\theta_2)\dot{\theta}_1\dot{\theta}_2 \\ &\quad - m_2 d_1 d_2 \sin(\theta_2)\dot{\theta}_2^2 + (m_1 + m_2) g d_1 \sin(\theta_1) + m_2 g d_2 \sin(\theta_1 + \theta_2) \end{aligned}$$

$$\frac{\partial L}{\partial \theta_2} = -m_2 d_1 d_2 \sin(\theta_2)(\dot{\theta}_1^2 + \dot{\theta}_1\dot{\theta}_2) - m_2 g d_2 \sin(\theta_1 + \theta_2)$$

$$\frac{\partial L}{\partial \dot{\theta}_2} = m_2 d_2^2 (\dot{\theta}_1 + \dot{\theta}_2) + m_2 d_1 d_2 \cos(\theta_2)\dot{\theta}_1 - m_2 d_1 d_2 \sin(\theta_2)\dot{\theta}_1\theta_2$$

$$\frac{d}{dt}\frac{\partial L}{\partial \dot{\theta}_2} = m_2 d_2^2 (\ddot{\theta}_1 + \ddot{\theta}_2) + m_2 d_1 d_2 \cos(\theta_2)\ddot{\theta}_1 - m_2 d_1 d_2 \sin(\theta_2)\ddot{\theta}_1\ddot{\theta}_2$$

Torque T_2

$$T_2 = \frac{d}{dt}\frac{\partial L}{\partial \dot{\theta}_2} - \frac{\partial L}{\partial \theta_2}$$

$$= m_2 d_2^2(\ddot{\theta}_1) + m_2 d_2^2(\ddot{\theta}_2) + m_2 d_1 d_2 \cos(\theta_2)\ddot{\theta}_1 - m_2 d_1 d_2 \sin(\theta_2)\dot{\theta}_1\dot{\theta}_2$$
$$+ m_2 d_1 d_2 \sin(\theta_2)\dot{\theta}_1^2 + m_2 d_1 d_2 \sin(\theta_2)\dot{\theta}_1\dot{\theta}_2 + m_2 g d_2 \sin(\theta_1 + \theta_2)$$

$$T_2 = [m_2 d_2^2 + m_2 d_1 d_2 \cos(\theta_2)]\ddot{\theta}_1 + m_2 d_2^2(\ddot{\theta}_2)$$
$$+ m_2 d_1 d_2 \sin(\theta_2)\dot{\theta}_1^2 + m_2 g d_2 \sin(\theta_1 + \theta_2)$$

3.20 On Roboanalyzer

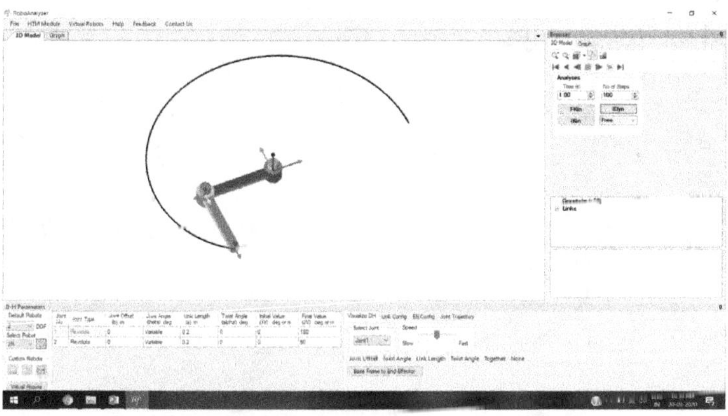

Graphs on Roboanalyzer

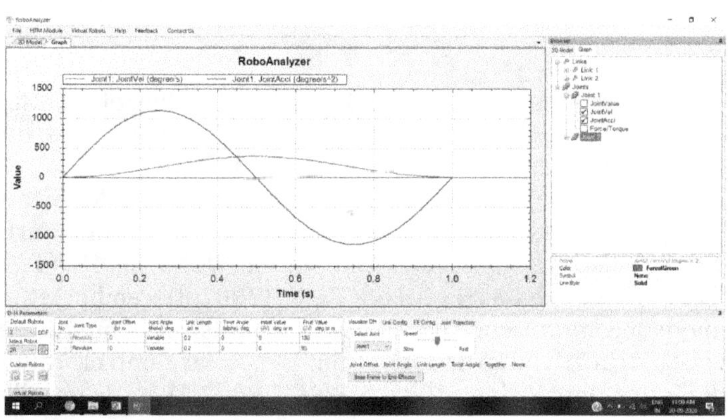

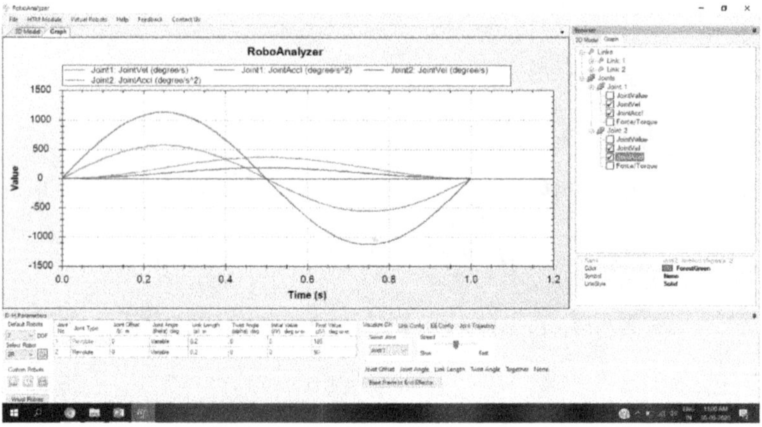

Free Fall on Roboanalyzer

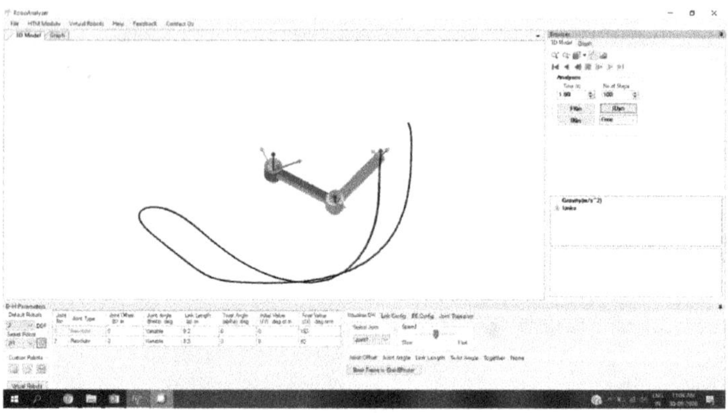

References

1. Singh, A., & Singla, A. (2017). Kinematic modeling of robotic manipulators. Proceedings of the national academy of sciences, India section A: physical sciences, 87(3), 303-319.
2. Everett, L., Driels, M., & Mooring, B. (1987, March). Kinematic modelling for robot calibration. In Proceedings. 1987 IEEE International Conference on Robotics and Automation (Vol. 4, pp. 183-189). IEEE.
3. Leow, Y. P., Low, K. H., & Loh, W. K. (2002, December). Kinematic modelling and analysis of mobile robots with omni-directional wheels. In 7th International Conference on Control, Automation, Robotics and Vision, 2002. ICARCV 2002. (Vol. 2, pp. 820-825). IEEE.

4. Kucuk, S., & Bingul, Z. (2006). Robot kinematics: Forward and inverse kinematics. INTECH Open.
5. Cursi, F., Mylonas, G. P., & Kormushev, P. (2020). Adaptive kinematic modelling for multiobjective control of a redundant surgical robotic tool. Robotics, 9(3), 68.
6. Deshpande, V., & George, P. M. (2014). Kinematic modelling and analysis of 5 DOF robotic arm. International Journal of Robotics Research and Development (IJRRD), 4(2), 17-24.
7. Gouasmi, M., Ouali, M., Fernini, B., & Meghatria, M. H. (2012). Kinematic modelling and simulation of a 2-R robot using solidworks and verification by MATLAB/Simulink. International Journal of Advanced Robotic Systems, 9(6), 245.
8. Betourne, A., & Campion, G. (1996, April). Kinematic modelling of a class of omnidirectional mobile robots. In Proceedings of IEEE International Conference on Robotics and Automation (Vol. 4, pp. 3631-3636). IEEE.
9. Larsen, T. D., Hansen, K. L., Andersen, N. A., & Ravn, O. (1999, August). Design of Kalman filters for mobile robots; evaluation of the kinematic and odometric approach. In Proceedings of the 1999 IEEE international conference on control applications (Cat. No. 99CH36328) (Vol. 2, pp. 1021-1026). IEEE.
10. Kebria, P. M., Al-Wais, S., Abdi, H., & Nahavandi, S. (2016, October). Kinematic and dynamic modelling of UR5 manipulator. In 2016 IEEE international conference on systems, man, and cybernetics (SMC) (pp. 004229-004234). IEEE.
11. Schröer, K., Albright, S. L., & Grethlein, M. (1997). Complete, minimal and model-continuous kinematic models for robot calibration. Robotics and Computer-Integrated Manufacturing, 13(1), 73-85.
12. Pisla, D., Gherman, B., Vaida, C., & Plitea, N. (2012). Kinematic modelling of a 5-DOF hybrid parallel robot for laparoscopic surgery. Robotica, 30(7), 1095-1107.
13. Tondu, B. (2009, June). Kinematic modelling of anthropomorphic robot upper limb with human-like hands. In 2009 International Conference on Advanced Robotics (pp. 1-9). IEEE.
14. Leena, N., & Saju, K. K. (2016). Modelling and trajectory tracking of wheeled mobile robots. Procedia technology, 24, 538-545.
15. Singh, S., Singla, A., Singh, A., Soni, S., & Verma, S. (2016). Kinematic modelling of a five-DOFs spatial manipulator used in robot-assisted surgery. Perspectives in Science, 8, 550-553.
16. Barrientos-Diez, J., Dong, X., Axinte, D., & Kell, J. (2021). Real-time kinematics of continuum robots: modelling and validation. Robotics and Computer-Integrated Manufacturing, 67, 102019.
17. Xin, G., Deng, H., Zhong, G., & Wang, H. (2015). Hierarchical kinematic modelling and optimal design of a novel hexapod robot with integrated limb mechanism. International Journal of Advanced Ro-

botic Systems, 12(9), 123.

18. Martínez, J. L., Mandow, A., Morales, J., Garcia-Cerezo, A., & Pedraza, S. (2004, September). Kinematic modelling of tracked vehicles by experimental identification. In 2004 IEEE/RSJ International Conference on Intelligent Robots and Systems (IROS) (IEEE Cat. No. 04CH37566) (Vol. 2, pp. 1487-1492). IEEE.
19. Pan, X., Wang, H., Jiang, Y., & Xiao, J. (2013). Automatic kinematic modelling of a modular reconfigurable robot. Transactions of the Institute of Measurement and Control, 35(7), 922-932.
20. Singh, T. P., Suresh, P., & Chandan, S. (2017). Forward and inverse kinematic analysis of robotic manipulators. International Research Journal of Engineering and Technology (IRJET), 4(2), 1459-1468.

4

Manipulator Dynamic Modelling in Robotics via Different Methods

Soumya Ranjan Mangaraj[1], Dillip Ranjan Nayak[2], Ashutosh Khuntia[3], Sandeep Bhoi[4], and Kishor Chandra Panda[5]

[1]*Department of Mechanical Engineering, Government College of Engineering, Kalahandi, Bhawanipatna, Odisha, India*
[2]*Department of Computer Science and Engineering, Government College of Engineering, Kalahandi, Bhawanipatna, Odisha, India*
[3]*Department of Electrical Engineering, College of Engineering and Technology, Bhubaneswar, Odisha, India*
[4]*Department of Mechanical Engineering, Parala Maharaja Engineering College, Berhampur Odisha, India*
[5]*Department of Civil Engineering, Government College of Engineering, Kalahandi, Bhawanipatna, Odisha, India*

4.1 Newton-Euler Formulation

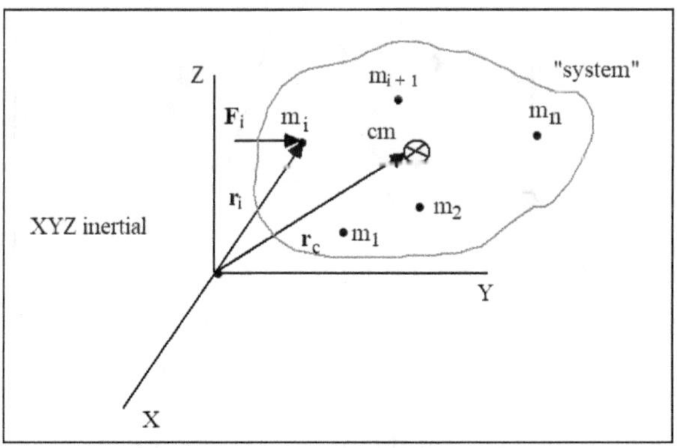

Figure 4.1 Newton-Euler formulation.

Given a system of particles translating through space, each particle i being acted upon by external force Fi, and each particle located relative to an inertial reference frame, the governing equations are [1,2]:

$$F_c = m\, a_c \quad \textbf{(1)}$$

$$M = H \quad \textbf{(2)}$$

$$M_c = H_c \tag{3}$$

where,
m = total mass (sum over all mass particles)
a_c = acceleration of center of mass (cm) of all mass particles
F_c = sum of external forces applied to system of particles as if applied at cm
$H_i = r_i \times m_i v_i$ = angular momentum of particle i (also called moment of momentum)
H, H_c = angular momentum summed over all particles, measured about inertial point, cm point, respectively
M, M_c = moment of all external forces applied to system of particles, measured about inertial point, cm point, respectively.

4.2 Rigid Bodies in General Motion (Translating and Rotating)

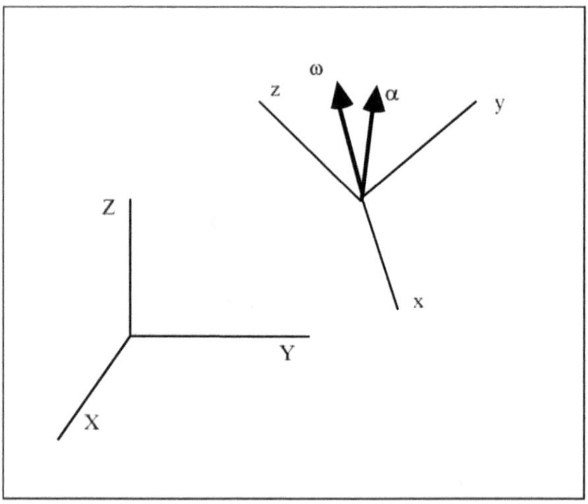

Figure 4.2 Rigid bodies in general motion (translating and rotating).

The time rate of change of any vector V capable of being viewed in either XYZ or xyz is

$$\left[\frac{dV}{dt}\right]_{XYZ} = \left[\frac{dV}{dt}\right]_{xyz} + \mathbf{w} \times \mathbf{V} \tag{4}$$

where w is the angular velocity of a secondary translating, rotating reference frame (xyz). Another common form of the equation is:

$$V = V_r + \mathbf{w} \times \mathbf{V} \quad \textbf{(5)}$$

and when applied to rate of change of angular momentum becomes

$$\mathbf{M} = H = H_r + \mathbf{w} \times \mathbf{H} \quad \textbf{(6)}$$

which is referred to as Euler's equation.

4.3 Rotating Rigid Body

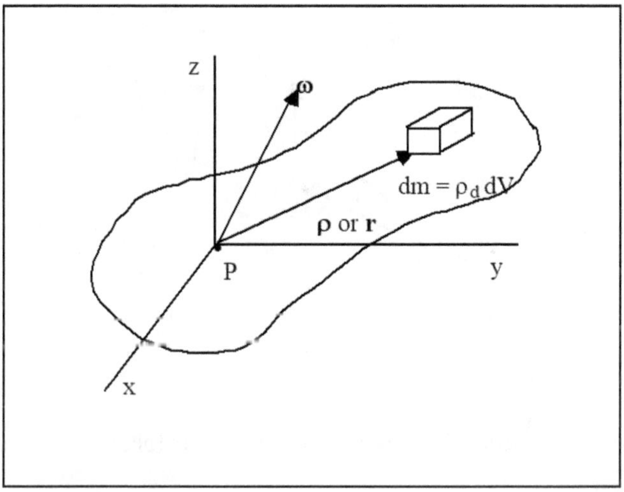

Figure 4.3 Rotating rigid body.

in matrix form

$$H = J\,w$$

where J = inertia matrix

By integrating the motion over the rigid body, we can express the angular momentum relative to the xyz axes as

$$\begin{aligned} H &= H_x\, i + H_y\, j + H_z\, k \\ &= (J_{xx}\, w_x + J_{xy}\, w_y + J_{xz}\, w_z)\, i \\ &\quad + (J_{yx}\, w_x + J_{yy}\, w_y + J_{yz}\, w_z)\, j \\ &\quad + (J_{zx}\, w_x + J_{zy}\, w_y + J_{zz}\, w_z)\, k \end{aligned} \quad \textbf{(7)}$$

Taking the derivative of (7) and substituting into (6), also assuming the body axes to be aligned with the *principal axes*, we get Euler's moment equations [3-7]:

$$M_x = J_{xx}w_x + (J_{zz} - J_{yy})\, w_y\, w_z \quad \textbf{(8)}$$
$$M_y = J_{yy}w_y + (J_{xx} - J_{zz})\, w_x\, w_z \quad \textbf{(9)}$$
$$M_z = J_{zz}w_z + (J_{yy} - J_{xx})\, w_x\, w_y \quad \textbf{(10)}$$

4.4 Acceleration Relative to a Non-Inertial Reference Frame

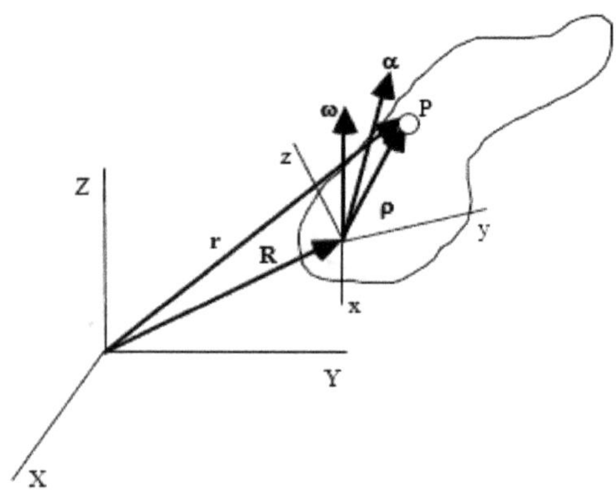

Figure 4.4 Acceleration relative to a non-inertial reference frame.

By taking two derivatives and applying appropriately, the absolute acceleration of point P can be shown to be

$$a = r = R + a \times r + w \times (w \times r) + \rho + 2\, w \times \quad \textbf{(11)}$$

where,
R = acceleration of xyz origin
a x r = tangential acceleration
w x (w x r) = centripetal acceleration
ρ= acceleration of P relative to xyz
2 w x ρ = Coriolis acceleration

For the special case of xyz fixed to rigid body and P a point in the body, and (11) reduces to

$$a = r = R + a \times r + w \times (w \times r) \quad \textbf{(12)}$$

If P at cm, then

a = r=R + a x r_c + w x (w x r_c) **(13)**

4.5 Recursive Newton-Euler Equations (Forward Recursion for Motion)

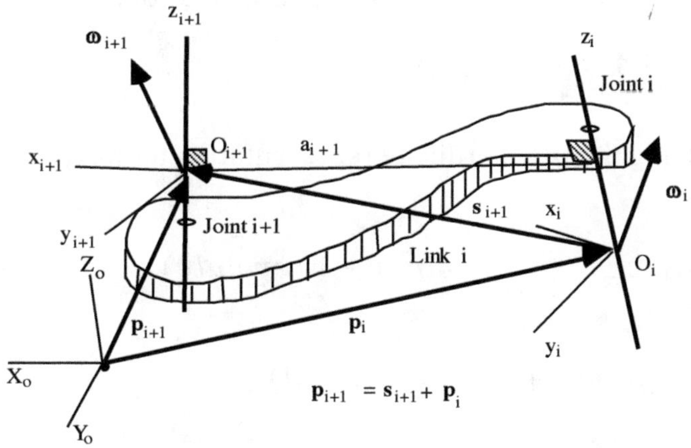

Figure 4.5 Recursive Newton-Euler Equations (forward recursion for motion).

If v_i = $_i$ and w_i is defined to be the angular velocity of the ith joint frame x_i y_i z_i with respect to base coordinates, then

$$\boldsymbol{V_{i+1}} = \boldsymbol{V_i} + \frac{\boldsymbol{d}}{\boldsymbol{dt}}(\boldsymbol{s_{i+1}})$$

$$\boldsymbol{V_{i+1}} = \boldsymbol{V_i} + \boldsymbol{\dot{S}^*_{i+1}} + \boldsymbol{\omega_i x s_{i+1}} \quad \textbf{(14)}$$

where describes the velocity of x_{i+1}, y_{i+1}, z_{i+1} relative to an observer in frame x_i, y_i, z_i.

Likewise, the acceleration becomes

$$\dot{v}_{i+1} = \dot{v}_i + \ddot{s}^*_{i+1} + \dot{\omega}_i x s_{i+1} + 2\omega_i x \dot{s}^*_{i+1} + \omega_i x(\ \omega_i\ x s_{i+1}) \quad \textbf{(15)}$$

Defining w_{i+1} to be the absolute angular velocity of the i+1 frame and to be the angular velocity of the i+1 frame relative to the ith frame [2, 8,9]:

$$\omega_{i+1} = \omega_{i+1}^{*} + \omega_i \tag{16}$$

Taking one more derivative for angular acceleration:

$$\dot{\omega}_{i+1} = \dot{\omega}_i + \dot{\omega}_{i+1}^{*} + \dot{\omega}_i X \omega_{i+1}^{*} \tag{17}$$

Now applying the DH coordinate representation for manipulators:

$$\omega_{i+1}^{*} = \dot{\theta}_{i+1} z_{i+1} \qquad (joint\ i+1\ revolute) \tag{18}$$

$$\omega_{i+1}^{*} = 0 \quad (joint\ i+1\ translational) \tag{19}$$

$$\dot{\omega}_{i+1}^{*} = \ddot{\theta}_{i+1} z_{i+1} \quad (joint\ i+1\ revolute) \tag{20}$$

$$\dot{\omega}_{i+1}^{*} = 0 \qquad (joint\ i+1 translational) \tag{21}$$

Using the previous equations, we can generate the angular motion recursive equations:

$$\omega_{i+1} = \omega_i + \dot{\theta}_{i+1} z_{i+1} \qquad (revolute) \tag{22}$$

$$\omega_{i+1} = \omega_i \qquad (translational) \tag{23}$$

$$\dot{\omega}_{i+1} = \dot{\omega}_i + \dot{\theta}_{i+1} z_{i+1} + \omega_i X \left(\dot{\theta}_{i+1} z_{i+1}\right) (revolute) \tag{24}$$

$$\dot{\omega}_{i+1} = \dot{\omega}_i \qquad (translational) \tag{25}$$

The linear velocity and acceleration equations use the D-H forms:

$$\dot{\boldsymbol{s}}^*_{i+1} = \mathbf{0} \qquad (\boldsymbol{revolute}) \tag{26}$$

$$\dot{\boldsymbol{S}}^*_{i+1} = \boldsymbol{d}_{1+1}\boldsymbol{z}_{i+1} \qquad (\boldsymbol{translation}) \tag{27}$$

$$\ddot{\boldsymbol{s}}^*_{i+1} = \mathbf{0} \qquad (\boldsymbol{revolute}) \tag{28}$$

$$\ddot{\boldsymbol{S}}^*_{i+1} = \ddot{\boldsymbol{d}}_{i+1}\boldsymbol{z}_{i+1} \qquad (\boldsymbol{translational}) \tag{29}$$

where $\dot{\mathbf{d}}_{i+1}$ is the translational velocity of x_{i+1}, y_{i+1}, z_{i+1} relative to x_i, y_i, z_i

Substituting (26) – (29), we get the velocity and acceleration recursion equations:

$$v_{i+1} = v_i + \omega_i x s_{i+1} \qquad (revolute) \tag{30}$$

$$v_{i+1} = v_i + \dot{d}_{i+1} z_{i+1} + \omega_{i+1} x s_{i+1} \quad (translational) \tag{31}$$

$$\dot{v}_{i+1} = \dot{v}_i + \dot{\omega} x(\omega_{i+1} x s_{i+1}) \qquad (revolute) \tag{32}$$

$$\dot{v}_{i+1} = \dot{v}_i + \ddot{d}_{i+1} z_{i+1} + \dot{\omega}_i + 1 X s_{i+1} + 2\omega_{i+1} X(\dot{d}_{i+1} z_{i+1}) + \omega_{i+1} x(\omega_{i+1} x s_{i+1}) \qquad (translational) \tag{33}$$

Note that $w_{i+1} = w_i$ for translational link i+1 [2, 10-13].

4.6 Recursive Newton-Euler Equations (Backward Recursion for Forces and Torques)

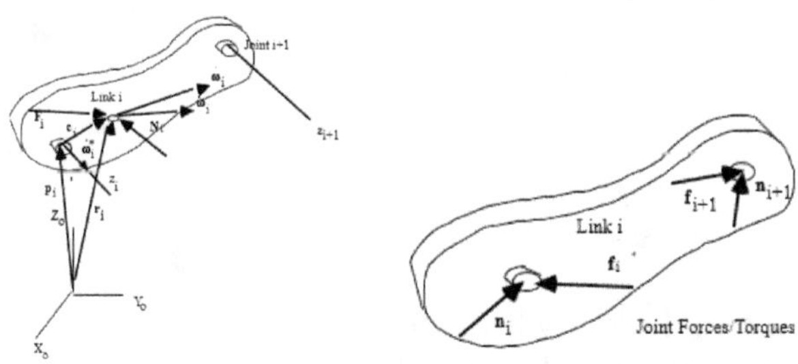

Figure 4.6 Recursive Newton-Euler Equations (backward recursion for forces and torques).

Define the terms:
m_i = mass of link i
r_i = position of link i cm with respect to base coordinates
F_i = total force exerted on link i
N_i = total moment exerted on link i
J_i = inertia matrix of link i about its cm determined in the X_o Y_o Z_o axes
f_i = force exerted on link i by link i-1
n_i = moment exerted on link i by link i-1

For each link we must apply the N-E equations:

$$\boldsymbol{F_i = m_i \ddot{r}_i} \quad \textbf{(34)}$$

$$\boldsymbol{N_i = J_i \dot{\omega}_i + \omega_i X (J_i \omega_i)} \quad \textbf{(35)}$$

The gravitational acceleration and damping torque will be added to the equations of motion later. Now $_i$ is easily calculated by knowing the acceleration of the origin of the ith frame attached to link i at joint i. We locate link i cm with respect to x_i y_i z_i by c_i such that $r_i = c_i + p_i$.

$$\boldsymbol{\ddot{r} = \dot{v}_i + \dot{\omega}_i X c_i + \omega_i X (\omega_i X C_i)} \quad \textbf{(36)}$$

The velocity of the cm of link i is obviously

$$\boldsymbol{\dot{r}_i = v_i + \omega_i X c_i} \quad \textbf{(37)}$$

To determine F_i and N_i define

f_i = force exerted on link i by link i-1
n_i = moment exerted on link i by link i-1

Then

$F_i = f_i - f_{i+1}$ **(38)**

and

$N_i = n_i - n_{i+1} + (p_i - r_i) \times f_i - (p_{i+1} - r_i) \times f_{i+1}$ **(39)**

$= n_i - n_{i+1} - c_i \times F_i - s_{i+1} \times f_{i+1}$

The previous equations can be placed in the backwards recursion form to work from the forces/moments exerted on the hand backwards to the joint torques necessary to react to these hand interactions and move the manipulator:

$\mathbf{f_i = f_{i+1} + F_i}$ **(40)**

$\mathbf{n_i = n_{i+1} + c_i \times F_i + s_{i+1} \times f_{i+1} + N_i}$ **(41)**

The motor torque t_i required at joint i is the sum of the joint torque n_i resolved along the revolute axis plus the damping torque,

$t_i = \mathbf{n_i} \cdot \mathbf{z_i} + b_i\ \boldsymbol{\theta_i}$ **(revolute)** **(42)**

where b_i is the damping coefficient.

For a translational joint

$t_i = l_i\ \mathbf{f_i} \cdot \mathbf{z_i} + b_i\ d_i$ **(translational)** **(43)**

where l_i is the torque arm for motor i.

What about gravity?

The effect of gravity on each link is accounted for by applying a base acceleration equal to gravity to the base frame of the robot: $_o = g\ z_o$ with z_o vertical. $_o$ is applied to the base link in equations (36) and (37) for i = 0 and this serves to transmit the acceleration of gravity to each link by recursion.

There are two basic problems with the derivation so far. What are they?

Problem 1 – Find J_i when resolved into base coordinates is a function of manipulator configuration. To avoid this unnecessary complexity, we apply the equations at the cm of each link where J_i is constant.

Problem 2 – The recursive relations have not resolved the various vectors from one joint frame to the next. We must adjust the equations accordingly.

Do we use the full homogeneous transformation in the recursive equations?

We resolve the free vectors by applying the rotational sub-matrix of the D-H transformations for each joint frame to the recursive vectors, using the Craig/Red D-H representation. Let us also use Tsai's notation [2, 8].

joint frame i+1 relative to joint frame i: $^{i}R_{i+1}$

joint frame i relative to joint frame i+1: $^{i+1}R_{i}$

4.7 Revised Angular Motion Equations

$$\dot{\omega}_{i+1} = {}^{i+1}R_i\omega_i + \dot{\theta}_{i+1}z_{i+1} \tag{44}$$

$$\omega_{i+1} = {}^{i+1}R_i\,\omega_i \tag{45}$$

$$\dot{\omega}_{i+1} = {}^{i+1}R_i\dot{\omega}_i + \ddot{\theta}_{i+1} + z_{i+1} + {}^{i+1}R_i\,\omega_i X(\dot{\theta}_{i+1}z_{i+1}) \tag{46}$$

$$\dot{\omega}_{i+1} = {}^{i+1}R_i\dot{\omega}_i \tag{47}$$

$$v_{i+1} = {}^{i+1}\mathbf{R}_i(v_i + \omega_i X s_{i+1}) \qquad \text{(revolute)} \qquad \mathbf{(48)}$$

$$v_{i+1} = {}^{i+1}\mathbf{R}_i + \dot{d}_{i+1} z_{i+1} + \dot{\omega}_{i+1} X \, {}^{i+1}\mathbf{R}_i \boldsymbol{s}_{i+1} \qquad \text{(translational)} \qquad \mathbf{(49)}$$

$$\dot{v}_{i+1} = {}^{i+1}\mathbf{R}_i(v_i + \dot{\omega} X s_{i+1} + \omega_i X(\dot{\omega}_X s_{i+1})) \quad \text{(revolute)} \qquad \mathbf{(50)}$$

$$\dot{v}_{i+1} = {}^{i+1}\mathbf{R}_i + \dot{d}_{i+1} z_{i+1} + \dot{\omega}_{i+1} X \, {}^{i+1}\mathbf{R}_i \boldsymbol{s}_{i+1} + \boldsymbol{2\omega_{i+1} X(\dot{d}_{i+1} z_{i+1})} + \boldsymbol{\omega_{i+1} X(\omega_{i+1}} \, {}^{i+1}\mathbf{R}_i \boldsymbol{s}_{i+1})\text{(translational)} \qquad \mathbf{(51)}$$

$$\ddot{r}_i = \dot{v}_i + \dot{\omega} X c_i + \omega_i x(\omega_i X c_i) \qquad \mathbf{(52)}$$

$$F_i = m_i \ddot{r}_i \qquad \mathbf{(53)}$$

$$N_i = J_i \dot{\omega}_i + \omega_i X(J_i \omega_i) \qquad \mathbf{(54)}$$

$$f_i = {}^{i}\boldsymbol{R}_{i+1} \boldsymbol{f}_{i+1} + \boldsymbol{F}_i \qquad \mathbf{(55)}$$

$$\boldsymbol{n}_i = {}^{i}\boldsymbol{R}_{i+1} \boldsymbol{n}_{i+1} + \boldsymbol{c}_i \boldsymbol{x} \boldsymbol{F}_i + \boldsymbol{s}_{i+1} \boldsymbol{x}({}^{i}\boldsymbol{R}_{i+1} \boldsymbol{f}_{i+1} + \boldsymbol{N}_i) \qquad \mathbf{(56)}$$

$$\tau_i = n_i T z_i + b_i \dot{\theta}_i \qquad \text{(revolute)} \qquad \mathbf{(57)}$$

$$\tau_i = I_i f_i T z_i + b_i \dot{d}_i \qquad \text{(translational)} \qquad \mathbf{(58)}$$

4.8 Dynamic Equation of a Rigid Body

Two methods to form dynamic equation of a rigid body:

Lagrangian equation (Energy-based)
Newton-Euler equation (Force-balance)

Summary of the steps to form Lagrangian equation of n-link manipulators:

Computing the velocity of any given point;
Computing total kinetic energy;
Computing total potential energy;
Forming Lagrangian function of the system;
Forming Dynamic Equation through Lagrangian equation [12].

4.9 Dynamics Summary

The N-E equations are applied recursively to generate the forces and torques at each joint motor. We first apply forward recursion to get the motion state for each link. We then use this motion state to propagate the forces and torques in backward recursion to each joint. The rotational sub-matrix of the D-H transformations must be applied to resolve the vectors correctly into each link's joint frame [2].

4.10 Overview of Sensors for Robotics

What makes a machine a robot?

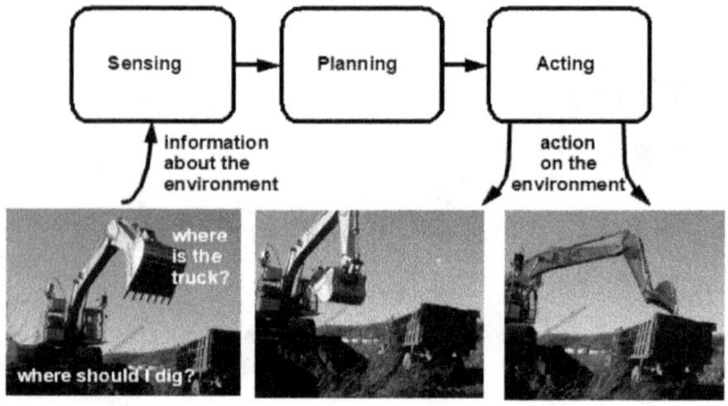

Figure 4.7 Machine as a robot.

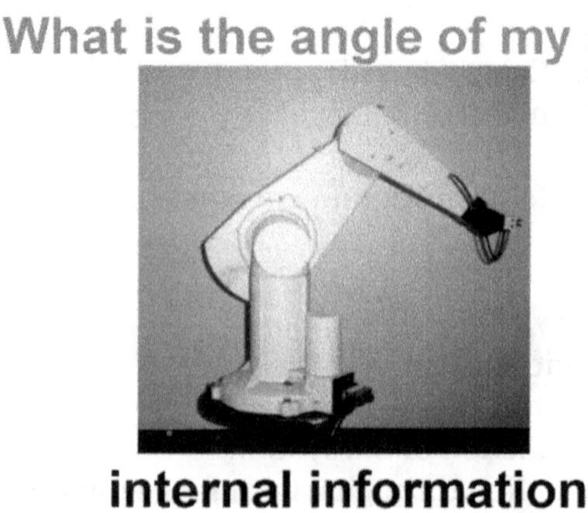

Figure 4.8 Simple robotic arm.

What is Sensing?

- Collect information about the world.
- Sensor - an electrical/mechanical/chemical device that maps an environmental attribute to a quantitative measurement
    - attribute mixtures - often no one to one map
    - hidden state in environment
    - Each sensor is based on a *transduction principle* - conversion of energy from one form to another
- Also known as transducers [1].

Why do robots need sensors?

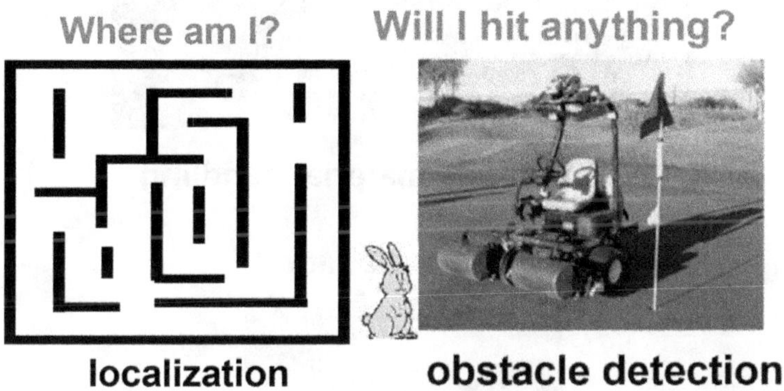

Figure 4.9 Location finder sensor.

Sensing for Specific Tasks

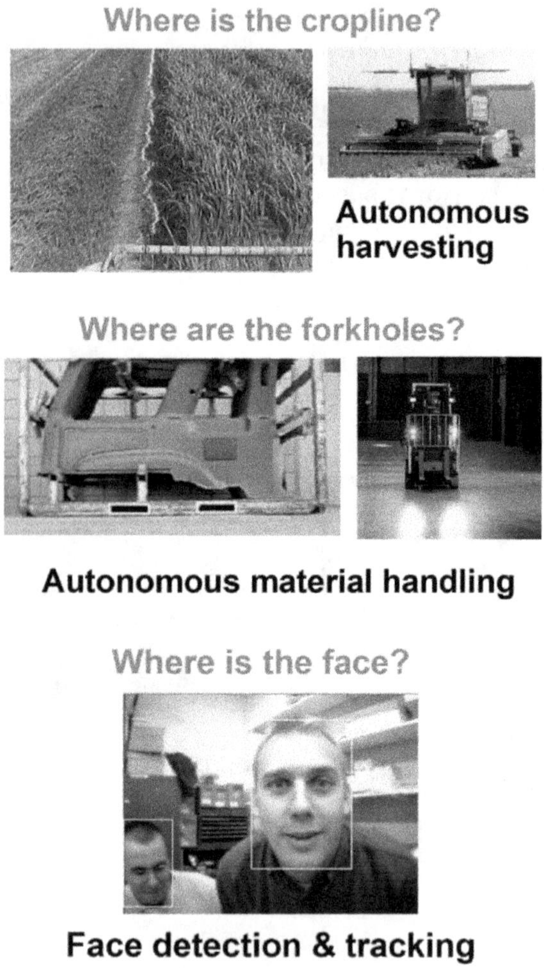

Figure 4.10 Sensing for specific tasks.

4.11 Types of Sensors

Active

- send signal into environment and measure interaction of signal w/ environment
- e.g., radar, sonar

Passive

- record signals already present in environment
- e.g., video cameras

4.12 Classification by Medium Used

- based on electromagnetic radiation of various wavelengths
- vibrations in a medium
- concentration of chemicals in environment
- by physical contact

4.13 Interoceptive

- what is my thirst level? (biochemical)
- what is my battery charge? (voltmeter)

4.14 Simple Practical Sensors We Can Purchase

- Touch sensors
- Tilt sensors
- Encoders
- Bend sensors
- Light sensors
- Temperature sensors
- Potentiometers
- Laser rangefinders
- Cameras

Touch Sensors

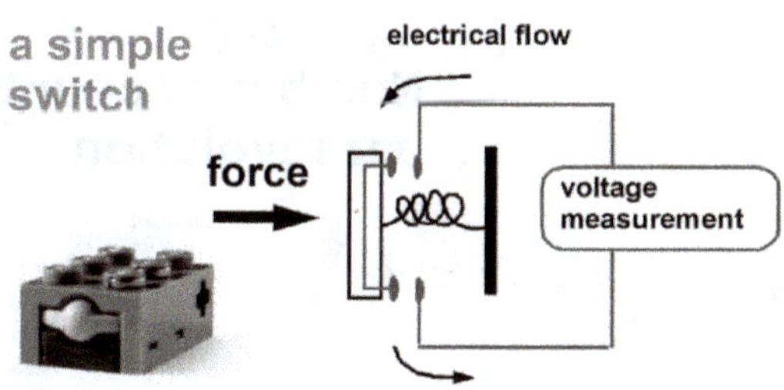

Figure 4.11 Touch sensors.

Tilt Sensors

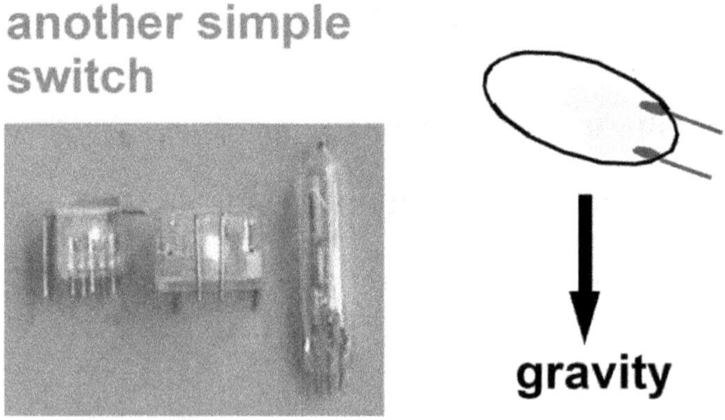

Figure 4.12 Tilt sensors.

Encoders

- Encoders measure rotational motion.
- They can be used to measure the rotation of a wheel.
- Servo motors: Used in conjunction with an electric motor to measure the motor's position and, in turn, control its position [1, 2].

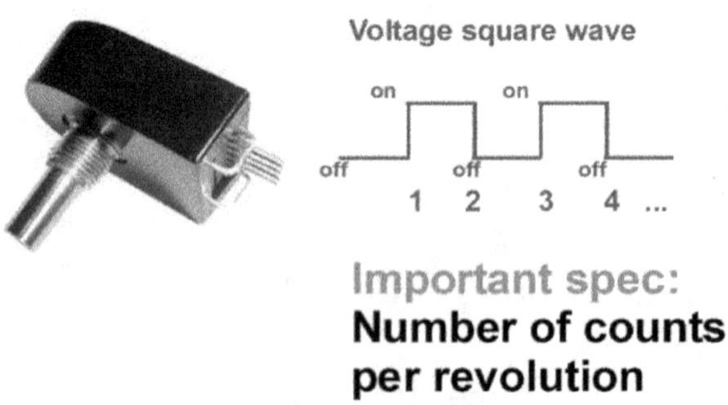

Figure 4.13 Encoders.

Sensor Analysis

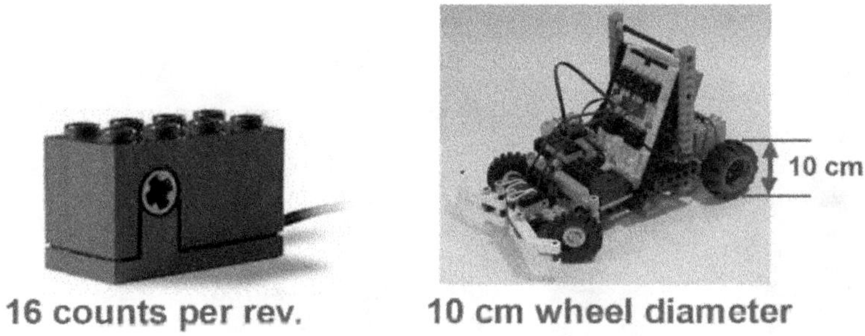

Figure 4.14 Sensor analysis.

- How far does the wheel travel for 1 encoder count?
- What happens if we change the wheel diameter?
- How many counts are there per meter of travel?

Electrical Analogy

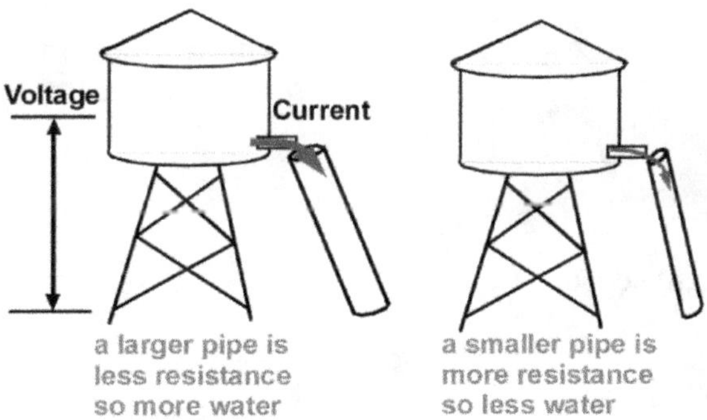

Figure 4.15 Electrical analogy.

Bend Sensor

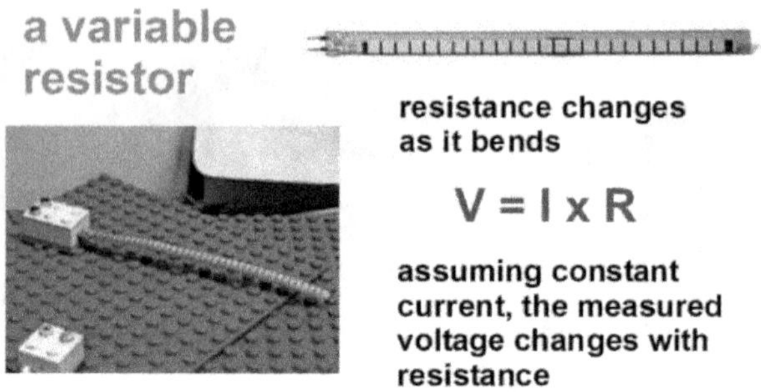

Figure 4.16 Bend sensor.

Light Sensor

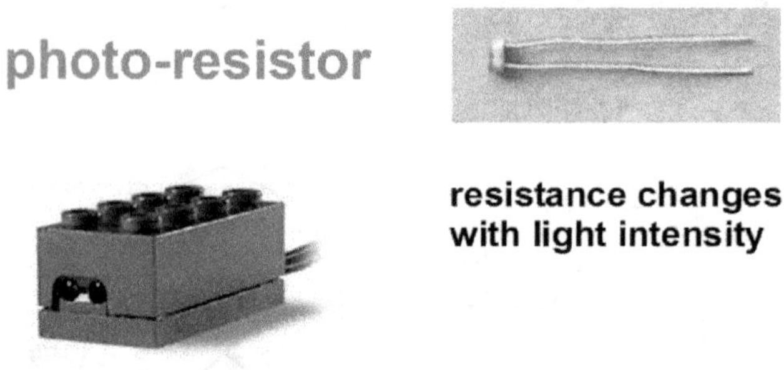

Figure 4.17 Light sensor.

Temperature Sensor

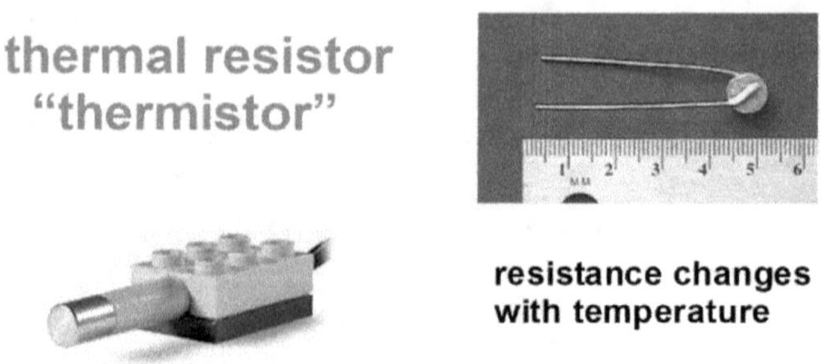

Figure 4.18 Temperature sensor.

Potentiometer

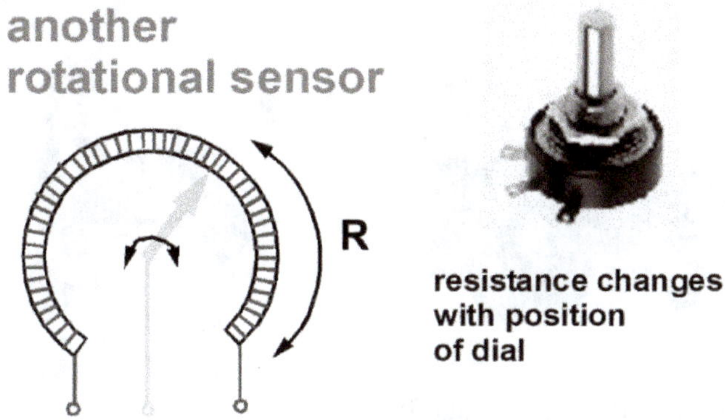

Figure 4.19 Potentiometer.

Sensors Based on EM Spectrum

- Basically used for ranging
- Light sensitive
    - eyes, cameras, photocells etc.
- Operating principle
    - CCD - charge coupled devices
    - photoelectric effect
- IR sensitive - FLIR
    - sense heat differences and construct images
    - night vision application
- Radio and Microwave
    - RADAR: **Ra**dio **D**etection **a**nd **R**anging
    - Microwave radar: insensitive to clouds
- Coherent light
    - all photons have same phase and wavelength
    - LASER: **L**ight **A**mplification by **S**timulated **E**mission of **Ra**-diation
    - LASER RADAR: LADAR - accurate ranging

The SICK Laser Rangefinder

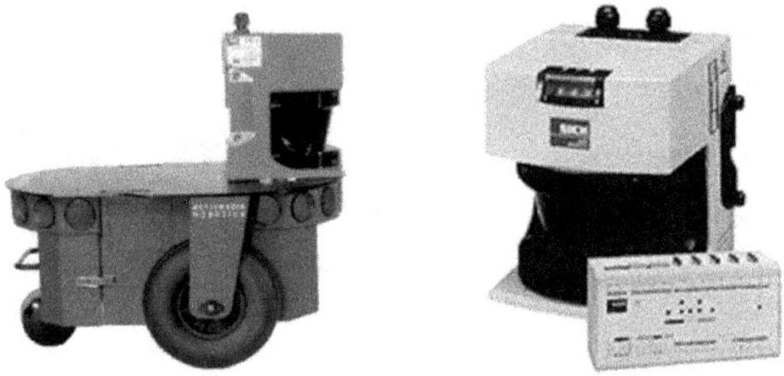

Figure 4.20 The SICK laser rangefinder.

EM Spectrum

- Nuclear Magnetic Resonance (NMR)
    - heavy duty magnetic field lines up lines up atoms in a body
    - now expose body to radio signals
    - different nuclei resonate at different frequencies which can be measured leading to an image

Local Proximity Sensing in EM

- Infrared LEDs
    - cheap, active sensing
    - usually low resolution - normally used for presence/absence of obstacles rather than ranging
    - operate over small range

4.15 Sensors Based on Sound

- SONAR: **So**und **N**avigation **a**nd **R**anging
    - bounce sound off of something
    - measure time for reflection to be heard - gives a range measurement
    - measure change in frequency - gives the relative speed of the object (Doppler effect)
    - bats and dolphins use it with amazing results

- robots use it w/ less than amazing results

Sonar and IR Proxmity

Figure 4.21 Sonar and IR proxmity.

Odour Sensors

- Detection of chemical compounds and their density in an area
    - spectroscopy - mostly lab restricted
    - fibre-optic techniques - recently developed
    - chemical detection - sniffers aand electronic noses via "wet chemistry on a chip"
    - No major penetration in robotics yet applications are vast (e.g., mine detection)

Touch Sensors

- Whiskers, bumpers etc.
    - mechanical contact leads to
        - closing/opening of a switch
        - change in resistance of some element
        - change in capacitance of some element
        - change in spring tension

Proprioceptive Sensors

- Encoders, Potentiometers
    - measure angle of turn via change in resistance or by counting optical pulses
- Gyroscopes
    - measure rate of change of angles
    - fiber-optic (newer, better), magnetic (older)

- Compass
    - measure which way is north
- GPS: measure location relative to globe

Figure 4.22 Proprioceptive sensors.

4.16 What Sensors to Employ?

- E.g., mapping
    - ranging - laser, sonar, IR, stereo camera pair
    - salient feature detection - doors using color
- Factors
    - Accuracy, cost, information needed etc.

4.17 Where Do You Put Them?

- On/off board (e.g., localization using odometry vs. localization using beacons)
- If onboard - where?
- Reasonable arrangements - heuristic
- Optimal arrangements - mathematically rigorous [1, 3, 6, 14-23]

References

1. Montazeri, A., West, C., Monk, S. D., & Taylor, C. J. (2017). Dynamic modelling and parameter estimation of a hydraulic robot manipulator using a multi-objective genetic algorithm. International Journal of Control, 90(4), 661-683.

2. Tang, L., Gouttefarde, M., Sun, H., Yin, L., & Zhou, C. (2021). Dynamic modelling and vibration suppression of a single-link flexible manipulator with two cables. Mechanism and Machine Theory, 162, 104347.
3. Chen, M. W., & Zalzala, A. M. S. (1997). Dynamic modelling and genetic-based trajectory generation for non-holonomic mobile manipulators. Control Engineering Practice, 5(1), 39-48.
4. Martins, J. M., Mohamed, Z., Tokhi, M. O., Da Costa, J. S., & Botto, M. A. (2003). Approaches for dynamic modelling of flexible manipulator systems. IEE Proceedings-Control Theory and Applications, 150(4), 401-411.
5. Codourey, A. (1996, November). Dynamic modelling and mass matrix evaluation of the DELTA parallel robot for axes decoupling control. In Proceedings of IEEE/RSJ international conference on intelligent robots and systems. IROS'96 (Vol. 3, pp. 1211-1218). IEEE.
6. Serrezuela, R. R., Chavarro, A. F. C., Cardozo, M. A. T., Toquica, A. L., & Martinez, L. F. O. (2017). Kinematic modelling of a robotic arm manipulator using Matlab. ARPN Journal of Engineering and Applied Sciences, 12(7), 2037-2045.
7. Tarn, T. J., Shoults, G. A., & Yang, S. P. (1996). A dynamic model of an underwater vehicle with a robotic manipulator using Kane's method. Autonomous robots, 3(2), 269-283.
8. West, C., Montazeri, A., Monk, S. D., & Taylor, C. J. (2016). A genetic algorithm approach for parameter optimization of a 7DOF robotic manipulator. IFAC-PapersOnLine, 49(12), 1261-1266.
9. Terry, J. S., Rupert, L., & Killpack, M. D. (2017, November). Comparison of linearized dynamic robot manipulator models for model predictive control. In 2017 IEEE-RAS 17th International Conference on Humanoid Robotics (Humanoids) (pp. 205-212). IEEE.
10. Chen, M. W., & Zalzala, A. M. (1995). Dynamic modelling and genetic-based motion planning of mobile manipulator systems with nonholonomic constraints.
11. Tokhi, M. O., & Azad, A. K. (Eds.). (2008). Flexible robot manipulators: modelling, simulation and control (Vol. 68).
12. Dwivedy, S. K., & Eberhard, P. (2006). Dynamic analysis of flexible manipulators, a literature review. Mechanism and machine theory, 41(7), 749-777.
13. Rahimi, H. N., & Nazemizadeh, M. (2014). Dynamic analysis and intelligent control techniques for flexible manipulators: a review. Advanced Robotics, 28(2), 63-76.
14. Rahimi, H. N., & Nazemizadeh, M. (2014). Dynamic analysis and intelligent control techniques for flexible manipulators: a review. Advanced Robotics, 28(2), 63-76.
15. Fernandes, J. J. (2018). Kinematic and dynamic analysis of 3PUU parallel manipulator for medical applications. Procedia Computer

Science, 133, 604-611.

16. Li, K., Zhang, Y., & Hu, Q. (2019). Dynamic modelling and control of a tendon-actuated lightweight space manipulator. Aerospace Science and Technology, 84, 1150-1163.
17. Dixon, W. E. (2007). Adaptive regulation of amplitude limited robot manipulators with uncertain kinematics and dynamics. IEEE Transactions on Automatic Control, 52(3), 488-493.
18. Mishra, N., Singh, S. P., & Nakra, B. C. (2015). Dynamic modelling of two link flexible manipulator using lagrangian assumed modes method. Global Journal of Multidisciplinary Studies, 4(12), 93-105.
19. Kebria, P. M., Al-Wais, S., Abdi, H., & Nahavandi, S. (2016, October). Kinematic and dynamic modelling of UR5 manipulator. In 2016 IEEE international conference on systems, man, and cybernetics (SMC) (pp. 004229-004234). IEEE.
20. Kozlowski, K. R. (2012). Modelling and identification in robotics. Springer Science & Business Media.
21. Subudhi, B. A. M. S., & Morris, A. S. (2002). Dynamic modelling, simulation and control of a manipulator with flexible links and joints. Robotics and Autonomous Systems, 41(4), 257-270.
22. Tatlicioglu, E., Walker, I. D., & Dawson, D. M. (2007, April). Dynamic modelling for planar extensible continuum robot manipulators. In Proceedings 2007 IEEE International Conference on Robotics and Automation (pp. 1357-1362). IEEE.
23. Sciavicco, L., & Siciliano, B. (2001). Modelling and control of robot manipulators. Springer Science & Business Media.

5

Study of Robotic Sensors and Actuators with Their Performance Characteristics

Ajit Kumar Pattanaik[1], Gopal Charan Behera[2], Asish Kumar Bagal[3], Rajendra Prasad Nayak[4], and Pulkit Kumar Agrawal[1]

[1]*Department of Mechanical Engineering, Government College of Engineering, Kalahandi, Bhawanipatna, Odisha, India*
[2]*Department of Civil Engineering, Government College of Engineering, Kalahandi, Bhawanipatna, Odisha, India*
[3]*Department of Civil Engineering, College of Engineering and Technology, Bhubaneswar, Odisha, India*
[4]*Department of Computer Science and Engineering, Government College of Engineering, Kalahandi, Bhawanipatna, Odisha, India*

5.1 Sensors in Robotics

Sensory input is required by industrial robots in order to: locate randomly placed items; accommodate changes in the form of objects; and protect against potentially hazardous and unexpected circumstances. Especially if the robot is required to operate in close proximity to humans: Allow for "intelligent" recovery from error situations, as well as the performance of quality control. The primary goal of integrating sensors into robotic systems is to allow robots to operate in surroundings that are nonstructural and unpredictable. Robots will become more intelligent as a result of the use of sensors. However, the robotic software that is connected with the sensors must be capable of receiving data from the sensors and processing the essential real-time information and instructions that are required for decision-making [1-4].

What is Sensing?

Sensing is the process of gathering information about the surrounding environment. The term "sensor" refers to an electrical, mechanical, or chemical instrument that converts an environmental characteristic into a numerical measurement. Each sensor is based on the transduction principle - the transfer of energy from one form to another – which is the basis of its operation [5-7].

Transduction to Electronics

- Thermistor: temperature-to-resistance
- Electrochemical: chemistry-to-voltage
- Photocurrent: light intensity-to-current
- Pyroelectric: thermal radiation-to-voltage
- Humidity: humidity-to-capacitance
- Length (LVDT: Linear variable differential transformers): position-to-inductance
- Microphone: sound pressure-to-anything

Human Sensing and Organs

- Vision: eyes (optics, light)
- Hearing: ears (acoustics, sound)
- Touch: skin (mechanics, heat)
- Odor: nose (vapor-phase chemistry)
- Taste: tongue (liquid-phase chemistry)

Extended Ranges and Modalities

- Vision outside the RGB spectrum
    - Infrared Camera, see at night
- Active vision
    - Radar and optical (laser) range measurement
- Hearing outside the 20 Hz – 20 kHz range
    - Ultrasonic range measurement
- Chemical analysis beyond taste and smell
- Radiation: alpha, beta, and gamma-rays, neutrons, etc. [4].

Electromagnetic Spectrum

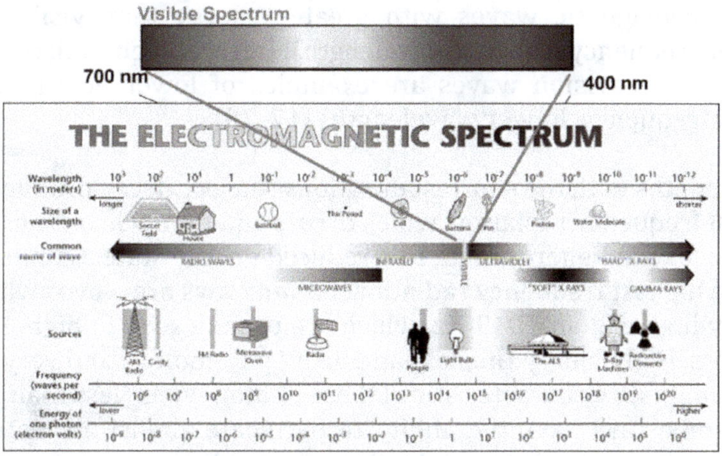

Figure 5.1 Electromagnetic spectrum.

The energy carried by an electromagnetic wave compensates for the lack of mass carried by it. Moreover, it possesses momentum and has the capability of exerting pressure (known as radiation pressure). When comet tails point away from the Sun, it's because of the radiation pressure exerted on the tail by the light (and other forms of radiation) emitted by the Sun. It is proportional to the frequency of an electromagnetic wave how much energy it can carry. A connection exists between the wave's wavelength and frequency thanks to the speed of light [1, 8-19]:

$$C=fl \tag{1}$$

In accordance with their frequency, electromagnetic waves are divided into many types (or, equivalently, on their wavelength). We may put it another way: we divide the electromagnetic spectrum into bands depending on their frequency of occurrence. When it comes to visibility, the spectrum spans from violet to red, for instance. Approximately 400 nanometers (nm) of violet light have been measured, and the frequency of this light is 7.5 x 1014 hertz (Hz). RED light has a wavelength of 700 nm and a frequency of 4.31 x 1014 Hz, making it the most energetic colour in the visible spectrum. In between the two extremes, humans can perceive any electromagnetic wave with a frequency (or wavelength) in the middle. In comparison to the rest of the electromagnetic spectrum, visible light represents a relatively tiny fraction. Ultraviolet light, X-rays, and gamma rays are examples of electromagnetic waves with greater energy than visible light (higher frequency, shorter wavelength). Infrared light, microwaves, radio and television waves are examples of lower energy waves (lower frequency, longer wavelength) [17, 18].

Wavelengths with the shortest durations Nuclear decay produces the highest frequency radiation, which occurs when a nucleus transitions from an excited energy state to a reduced energy state. Gamma rays are the highest frequency radiation. Gamma rays are waves with frequency higher than 1019 Hz, which are usually seen in high-energy particles. High energy (more than 104 eV per photon) and very short wavelengths are characteristic of these photon types (less than 10-14 m). Gamba rays have the ability to permeate almost all materials, making them difficult to detect. Gamma rays have mostly been seen in space-based activity such as those of the Crab Nebula and the Vela Pulsar, among other locations. Gamma radiation has been discovered at frequencies up to 1030Hz, which has been measured from diffuse gamma radiation emissions.

In terms of frequency, the longest wavelength is the lowest. Many textbooks state that the lowest electromagnetic waves have a frequency on the order of 102 Hertz (Hz), and that these waves are classed as Extremely Low Frequency (ELF) waves (ELF). According to recent discoveries, the lowest electromagnetic waves have frequencies on the order of 10-3 Hz (millihertz or mHz) and are known as micro-pulsations because of their very low frequencies. Changes in the magnetosphere cause micro-pulsations, also known as geomagnetic pulsations. Known as a magnetosphere, it is a cavity in the solar wind created by the geomagnetic field (the earth's magnetic field) preventing the direct entrance of ionised gas (plasma) from the solar wind into the cavity. When Balfour Steward discovered and published micro-pulses in 1861, it was a watershed moment for science. In his description, pulsations with frequency ranging from three millihertz to thirty millihertz were mentioned. Geographical pulsations occur at frequencies ranging from one millihertz to one hertz in the present day and age. There are two types of pulsations: continuous and irregular. Each kind of pulsation is further subdivided based on the length of time that the pulsations last. Its name comes from the term "crab," which means "cloud," and refers to the cloudlike masses of gases or dust that exist between stars. Micro-pulsations produced by the Vela Pulsar [1].

5.2 Sensors used in Robots

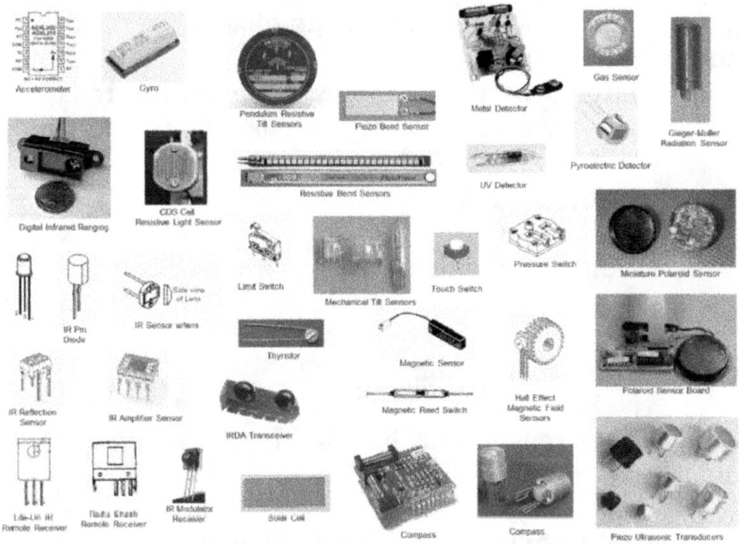

Figure 5.2 Different types of sensors used in robots.

Sensors used in Robot Navigation

- Resistive sensors: bend sensors, potentiometers, resistive photocells, and other similar devices.
- Tactile sensors: contact switch, bumpers, and other similar devices
- Infrared sensors, including reflective, proximity, and distance sensors
- Ultrasonic Distance Sensor (optional).
- Inertial Sensors (which measure the second derivatives of position): Accelerometers, Gyroscopes.
- Orientation Sensors (which measure the second derivatives of orientation): Compass, Inclinometer.
- Laser range sensors are used in a variety of applications.
- Vision Sensors (sometimes referred to as vision sensors)
- The Global Positioning System (GPS).

Classification of Sensors

- Internal state (proprioception) vs. external state (exteroceptive)
    - feedback of robot internal parameters, e.g., battery level, wheel position, joint angle, etc.
    - observation of environments, objects
- Active vs. non-active
    - emitting energy into the environment, e.g., radar, sonar
    - passively receive energy to make observation, e.g., camera
- Contact vs. non-contact
- Visual vs. non-visual
    - vision-based sensing, image processing, video camera [1,4].

Robotic Sensor Classification

In general, robotic sensors may be split into two types: position sensors and motion sensors.

1. **Internal state sensors** - a device that measures the position, velocity, and acceleration of a robot joint and/or end-effector from inside the robot itself. Potentiometers, tachometers, synchros, resolvers, differential transformers, optical interrupters, optical encoders, and accelerometers are some of the devices in this category.

2. **External state sensors** are devices that are used to monitor the connection between the robot's kinematics and/or dynamics and its job, its environment, or the item that is being handled.

Sensor Selection/Sensing Taxonomy

There are many different types of robot sensors available and there are many different parameters measured by these sensors. The application process should be carried out in a top down manner, starting with task requirements, and going through several levels of analysis, eventually leading to the selection of a specific device. A taxonomy for sensing to aid this process consists of five levels of refinement leading to sensor selection:

1. **Specification of task requirements**: e.g., localization, slippage detection, size confirmation, inspection, defect testing.
2. **Choice of modality**: e.g., vision, force, tactile
3. **Specification on sensor attributes**: e.g., output, complexity, discrete or continuous variable, imaging or non-imaging, local or global
4. **Specification of operational parameters**: e.g., size, accuracy, cost
5. **Selection of mechanism**: e.g., switching devices, inductive sensors, CCD vision imaging [6].

Some tasks requirements features:

- Insertion Monitoring
- Assembly Verification
- Detection of Reject Parts
- Recognition of Part Types
- Assembly Test Operations
- Check Gripper/Tool Operation
- Location & Orientation of Parts
- Workspace Intrusion Detection
- Check Correct Manipulation of Parts
- Analysis of Spatial Relations Between Parts

Some typical sensor operational data

- Ultrasonic Effects
- Resistive Effects
- Capacitive Effects
- Piezo-Electric Effects
- Visible Light Imaging
- Photo-Electric & Infrared
- Mechanical Switching
- Inductive Effects
- Thermal Effects
- Hall Effects [8].

5.3 Primary Physical Mechanisms

Main primary physical mechanisms are employed in sensors such as:

- Cost
- Range
- Accuracy
- Repeatability
- Power Requirements
- Output Signal Specification
- Processing Requirements
- Sensitivity
- Reliability
- Weight
- Size

5.4 Sensors for Industrial Robots

- Proximity and Range Sensors
- Tactile Sensors
- Vision Sensors
- Miscellaneous Sensors

Proximity and Range Sensors

It is a technique of detecting the presence or absence of an object with electronic noncontact sensors. Typical application of proximity sensors includes:

- Object detection
- Collision avoidance
- Object verification & counting

Commonly available proximity sensors are:

1. Photoelectric/optical sensors
2. Inductive proximity sensors
3. Capacitive proximity sensors
4. Ultrasonic proximity sensors [8, 15]

Resistive Sensors

Bend Sensors

- Resistance = 10k to 35k
- As the strip is bent, resistance increases

Potentiometers

- Can be used as position sensors for sliding mechanisms or rotating shafts
- Easy to find, easy to mount

Light Sensor (Photocell)

- Good for detecting direction/presence of light
- Non-linear resistance
- Slow response to light changes [1]

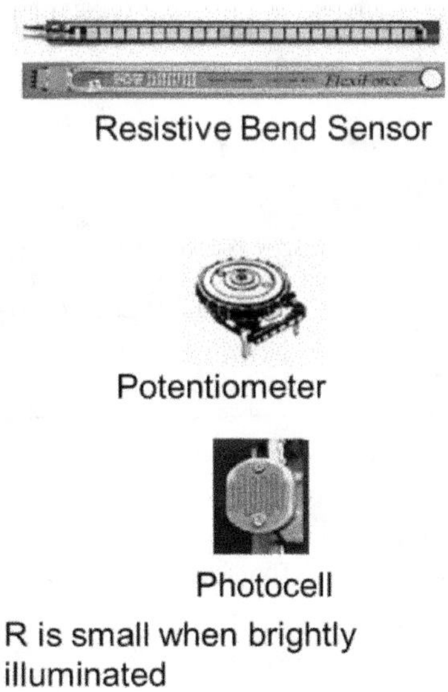

Figure 5.3 Resistive sensors.

Applications

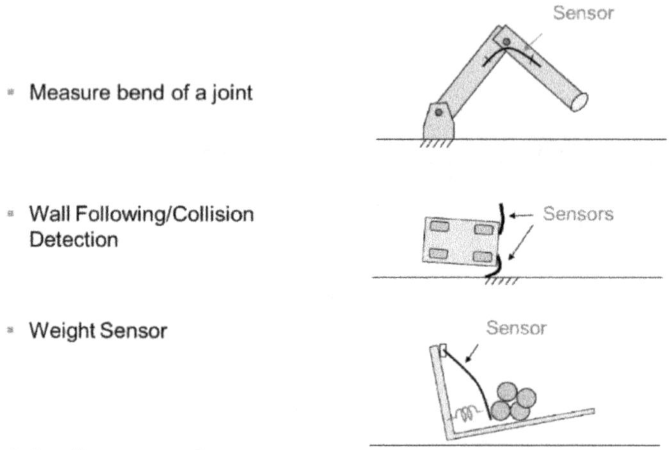

Figure 5.4 Applications of sensing.

Inputs for Resistive Sensors

Voltage divider: You have two resisters, one is fixed and the other varies, as well as a constant voltage

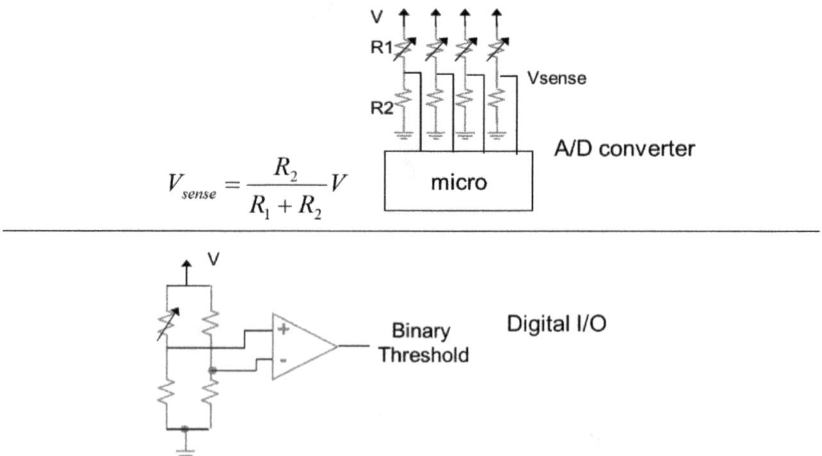

Figure 5.5 Inputs for resistive sensors.

Comparator: If voltage at + is greater than at -, digital high out

Infrared Sensors

- Intensity based infrared
  - Reflective sensors
  - Easy to implement
  - susceptible to ambient light
- Modulated Infrared
  - Proximity sensors
  - Requires modulated IR signal
  - Insensitive to ambient light
- Infrared Ranging
  - Distance sensors
  - Short range distance measurement
  - Impervious to ambient light, color and reflectivity of object

Intensity Based Infrared

- Easy to implement (few components)
- Works very well in controlled environments
- Sensitive to ambient light

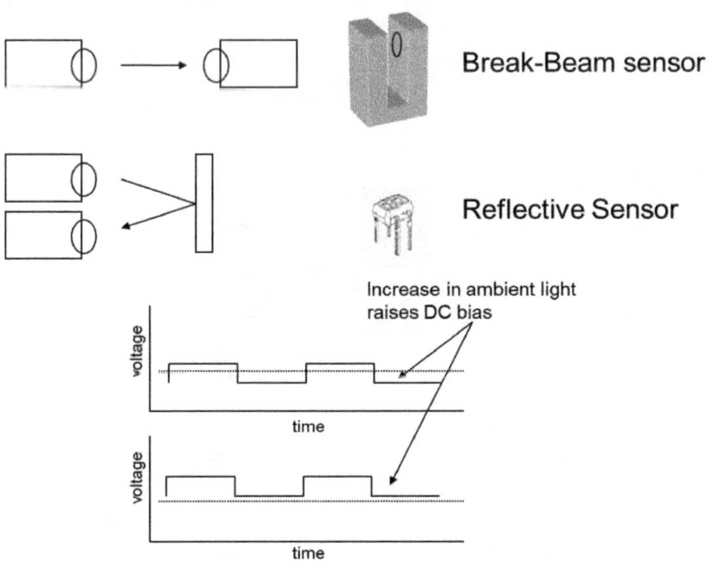

Figure 5.6 Infrared sensors.

IR Reflective Sensors

- Reflective Sensor
    - Emitter IR LED + detector photodiode/phototransistor
    - Phototransistor: the more light reaching the phototransistor, the more current passes through it
    - A beam of light is reflected off a surface and into a detector
    - Light usually in infrared spectrum, IR light is invisible
- Applications
    - Object detection,
    - Line following, Wall tracking
    - Optical encoder (Break-Beam sensor)
- Drawbacks
    - Susceptible to ambient lighting
    - Provide sheath to insulate the device from outside lighting
    - Susceptible to reflectivity of objects
    - Susceptible to the distance between sensor and the object

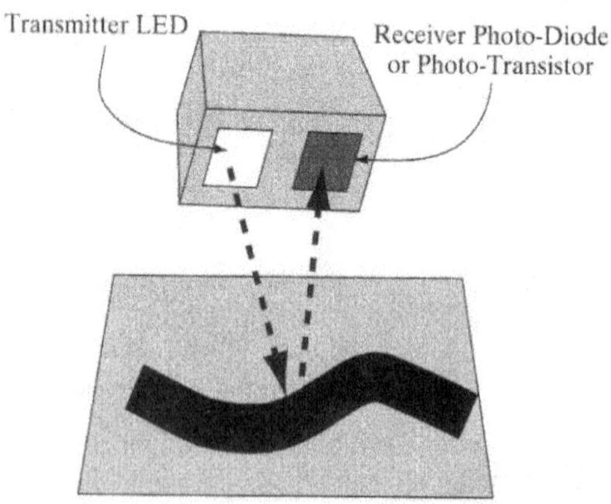

Figure 5.7 IR reflective sensors.

Modulated Infrared

- Modulation and Demodulation

- Flashing a light source at a particular frequency
- Demodulator is tuned to the specific frequency of light flashes. (32kHz~45kHz)
- Flashes of light can be detected even if they are very week
- Less susceptible to ambient lighting and reflectivity of objects
- Used in most IR remote control units, proximity sensors

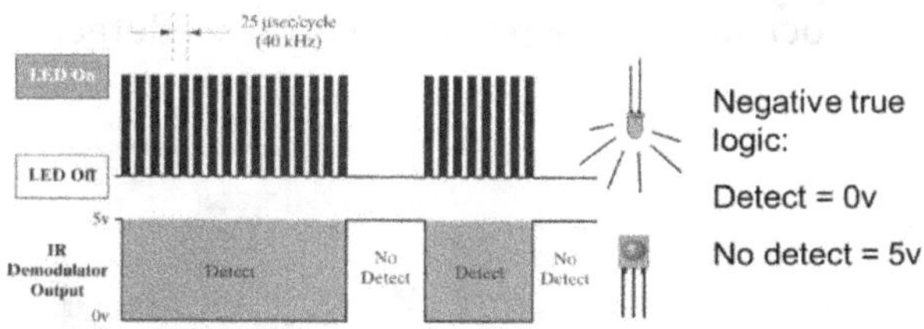

Figure 5.8 Modulated infrared.

IR Proximity Sensors

- Proximity Sensors:
  - Requires a modulated IR LED, a detector module with built-in modulation decoder
  - Current through the IR LED should be limited: adding a series resistor in LED driver circuit
  - Detection range: varies with different objects (shiny white card vs. dull black object)
  - Insensitive to ambient light
- Applications:
  - Rough distance measurement
  - Obstacle avoidance
  - Wall following, line following

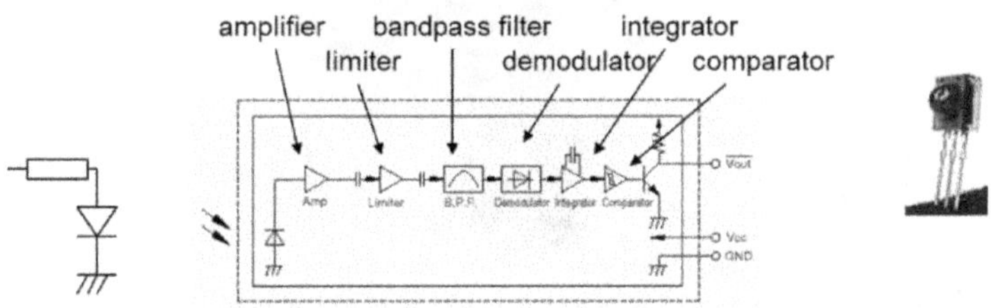

Figure 5.9 IR proximity sensors.

IR Distance Sensors

Location of the spot on the detector corresponds to the distance to the target surface, Optics to covert horizontal distance to vertical distance

Basic principle of operation:

- IR emitter + focusing lens + position-sensitive detector

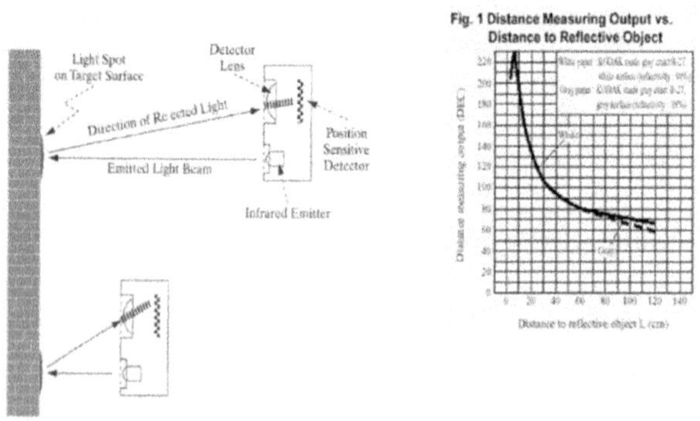

Figure 5.10 IR distance sensors.

- Sharp GP2D02 IR Ranger
    - Distance range: 10cm (4") ~ 80cm (30").
    - Moderately reliable for distance measurement
    - Immune to ambient light
    - Impervious to color and reflectivity of object
    - Applications: distance measurement, wall following,

Figure 5.11 Sharp GP2D02 IR ranger.

Range Finder

- Time of Flight: The measured pulses typically come from ultrasonic, RF and optical energy sources.
- **D = v * t** (2)
- Where D = round-trip distance, v = speed of wave propagation and t = elapsed time
- Velocity of Sound = 0.3 meters/msec
- RF/light = 0.3 meters / ns (Very difficult to measure short distances 1-100 meters)

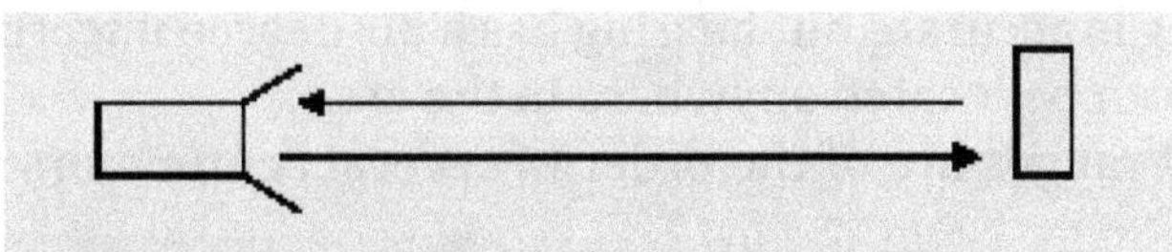

Figure 5.12 Range finder.

Ultrasonic Sensors

- Basic principle of operation:
    - Emit a quick burst of ultrasound (50kHz), (human hearing: 20Hz to 20kHz)
    - Measure the elapsed time until the receiver indicates that an echo is detected.
    - Determine how far away the nearest object is from the sensor

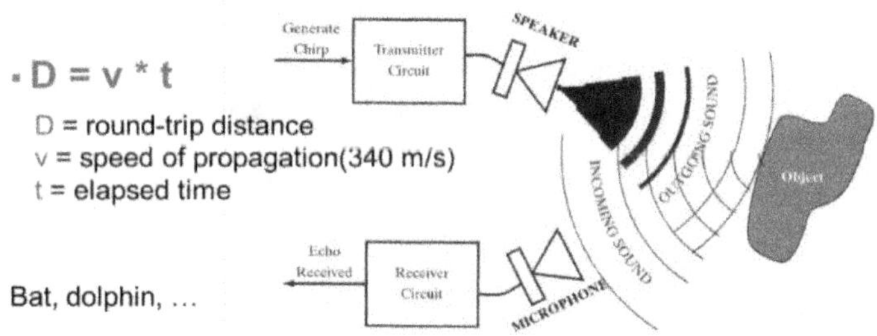

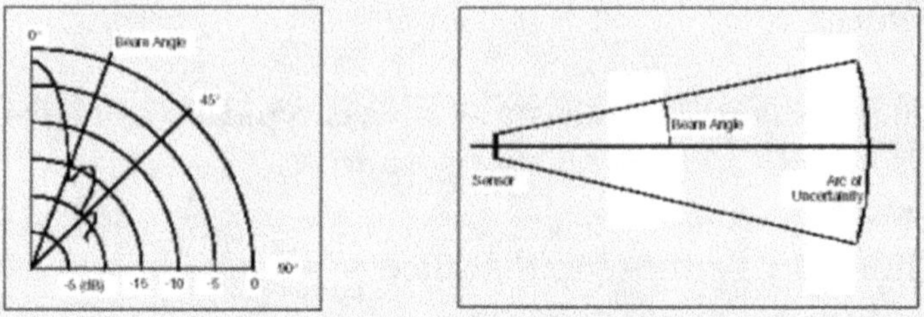

Sensor Specification

Sensor Model, angle = 15 degrees

Figure 5.13 Ultrasonic sensors.

- Ranging is accurate but bearing has a 30-degree uncertainty. The object can be located anywhere in the arc.
- Typical ranges are of the order of several centimeters to 30 meters.
- Another problem is the propagation time. The ultrasonic signal will take 200 msec to travel 60 meters. (30 meters roundtrip @ 340 m/s)
- Polaroid ultrasonic ranging system
    - It was developed for auto-focus of cameras.
    - Range: 6 inches to 35 feet

Transducer Ringing

- transmitter + receiver @ 50 KHz
- Residual vibrations or ringing may be interpreted as the echo signal
- Blanking signal to block any return signals for the first 2.38 ms after transmission

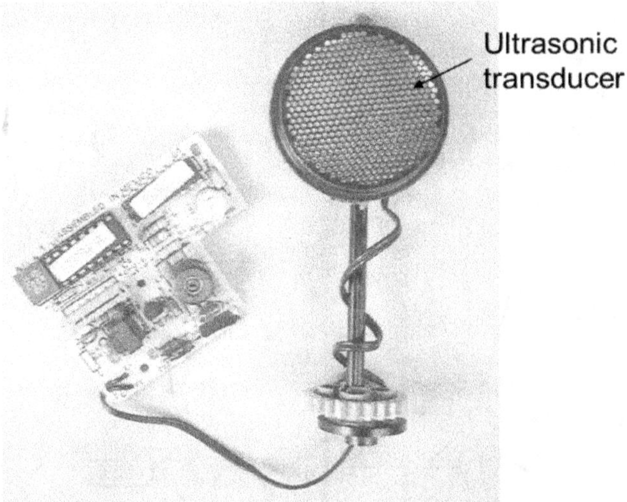

Figure 5.14 Transducer ringing.

Operation with Polaroid Ultrasonic

- The Electronic board supplied has the following I/0
  - **INIT**: trigger the sensor (16 pulses are transmitted)
  - **BLANKING**: goes high to avoid detection of own signal
  - **ECHO:** echo was detected.
  - **BINH**: goes high to end the blanking (reduce blanking time < 2.38 ms)
  - **BLNK**: to be generated if multiple echoes are required
  -

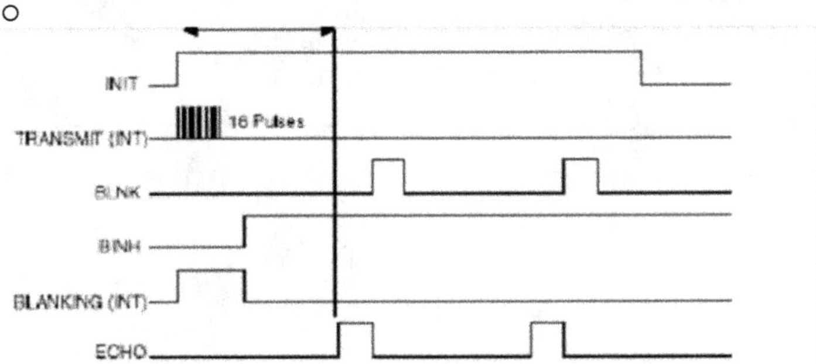

Figure 5.15 Different operations with polaroid ultrasonic.

Applications

- Distance Measurement
- Mapping: Rotating proximity scans (maps the proximity of objects surrounding the robot)

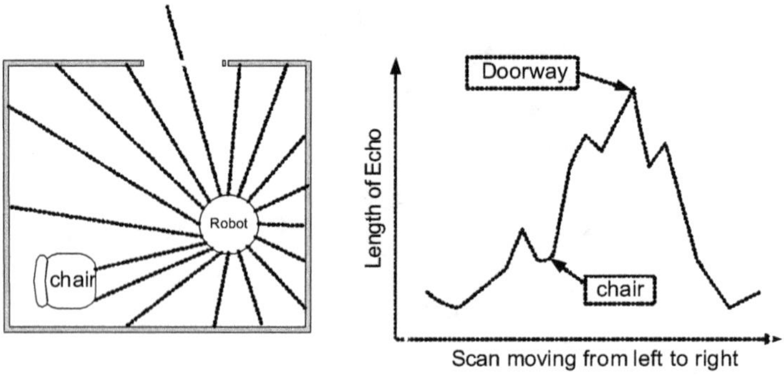

Figure 5.16 Distance measurement by scanning.

Scanning at an angle of 15º apart can achieve best results

Noise Issues

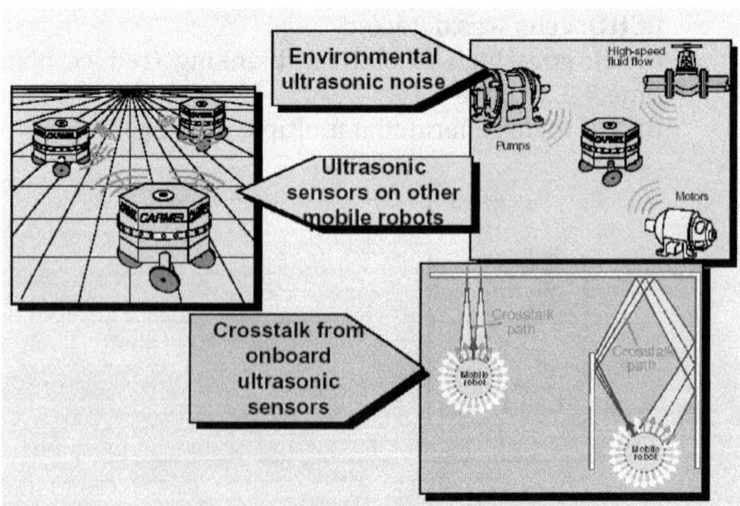

Figure 5.17 Noise issues.

Laser Ranger Finder

- Range 2-500 meters
- Resolution: 10 mm
- Field of view: 100 - 180 degrees
- Angular resolution: 0.25 degrees
- Scan time: 13 - 40 msec.
- These lasers are more immune to Dust and Fog [1, 2, 7].

Figure 5.18 Laser ranger finder.

Tactile Sensors

Tactile sensing is defined as any type of sensing that needs physical contact between the sensor and the item to be sensed in order to be effective. Using touch or tactile sensors is necessary in a wide range of robotic applications, ranging from picking oranges to loading machines. Current research indicates that the basic challenge of locating, identifying, and arranging pieces that need to be assembled is the most essential application at this time. Tactile sensor systems have the capability of detecting a variety of objects, including:

- Presence
- Part shape, location, orientation, contour examination
- Contact are pressure and pressure distribution
- Force magnitude, location, and direction
- Surface inspection: texture monitoring, joint checking, damage detection

- Object classification: recognition, discrimination
- Grasping: verification, error compensation (slip, position, orientation)
- Assembly monitoring

The following are the key components of a tactile/touch sensor system:

- A surface that can be touched
- The medium that converts local pressures or moments into electrical impulses is referred to as a transduction media.
- Structure
- Control/interface [6]

5.5 Method of Transduction

Resistive

The transduction approach is the one that has garnered the most attention in the development of touch sensors. It is concerned with the change in resistance of a conductive substance as a result of the application of pressure to the material. This approach includes measuring the resistance of a conductive elastomer either through or across the thickness of the material being tested. The majority of elastomers are composed of rubber that has been doped with carbon or silicon.

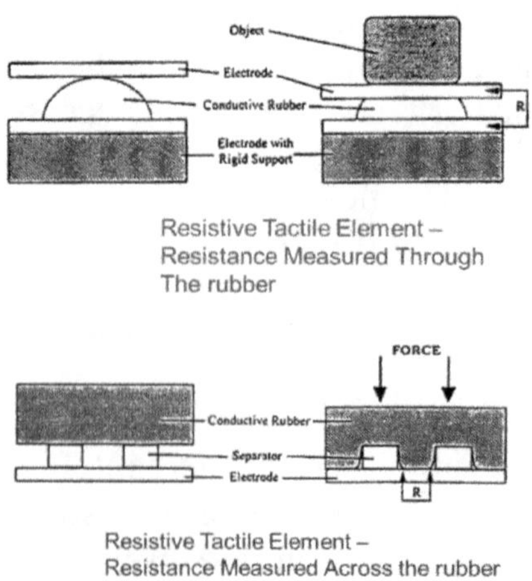

Figure 5.19 Resistive transduction.

Advantages:

- Wide dynamic range
- Durability
- Good overload tolerance
- Compatibility with integrated circuitry, particularly VLSI.

Disadvantages:

- Hysteresis in some designs.
- Elastomer needs to be optimized for both mechanical and electrical properties.
- Limited spatial resolution compared with vision sensors.
- Larger numbers of wires may have to be brought away from the sensor.
- Monotonic response but often not linear.

Piezoelectric & Pyroelectric Effects

When pressure is applied to a sensor element, the piezoelectric effect causes a voltage to be generated across the element. The voltage that is created is proportional to the amount of pressure that is applied. There is no requirement for an external voltage, and a continuous analogue output is provided by such a sensor. The production of a voltage when a sensing element is heated or cooled is referred to as the pyroelectric effect. Piezoelectric and pyroelectric characteristics of polymeric materials make them suitable for use with sensors.

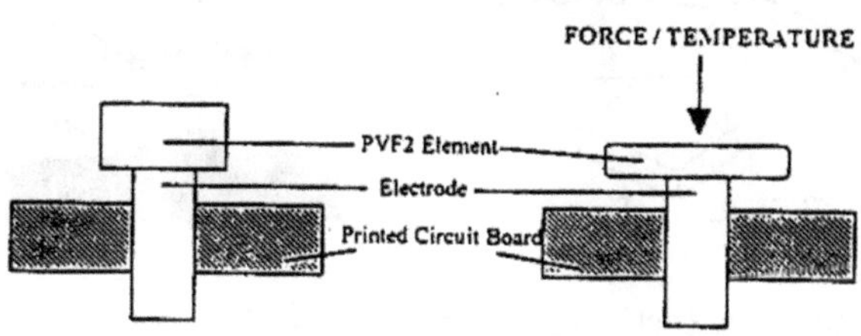

Figure 5.20 Piezoelectric/pyroelectric tactile element.

Advantages:

- Wide dynamic range
- Durability
- Good mechanical properties of piezoelectric from pyroelectric materials
- Temperature as well as force sensing capabilities

Disadvantages:

- Difficult of separating piezoelectric from pyroelectric effects
- Inherently dynamic - output decay to zero for constant load
- Difficult of scanning elements
- Good solution are complex

Capacitive Technique

Sensors in this category are focused with measuring capacitance, which is manufactured to change depending on the amount of load that they are exposed to. In a parallel plate capacitor, the capacitance varies depending on the distance between the plates and the area of the plates; thus, adding an elastomeric separator between the plates offers compliance, allowing the capacitance to change depending on the applied load.

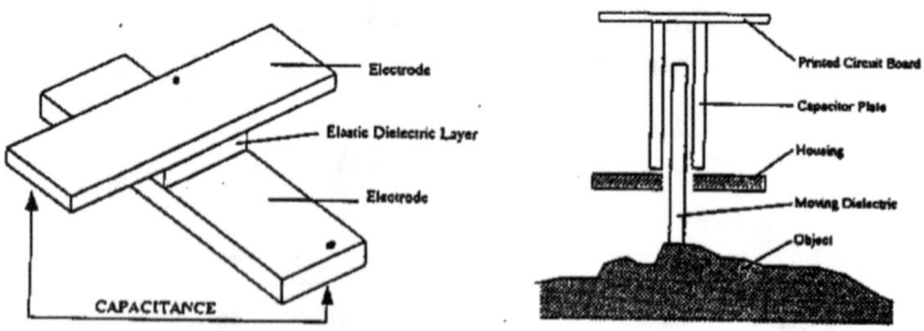

Figure 5.21 Capacitive tactile element.

Advantages:

- Wide dynamic range
- Linear response

- Robust

Disadvantages:

- Susceptible to noise
- Some dielectrics are temperature sensitive
- Capacitance decreases with physical size ultimately limiting spatial resolution.

Mechanical Transduction

- A linear potentiometer

Advantages:

- Well known Technology
- Good for probe application

Disadvantages:

- Limited spatial resolution
- Complex for array construction

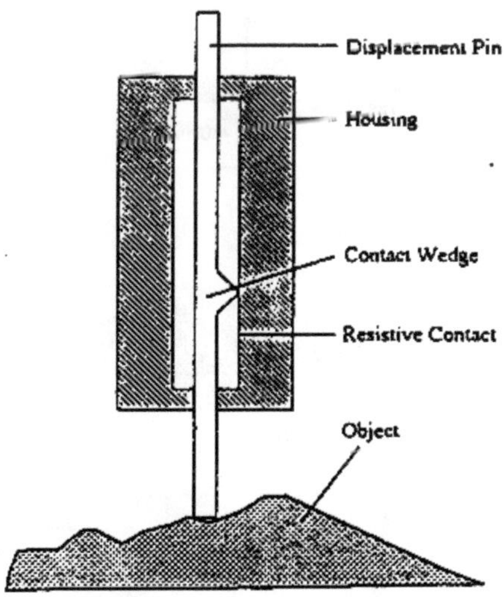

Figure 5.22 Mechanical transducer (a linear potentiometer).

Magnetic Transduction Methods

Sensors that operate on magnetic transduction may be classified into two main categories: passive and active. Sensors that utilize mechanical movement to generate changes in magnetic flux are grouped together in this category.

Advantages:

- Wide dynamic range
- Large displacements possible
- Simple

Disadvantages:

- Poor spatial resolution
- Mechanical problems when sensing on slopes.

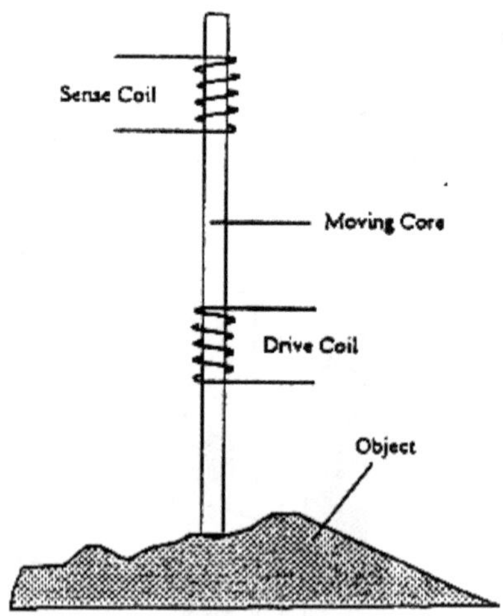

Figure 5.23 Magnetic tactile element.

Magnetic Tactile Element

In this case, the magneto elastic materials undergo a change in magnetic field when exposed to mechanical stress.

Advantages:

- Wide dynamic range
- Linear response
- Low hysteresis
- Robust

Disadvantages:

- Susceptible to stray field and noise.
- A.C. circuit required

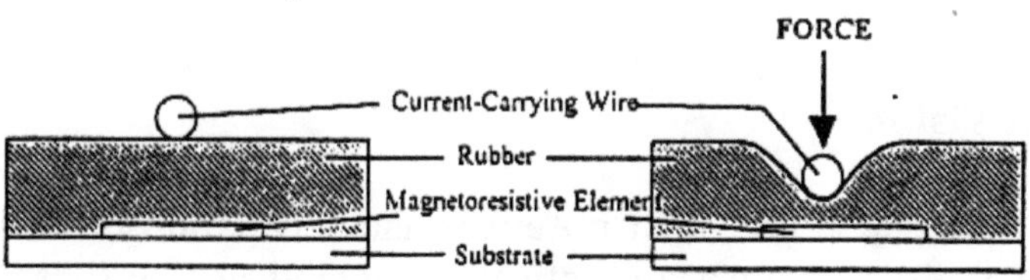

Figure 5.24 Magneto resistive tactile element.

Advantages:

- Very high resolution
- Compatible with vision sensing technology
- No electrical interference problems
- Processing electronics can be remote from sensor
- Low cabling requirements

Disadvantages:

- Dependence on elastomer in some designs – affects robustness
- Some hysteresis

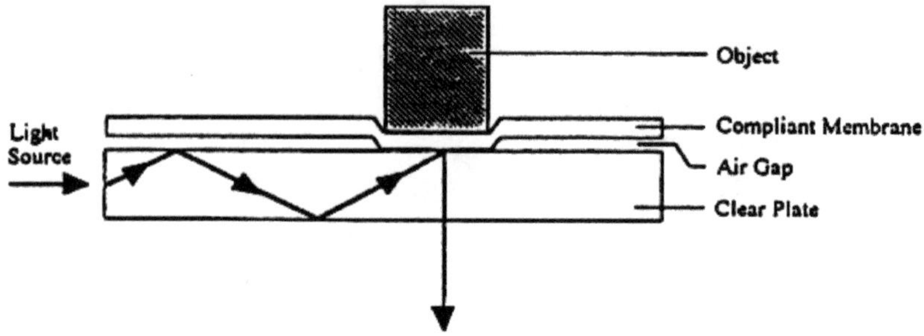

Figure 5.25 Optical tactile element (pressure to light transduction) [3, 20-23].

5.6 Robot Vision

The ability to see is the most potent sensory capability a robot can have. The ability to have a sophisticated sensing system that enables a robot to react intelligently and flexibly to its surroundings is enabled by this technology. As a result, machine vision is the most complicated sensor kind available. Robot vision may be described as the process of obtaining, characterising, and interpreting information from pictures of a three-dimensional environment using computer vision algorithms. This technique, which is also known as machine vision or computer vision, may be split into six major categories. These are the ones:

1. **Sensing**: the process that yields visual image
2. **Preprocessing**: deals with techniques such as noise reduction and enhancement of details
3. **Segmentation**: the process that partitions an image into objects of interest
4. **Description**: deals with that computation of features for example size or shape, suitable for differentiating one type of objects from another.
5. **Recognition**: the process that identifies these objects (for example wrench, bolt, engine block, etc.)
6. **Interpretation**: assigns meaning to an ensemble of recognized objects [5, 9].

Imaging Components

The first link in the visual chain is the imaging component, often known as the "eye" or sensor. A large number of sensors may be utilized to view the outside environment. Vision sensors or imaging components are classified into four categories:

a) Point sensors

These sensors are only capable of detecting light at a single point in space, which limits their use. These sensors are used in conjunction with a light source (such as an LED) and as a noncontact "feeler" for a variety of applications. It may also be utilized to generate a collection of visions with greater - spatial dimensions. In order to gather information, techniques such as an orthogonal set of scanning mirrors are used to scan over a field of vision.

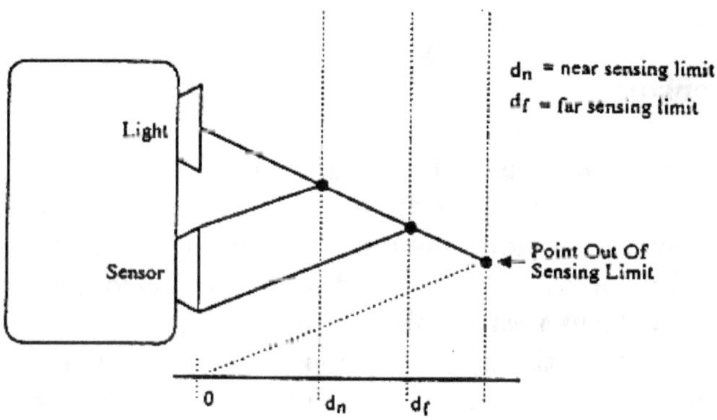

Figure 5.26 Noncontact feeler-point sensor.

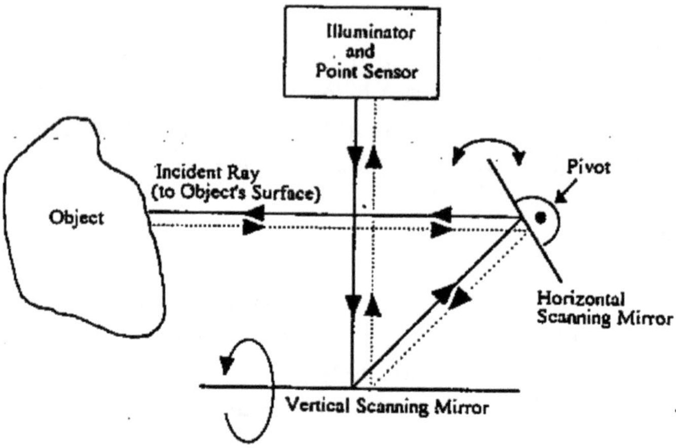

Figure 5.27 Image scanning using a point sensor and oscillating deflecting mirrors.

b) Line sensors

When used to gather visual information from a real scene in the real world, line sensors are one-dimensional devices that collect information in one dimension. The most commonly seen sensor configuration is a "line array" of photodiodes or charger-couple device components. It works in a way similar to that of an analogue shift register, generating a sequential, synchronized output of electrical signals that correspond to the intensity of light falling on an integrated light-collecting cell, as shown in the figure.

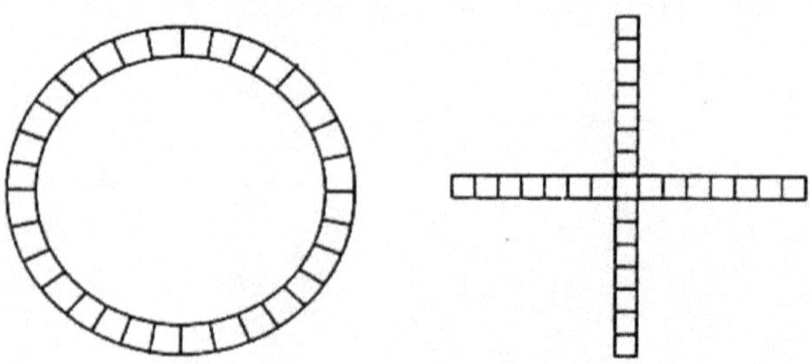

Figure 5.28 Circular and cross configurations of light sensors.

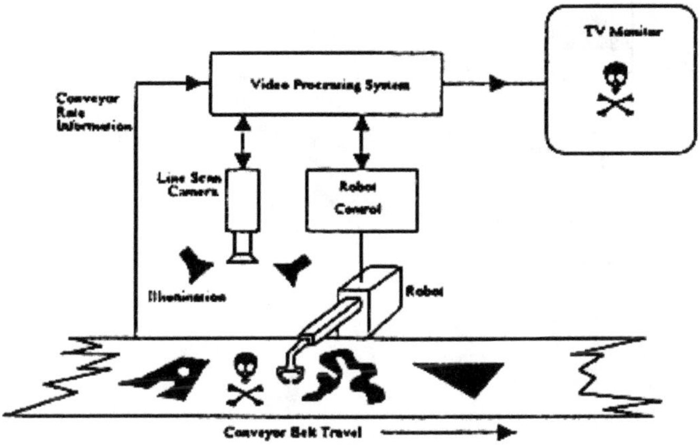

Figure 5.29 Light sensors mechanism.

Automated robot sorting system that generates two-dimensional pictures with the use of a line scan camera to represent a scene, a line array may be utilized. If one fixes the location of a straight-line sensor and moves an item orthogonally to the orientation of the array, one may scan the whole object of interest, as shown in the figure.

c) Planar sensors

The line-scan idea is represented as a two-dimensional arrangement. Scanning photomultipliers and solid-state sensors are two kinds of sensors that are often used nowadays and are classified as generic types. Photo synthesizers are represented by television cameras, the most widely used of which is the vidicon tube, which is basically a device that converts optical signals into electrical signals. In addition to vidicon tubes, there are many other kinds of solid-state cameras to choose from. Because of the weight and noise factor, solid-state sensors are used in a variety of applications (solid-state arrays are less noisy but more expensive). When placing a camera near or on the end-effector of a robot, this is particularly essential.

d) Volume sensors

The sensor must be able to give three-dimensional information. Sensors that use directional lasers or acoustic range finders may gather information about their surroundings.

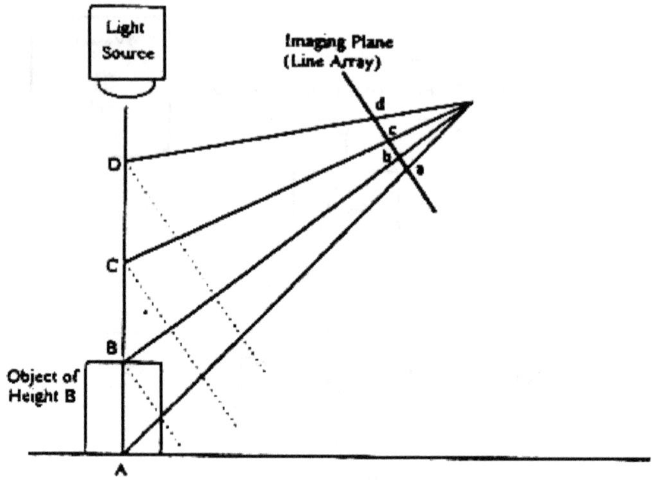

Figure 5.30 Schematic representation of a triangulation range finder: volume sensor.

Image Representation

According to the figure below. The two-dimensional picture produced by a television camera or other imaging equipment is denoted by the function F(x,y). The spatial coordinates (image plane) are denoted by the letters "x" and "y." The value of "f" at any location (x,y) is proportional to the brightness (intensity) of the picture at that position. An image function f(x,y) must be digitized both spatially and in amplitude in order to be represented in a format appropriate for computer processing (intensity). The digitization of spatial coordinates (x,y) will be referred to as picture sampling, while the digitization of amplitude will be referred to as intensity or grey-level quantization. A digital image is an array of (N, M) rows and columns in which each sample is sampled evenly and quantized in intensity, and which is also known as a digital image array. Each element in the array is referred to as an image element, picture element, or picture element (or pixel) [12, 14, 18].

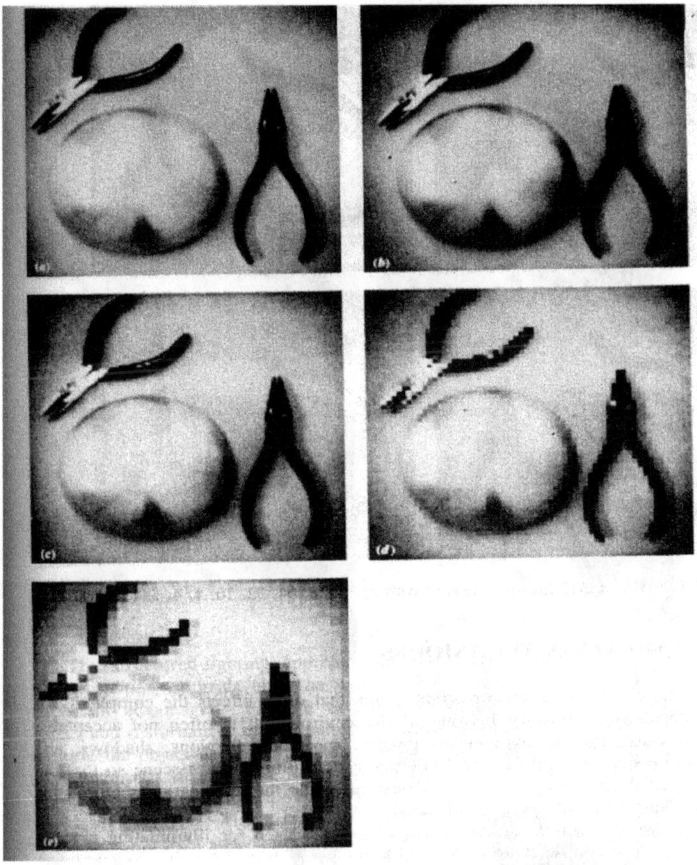

Figure 5.31 Effects of reducing sampling grid size with: 512x512, 256x256, 128x128, 64x64, 32x32.

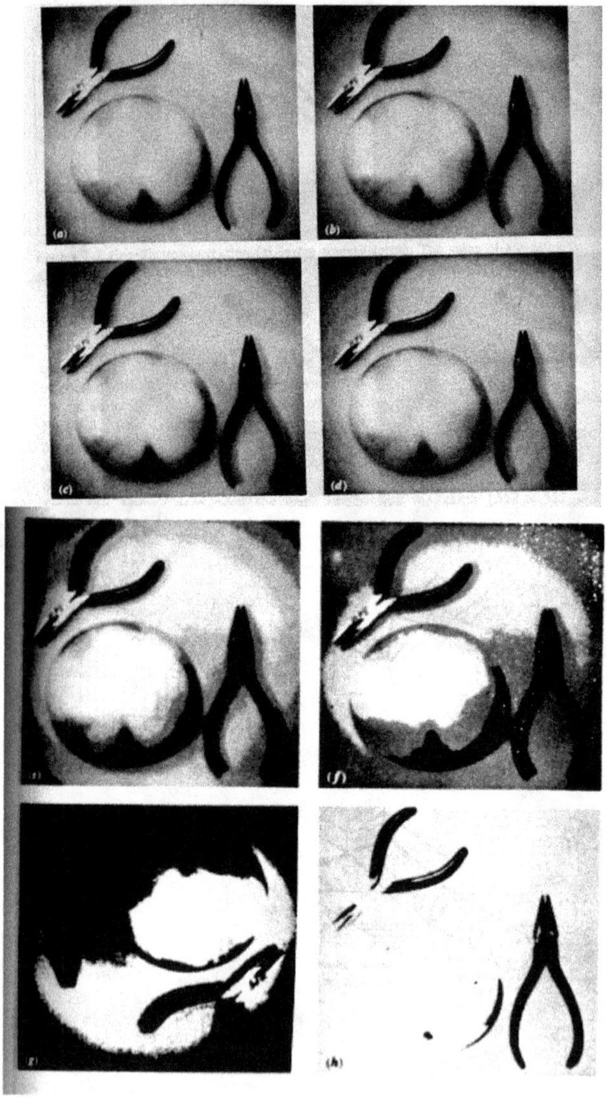

Figure 5.32 Spatial resolution constant at 512x512a) 256, b) 128, c) 64, d) 32, e) 16, f) 8, g) 4, and h) 2 levels.

Effect produced by reducing the number of intensity levels while maintaining the spatial resolution constant at 512x512. The 256-, 128- and 64-levels are of acceptable quality.

Illumination Techniques

The amount of illumination in a scene is a significant element that often influences the complexity of vision algorithms. Using a well-designed lighting system, a scene is illuminated in such a way that the complexity of the resultant picture is minimised while the amount of information needed for object identification and extraction is increased. Because it may result in low contrast pictures, specular reflections, shadows, and other superfluous features, arbitrary lighting of the surroundings is often not acceptable. For a robot work environment, there are four primary lighting methods to consider.

a) Diffuse-Lighting

- This technique is for smooth, regular surface object. It is used where surface characteristic are important [5, 9].
- Example:

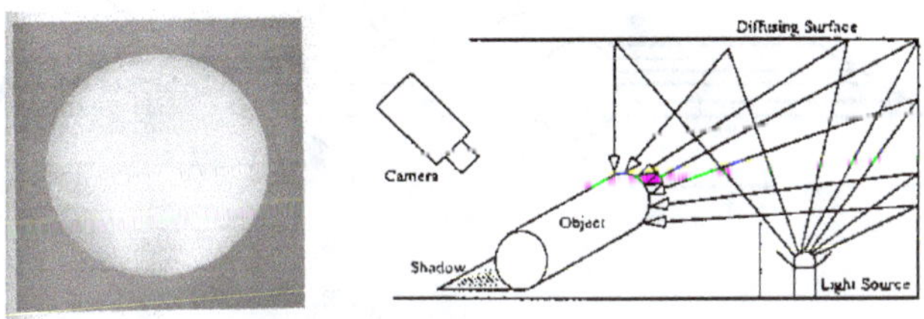

Figure 5.33 Diffuse-lighting technique.

b) Backlighting

- Create a picture in black and white. These techniques are most appropriate for situations in which object silhouettes are adequate for identification or other measuring purposes.
- Example:

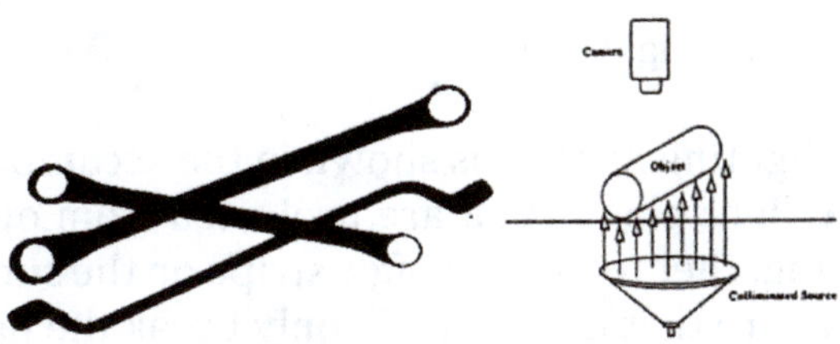

Figure 5.34 Backlighting technique.

c) Structured lighting

The process of projecting points, stripes, and grids onto a work area is called projection. There are two significant benefits to using this lighting technique: In the work area, it creates a recognised light pattern, and disruptions of this pattern signal the presence of an item, thus simplifying the object detection issues. It is possible to get insight into the three-dimensional features of an item by examining the manner in which the light pattern has been warped.

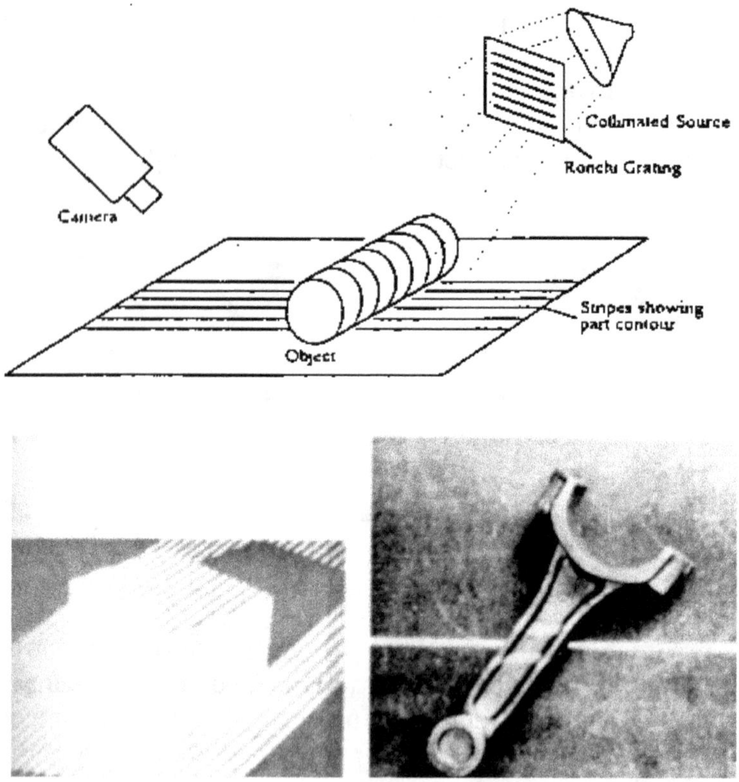

Figure 5.35 Structured lighting technique.

The structured lighting method is shown in the accompanying image, which uses two light planes that are projected from opposite directions yet come together to form a single stripe on the surface. The two light sources ensure that the item will only break the light stripe if it is immediately below the camera, which is the case in this case. This approach is appropriate for objects that are in motion.

Note: "While the line scan camera only observes the line on where the two light planes meet, two-dimensional information may be collected as the item moves past the camera".

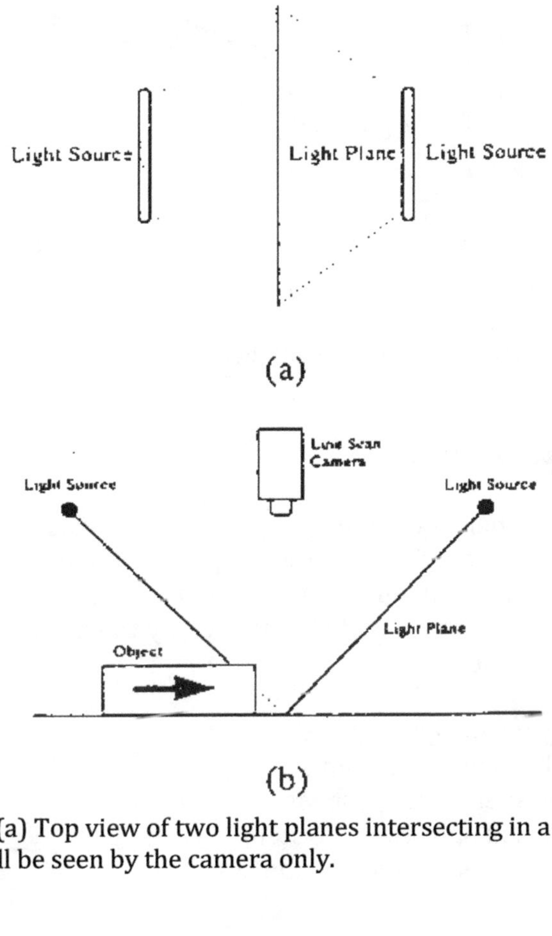

Figure 5.36 (a) Top view of two light planes intersecting in a line sight and (b) Object will be seen by the camera only.

d) Directional lighting

This technique is used to visually examine the surfaces of objects. A highly focused light beam (such as a laser beam) may be used to identify defects on the surface of a surface, such as scratches, by measuring the amount of scatter produced.

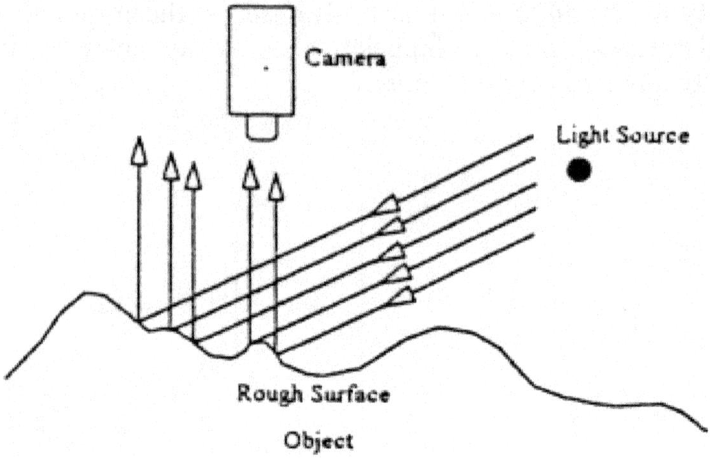

Figure 5.37 Directional lighting technique.

Robot Vision System

There are a number of commercial programs available for purchase that may be used for visual processing tasks. A common hardware setup is shown in the diagram below. There are three main kinds of robotic vision systems, each distinguished by the method that is employed to create them:

1. Binary vision systems
2. Gray-level vision systems
3. Ad hoc special-purpose vision systems
4. Structured light vision systems
5. Character recognition vision systems

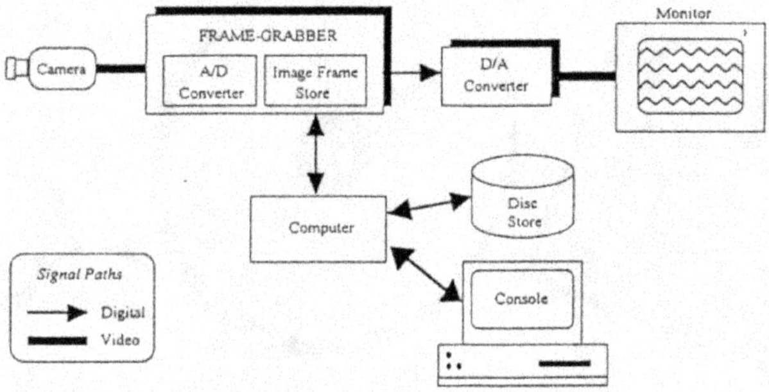

Figure 5.38 Vision system hardware.

For example, most systems will have remote camera control capabilities, and they may even include interfaces for remote lighting management as well. The major issue with commercial vision packages is that they must be general purpose in order to be usable in a wide range of scenarios. This presents a significant challenge. Because of this need, they are often deemed unsuitable or too complex for the specific robot job at hand. Visualization is not utilized in an exploratory manner in the realm of industrial robots; instead, it is used to validate, quantify, and improve previously collected data. Regardless of whatever commercial vision system one acquires, it is probable that it will be used for applications such as those mentioned in the section below.

Vision Dev. Tools: Survey

Commercial products

- Matrox: MIL, Inspector
- Coreco Imaging: Sapera, MVTools, WiT
- MVTec: Halcon
- Euresys: eVision, EasyAccess
- AAI: Aphlion

Free tools

- Intel: Open Source Computer Vision
- Microsoft: Vision SDK
- XMega Wave: XMega Wave
- UTHSCSA: Image Tool [5, 8].

Figure 5.39 Slide borrowed from CAIRO.

Vision Applications

1. OBJECT LOCATION
    - Used in object handling and processing: (a) Position and (b) Orientation

2. OBJECT PROPERTIES
    - Used in inspection, identification, measurement: (a) Size, (b) Area, (c) Shape, (d) Periphery length / area ratio, (e)Texture, (f) Repetition of pattern and (g) Properties of internal features

3. SPATIAL RELATIONS
    - Used in measurement and task verification: (a) Relative positions, (b) Relative orientations, (c) Occlusions, (d) Alignments and (e) Connectivity

4. ACTION MONITORING
    - Used in actuator control and verification: (a) Direct feedback, (b) Error measurement, (c) Action confirmation, (d) Inspection and (e) Collision avoidance planning.

5.7 Miscellaneous Sensors (Position, Velocity and Acceleration Sensors)

When it comes to determining the position of robot joints, there are many different types of sensors that may be utilised, including potentiometers, optical encoders, Linear Variable Differential Transformers (LVDT), Force & Torque Sensors, and others.

Potentiometer

- Potentiometer transducers can be used to measure both linear and angular displacement

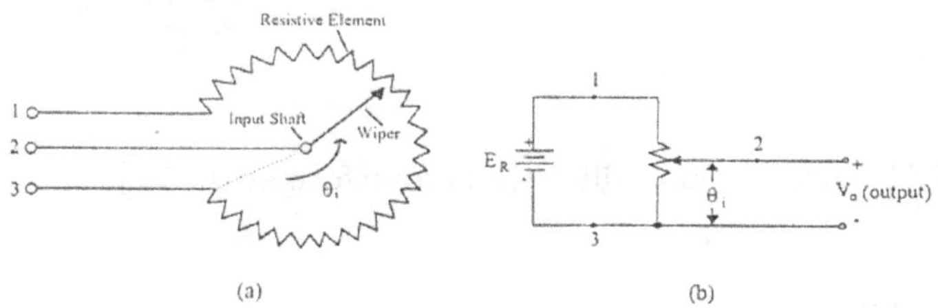

Figure 5.40 Schematic diagram of the potentiometer.

Linear Variable Differential Transformer (LVDT)

The LDVT is a rugged and accurate device that produces a voltage output proportional to the movement of a ferrous armature for the purpose of measuring the position of robot joints or end-effectors. It is much more costly than the potentiometer transducer, but it outperforms it.

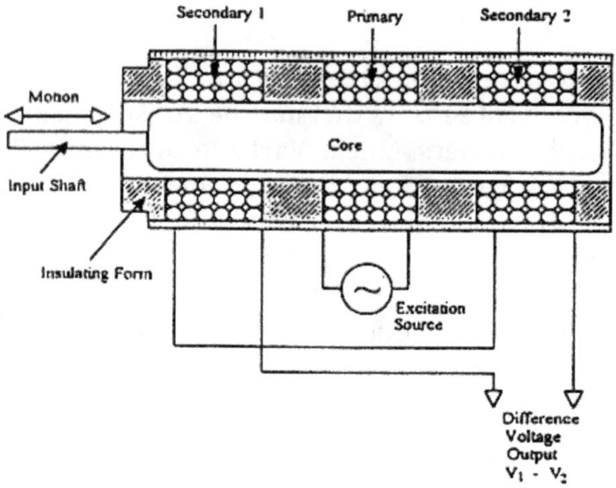

Figure 5.41 Linear variable differential transformer (LVDT).

Force and Torque Sensors

Displacement concepts are often used in the design of force transducers. There are several different types of force and torque transducers available on the market.

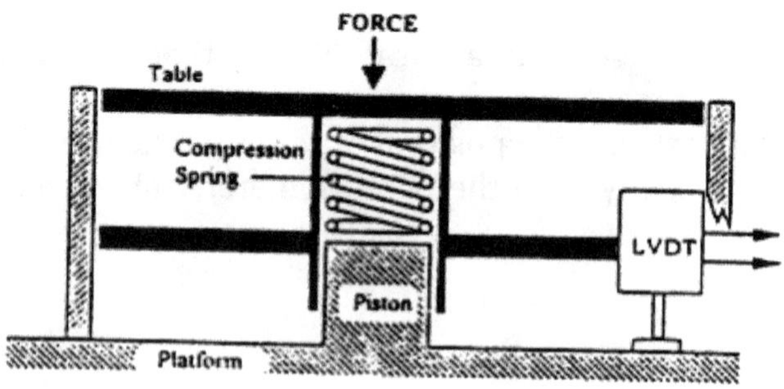

Figure 5.42 Force and torque sensors.

A compression spring and LDVT are used to create a force-measuring gadget.

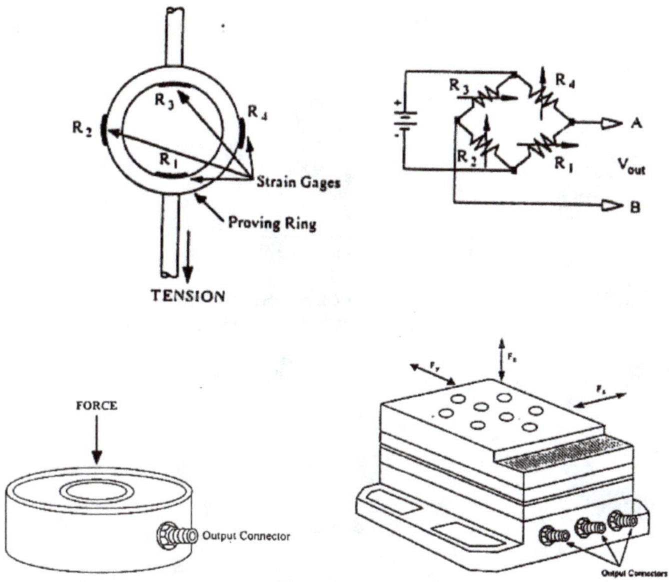

Figure 5.43 Tension load cell.

A tension load cell is shown in this illustration. It may be used to determine the amount of force needed to lift a large weight in the industrial setting. The piezoelectric concept allows for the measurement of force. The piezoelectric force transducer of the load washer type is shown in the illustration. It is intended for the measurement of axial forces. When it is produced, it is preloaded with tensile and compressive forces, and it is capable of measuring both.

An example of a tension load cell may be seen in this image. In the industrial context, it may be used to estimate the amount of force required to raise a big weight from the ground. The notion of piezoelectricity allows for the measuring of mechanical force. The piezoelectric force transducer of the load washer type is shown in the picture for your convenience. Its primary use is the measuring of axial forces. When it is manufactured, it is already preloaded with tensile and compressive forces, and it is capable of monitoring both of these forces simultaneously.

Incremental Optical Encoders

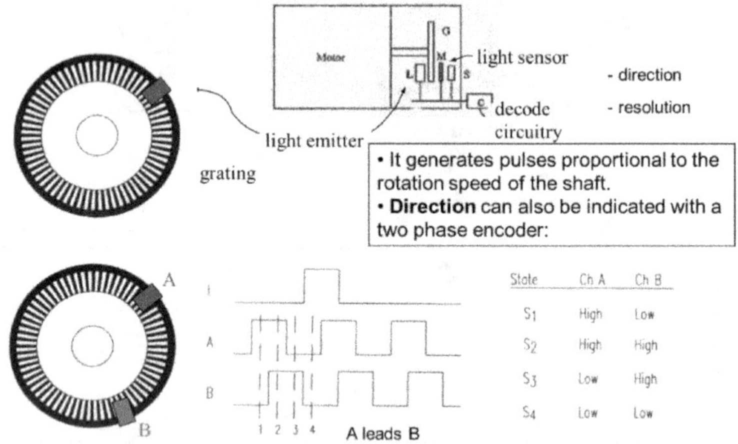

Figure 5.44 Incremental optical encoders.

Absolute Optical Encoders

When it is not feasible to lose a reference, this method is used. Gray codes are those in which only one bit changes at a time (less uncertainty). During the transmission of information, two streams of information are sent simultaneously (many wires are necessary).

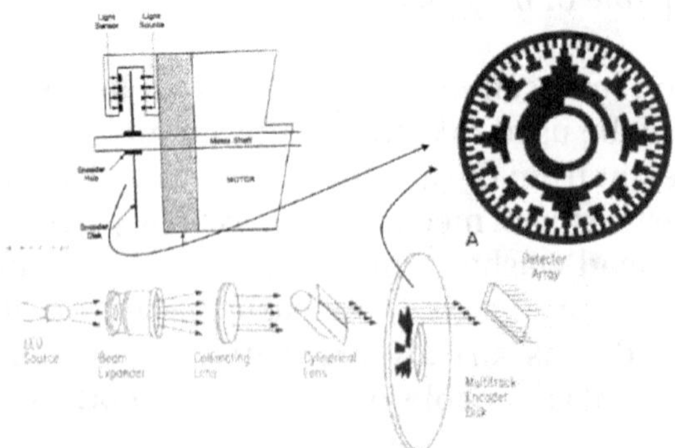

Binary	Gray Code
000	000
001	001
010	011
011	010
100	110
101	111
110	101
111	100

Figure 5.45 Absolute optical encoders.

Other Odometry Sensors

It features two stator windings that are oriented at 90 degrees to one another. Depending on the angle of the rotor, the output voltage will be proportional to the sine or cosine function of the angle. The rotor is made up of a third winding, denoted by the letter C.

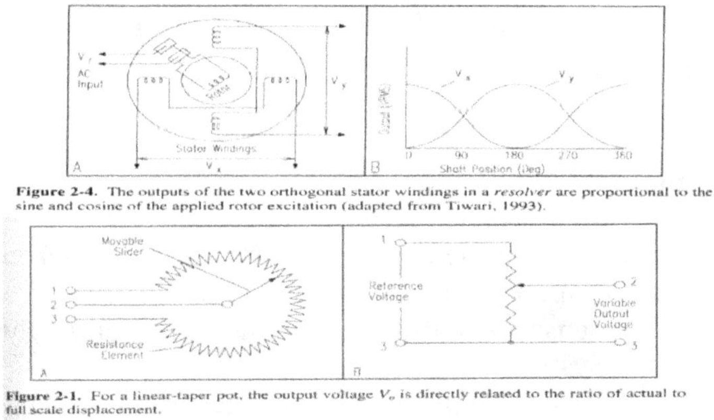

Figure 5.46 Other odometry sensors.

Inertial Sensors

Gyroscopes

- Gyroscopes Measure the rate of rotation independent of the coordinate frame
- Common applications:
  - Heading sensors, Full Inertial Navigation systems (INS)
  - These devices return a signal proportional to the rotational velocity.
  - There is a large variety of gyroscopes that are based on different principles

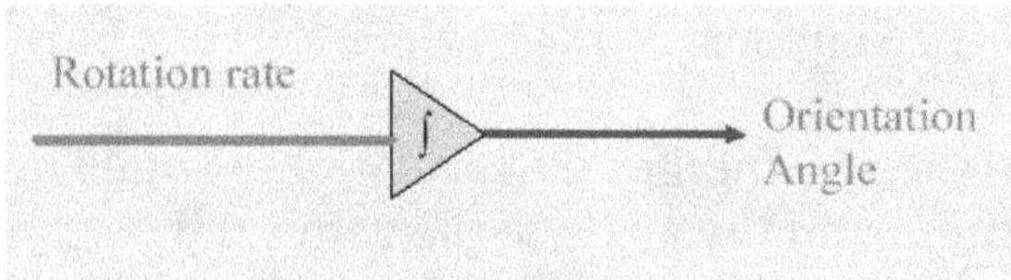

$$\theta(t) = \int \dot{\theta}(t)\, dt$$

Figure 5.47 Gyroscopes.

Accelerometers

- Measure accelerations with respect to an inertial frame
- Common applications:
    - Tilt sensor in static applications, Vibration Analysis, Full INS Systems
    - They measure the inertia force generated when a mass is affected by a change in velocity.
    - This force may change
        - The tension of a string
        - The deflection of a beam
        - The vibrating frequency of a mass
    - Main elements of an accelerometer:
    - Mass
    - Suspension mechanism
    - Sensing element

High quality accelerometers include a servo loop to improve the linearity of the sensor.

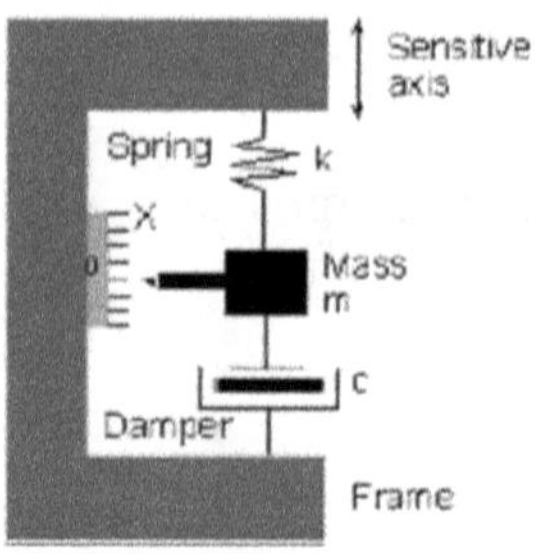

$$F = m\frac{d^2x}{d^2t} + c\frac{dx}{dt} + kx$$

Figure 5.48 Accelerometers.

Global Positioning System (GPS)

24 satellites (+several spares)
Broadcast time, identity, orbital parameters (latitude, longitude, altitude)

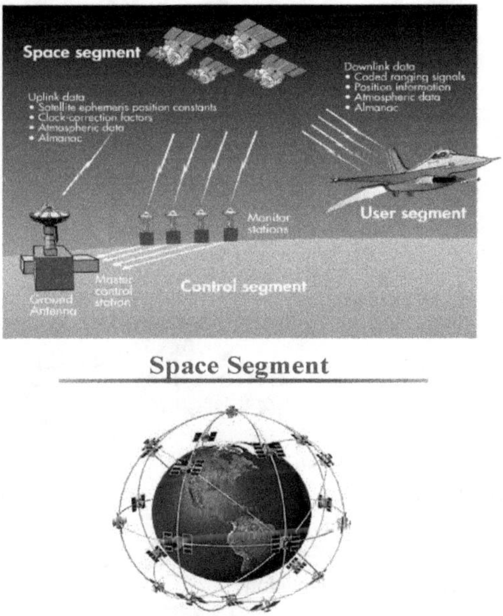

Figure 5.49 Global positioning system (GPS).

Noise Issues

- Real sensors are noisy
- Origins: natural phenomena + less-than-ideal engineering
- Consequences: limited accuracy and precision of measurements
- Filtering:
  - software: averaging, signal processing algorithm
  - hardware tricky: capacitor [7, 9].

Robot Actuators:

1. Hydraulic and pneumatic actuators, and
2. Electrical actuators

5.8 Robot as a "System"

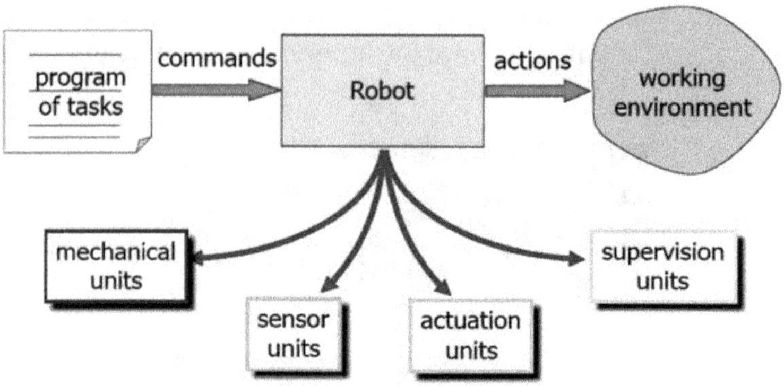

Figure 5.50 Robot as a "System".

Functional units of a robot

- mechanical units (robot arms)
  - rigid links connected through *rotational* or *prismatic* joints (each 1 dof)
  - mechanical subdivisions:
    - *supporting structure* (mobility), *wrist* (dexterity), *end-effector* (task execution, e.g., manipulation)
- sensor units
  - proprioceptive (internal robot state: position and velocity of the joints)
  - exteroceptive (external world: force and proximity, vision, ...)
- actuation units
  - motors (*electrical, hydraulic, pneumatic*)
  - motion control algorithms
- supervision units
  - task planning and control
  - artificial intelligence and reasoning

Actuators:

Actuation is the process of energy conversion into mechanical form. An instrument that performs this conversion is known as an actuator. The actor plays a very significant part in the execution of control. The controller gives the actuator the command signal for action. The control codes are intended to "derive the actuator when an event occurs".

Actuators for Robots:

Actuators for mechanical movement in robots are utilised. Actuators are robots' muscles. Depending on the weight involved, various kinds of actuators are available. The word load is linked with several characteristics, including strength, torque and operating speed, precision, and energy consumption.

Types of Actuators

- Electric Motor
- Servomotor
- Stepper Motor
- DC Motor
    - Direct Drive Motor
    - Hydraulic Actuators
    - Pneumatic Actuators
    - Magnetostrictive Actuators
    - Shape Memory Metal Actuators
- Electromechanical actuators convert electrical energy into mechanical energy. Magnetism is the basis of their principle of operation. They are DC, AC and stepper motors.
- DC motors require a direct current or voltage source as the input signals.
- AC motors require an alternating current or voltage source [1].

1. Stepper motors have capability of achieving precision angular rotation in both directions and are commonly employed to accommodate digital control technology.

2. Hydraulic and pneumatic actuators are under fluid power actuators. Fluid power refers to energy that is transmitted via a fluid under pressure. When a pressure is applied to a confined chamber containing a piston, the piston will exert a force causing a motion.

3. Materials which undergo some sort of transformations through physical interaction, are referred to as active materials. Piezoelectric (voltage-load), shape-memory alloys (react to heat), magnetostrictive are examples of these materials.

Characteristics of Actuating Systems

Weight, Power-to-weight Ratio, Operating pressure

1) Stepper motors are typically heavier for the same power as servo motors.
2) High electrical engine voltage, higher power-to-weight ratio.
3) The lowest power/weight ratio of pneumatic systems.
4) Hydraulic systems have the greatest power-to-weight ratio. Weight is really comprised of two parts in these systems. One is the hydraulic drive, the other the hydraulic drive (pump, cylinders, rams, reservoirs, filter, and electric motor). When the power unit also has to travel with the robot, the overall power to weight ratio is considerably lower.

Stiffness versus compliance

1) Stiffness is the material's resistance to distortion. The tighter the system, the greater the load to distort it. The more compatible the system, the easier it deforms under the strain.
2) Rigidity is directly linked to the material elasticity module. Hydraulic systems are very rigid and non-compliant, whereas pneumatic systems are compressed readily and are therefore compliant.
3) Stiff systems respond more quickly and are more precise to changing loads and pressures.
4) While steepness produces systems which are more sensitive and precise, it also presents risk if not always everything is flawless.

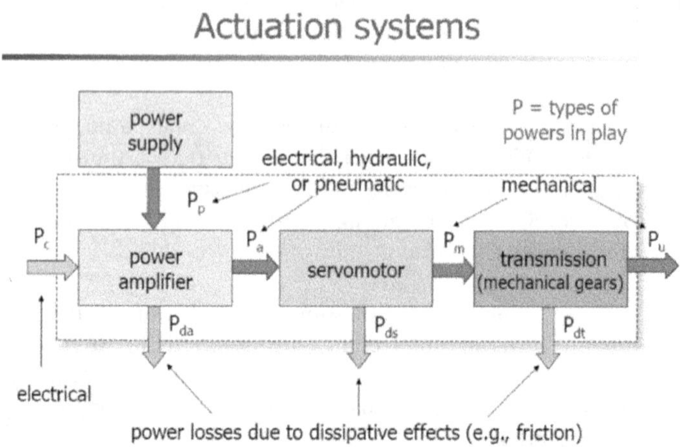

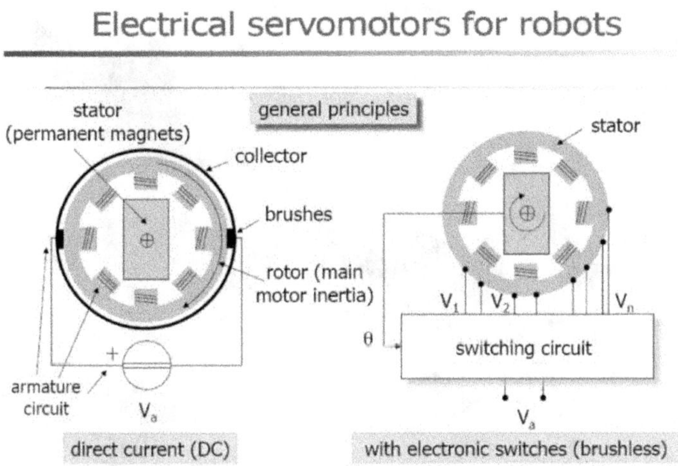

Figure 5.51 Actuators for robots.

Electrical Actuators

1) Easy to control
2) From W to MW
3) Normally high velocities 1000 - 10000 rpm
4) Several types
5) Accurate servo control
6) Ideal torque for driving
7) Excellent efficiency
8) Autonomous power system

Mainly rotating but also linearones are available. Linear movement with gear or with real linear motor.

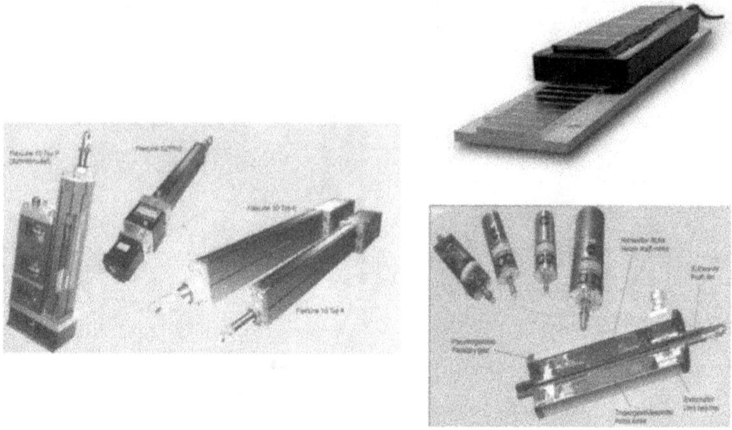

Figure 5.52 Electrical actuators.

Electrical Actuator Types

1. Servo Motor
2. DC-motors
3. Brushless DC-motors
4. Asynchronous motors
5. Synchronous motors
6. Reluctance motors
7. Stepper Motor
   a. Most of the time servo motor is used in the most of the robotics application.

Servo System

1. Servo is mechanism based on feedback control.
2. The controlled quantity is mechanical.

Properties of Servo Motor

1. High maximum torque/force allows high (de)acceleration
2. High zero speed torque/force
3. High bandwidth provides accurate and fast control
4. Works in all four quadrants
5. Robustness

Electrical servomotors

- advantages
  - power supply available everywhere
  - low cost
  - large variety of products
  - high power conversion efficiency
  - easy maintenance
  - no pollution in working environment
- disadvantages
  - overheating in static conditions (in the presence of gravity)
    - use of emergency brakes
  - need special protection in flammable environments

Stepper Motors

A (3 or more) pole sequence is triggered and the stator moves in tiny "steps." Very low speed / great angular accuracy may be achieved with numerous rotor teeth without reducing gears. Can also carry out a "micro step" if both spindles are activated at once. A Stepper engine is a digital actuator in which the input is programmed by the winding and output of a stator in the form of a discrete angular position. The magnetic contact between the rotor poles and poles of the sequentially powered stator winding rotation occurs [1].

Figure 5.53 Stepper motors.

Stepper Motor Selection

1. Permanent Magnet / Variable Reluctance
2. Unipolar vs. Bipolar
3. Number of Stacks
4. Number of Phases
5. Degrees Per Step
6. Micro-stepping
7. Pull-In/Pull-Out Torque
8. Detent Torque

Hydraulic Actuators

In general, a hydraulic system comprises the following parts:

1. A servo- or manual valve is used for controlling the hydraulic linear or rotating cylinders and rams to give the strength or torque required to move the joints.
2. A hydraulic pump to supply the system with high pressure fluid
3. Electric motor for hydraulic pump operation.
4. Heat removal cooling system (cooling fans, radiators, and cooled air).
5. Reservoir to maintain the system's fluid supply supplied.
6. Servo valve, which is a highly sensitive valve which regulates fluid quantity and rate on the cylinders. Generally, the servo valve is powered by a hydraulic servomotor.
7. Control sensors for cylinders motion (position, velocity, magnetic, touch...)
8. Connecting pressurised fluid hoses.
9. Valves for safety control, holding valves.

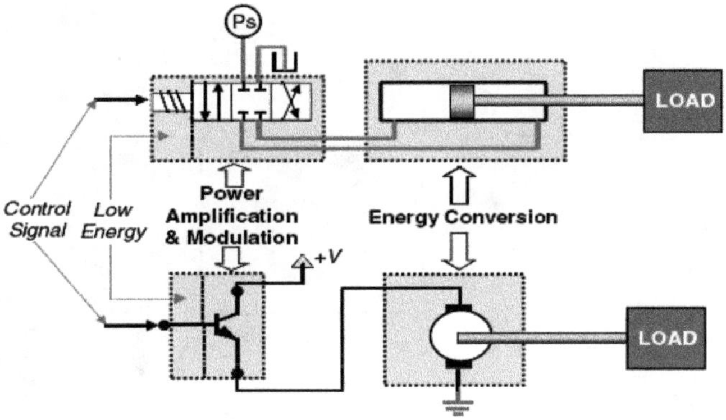

Electrohydraulic and electromechanical actuators.

Figure 5.54 Hydraulic actuators.

Pneumatic Actuators

Like hydraulic except power from compressed air.

Advantages:

- Fast on/off type tasks.
- Big forces with elasticity.
- No hydraulic oil leak problems.

Disadvantage:

Speed control is not possible because the air pressure depends on many variables that are out of control.

Other types of actuators

1. Piezoelectric
2. Magnetic
3. Ultrasound
4. Shape Memory Alloys [1, 11, 14-22].

Figure 5.55 Other types of actuators.

References

1. Dario, P., Laschi, C., & Guglielmelli, E. (1996). Sensors and actuators for'humanoid'robots. Advanced Robotics, 11(6), 567-584.
2. Ogenyi, U. E., Liu, J., Yang, C., Ju, Z., & Liu, H. (2019). Physical human-robot collaboration: Robotic systems, learning methods, collaborative strategies, sensors, and actuators. IEEE transactions on cybernetics.
3. Shahinpoor, M. (2003). Ionic polymer–conductor composites as biomimetic sensors, robotic actuators and artificial muscles—a review. Electrochimica Acta, 48(14-16), 2343-2353.
4. Schempf, H., & Yoerger, D. R. (1993). Study of dominant performance characteristics in robot transmissions.
5. Bilodeau, R. A., White, E. L., & Kramer, R. K. (2015, September). Monolithic fabrication of sensors and actuators in a soft robotic gripper. In 2015 IEEE/RSJ International Conference on Intelligent Robots and Systems (IROS) (pp. 2324-2329). IEEE.
6. Kim, B., Lee, M. G., Lee, Y. P., Kim, Y., & Lee, G. (2006). An earthworm-like micro robot using shape memory alloy actuator. Sensors and Actuators A: Physical, 125(2), 429-437.
7. Rao, S. S., & Sunar, M. (1993). Analysis of distributed thermopiezoelectric sensors and actuators in advanced intelligent structures. AIAA journal, 31(7), 1280-1286.
8. Liang, W., Liu, H., Wang, K., Qian, Z., Ren, L., & Ren, L. (2020). Comparative study of robotic artificial actuators and biological muscle. Advances in Mechanical Engineering, 12(6), 1687814020933409.
9. De Silva, C. W. (2015). Sensors and actuators: Engineering system instrumentation. CRC Press.

10. Ilas, C., Rosner, D., Tataroiu, R., & Surpateanu, A. (2012). Wireless sensors and actuators networks: Localization for medical and robotics applications. Journal of Control Engineering and Applied Informatics, 14(2), 34-39.
11. Jin, X., Feng, C., Ponnamma, D., Yi, Z., Parameswaranpillai, J., Thomas, S., & Salim, N. (2020). Review on exploration of graphene in the design and engineering of smart sensors, actuators and soft robotics. Chemical Engineering Journal Advances, 100034.
12. Wang, B., Wang, Z., Chen, T., & Zhao, X. (2020). Development of novel bioreactor control systems based on smart sensors and actuators. Frontiers in bioengineering and biotechnology, 8, 7.
13. Tzou, H. S., & Fu, H. Q. (1994). A study of segmentation of distributed piezoelectric sensors and actuators, part I: theoretical analysis. Journal of Sound and Vibration, 172(2), 247-259.
14. Cianchetti, M., Mattoli, V., Mazzolai, B., Laschi, C., & Dario, P. (2009). A new design methodology of electrostrictive actuators for bio-inspired robotics. Sensors and Actuators B: Chemical, 142(1), 288-297.
15. Dario, P., Laschi, C., & Guglielmelli, E. (1996). Sensors and actuators for'humanoid'robots. Advanced Robotics, 11(6), 567-584.
16. Ogenyi, U. E., Liu, J., Yang, C., Ju, Z., & Liu, H. (2019). Physical human-robot collaboration: Robotic systems, learning methods, collaborative strategies, sensors, and actuators. IEEE transactions on cybernetics.
17. Low, J. H., Khin, P. M., & Yeow, C. H. (2015, May). A pressure-redistributing insole using soft sensors and actuators. In 2015 IEEE International Conference on Robotics and Automation (ICRA) (pp. 2926-2930). IEEE.
18. Visone, C. (2008, November). Hysteresis modelling and compensation for smart sensors and actuators. In Journal of Physics: Conference Series (Vol. 138, No. 1, p. 012028). IOP Publishing.
19. Song, G., Qiao, P. Z., Binienda, W. K., & Zou, G. P. (2002). Active vibration damping of composite beam using smart sensors and actuators. Journal of aerospace engineering, 15(3), 97-103.
20. Giachino, J. M. (1986). Smart sensors. Sensors and actuators, 10(3-4), 239-248.
21. Tanaka, N., & Sanada, T. (2006). Modal control of a rectangular plate using smart sensors and smart actuators. Smart Materials and Structures, 16(1), 36.
22. Ogenyi, U. E., Liu, J., Yang, C., Ju, Z., & Liu, H. (2019). Physical human-robot collaboration: Robotic systems, learning methods, collaborative strategies, sensors, and actuators. IEEE transactions on cybernetics.
23. Dario, P., Laschi, C., & Guglielmelli, E. (1996). Sensors and actuators for 'humanoid' robots. Advanced Robotics, 11(6), 567-584.

6

Micro-actuator Study and Robotics Trajectory Planning

Bibekananda Naik[1], Nirmalya Parhi[2], Ashok Kumar Bhoi[3], Asini Kumar Baliarsingh[4], and Pradyut Kumar Muduli[5]

[1]CSIR - Institute of Minerals & Materials Technology, Bhubaneswar, Odisha, India
[2]Department of Mechanical Engineering, Government College of Engineering, Kalahandi, Bhawanipatna, Odisha, India
[3]Department of Computer Science and Engineering, Government College of Engineering, Kalahandi, Bhawanipatna, Odisha, India
[4]Department of Electrical Engineering, Government College of Engineering, Kalahandi, Bhawanipatna, Odisha, India
[5]Department of Civil Engineering, Government College of Engineering, Kalahandi, Bhawanipatna, Odisha, India

6.1 What are Actuators?

Input energy is used by actors and output energy is released in a regulated way. Mechanical actuators operate and move anything using strength or torque.

Many kinds of actuators are available:

- based on the kind of energy generated.
- based on the release of output energy.
- based on the energy consumed.

Actuators are also known as transducers. Transducers are devices that convert one type of energy into another.

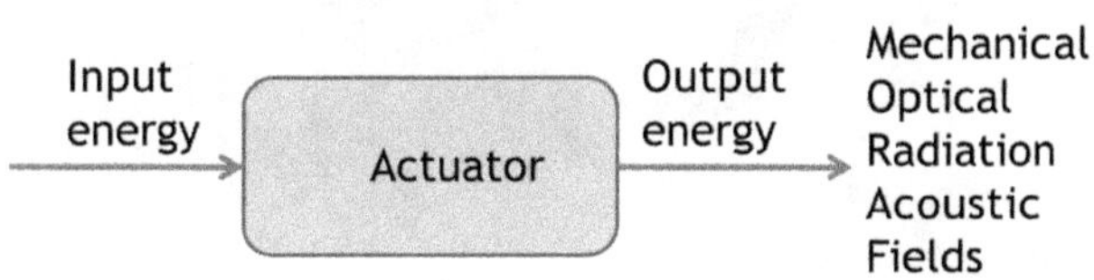

Figure 6.1 Function of transducers.

Actuators with different output energies

Mechanical actuators

- Produce motion
- Motors, engines, pumps

Acoustic actuators

- Produce sound

Optical actuators

- Produce light
- LEDs, solid-state lasers

Others

- Radiation
- Fields

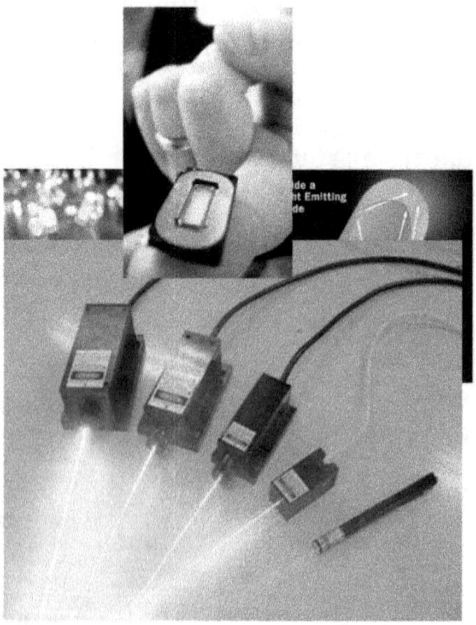

Figure 6.2 Optical actuators.

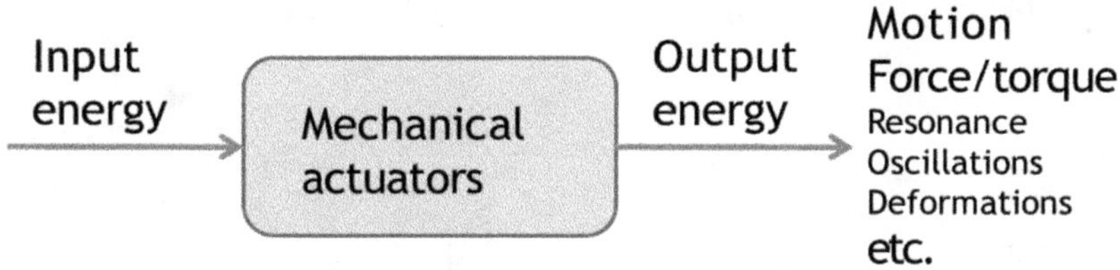

Figure 6.3 Mechanical actuator.

The output of an actuator is often mechanical in nature. In most cases, actuators convert the energy received into regulated mechanical energy [1, 2].

6.2 Micro-actuators

- Produce motions over small distances
  - of the order of microns to mm.
- Produce small forces
  - of the order of pN to mN.
- Produce motion and force in entities of small sizes.

Micro-actuator universe: displacements

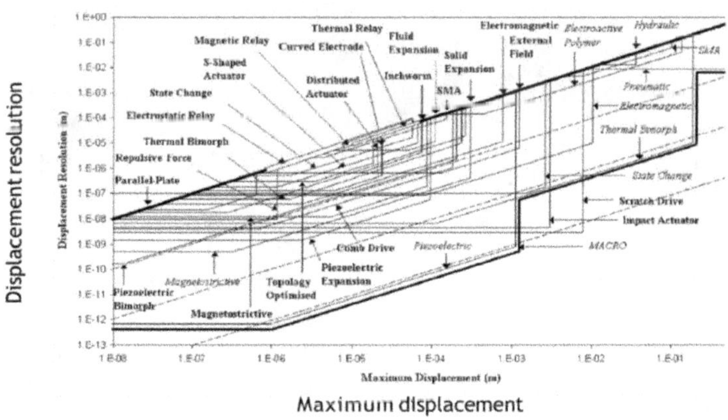

Figure 6.4 Micro-actuator universe: displacements.

Micro-actuator universe: forces

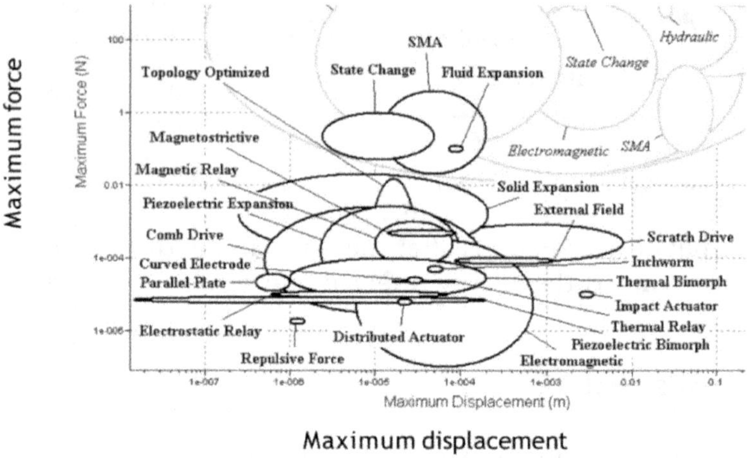

Figure 6.5 Micro-actuator universe: forces.

Biological micro-actuators: displacements

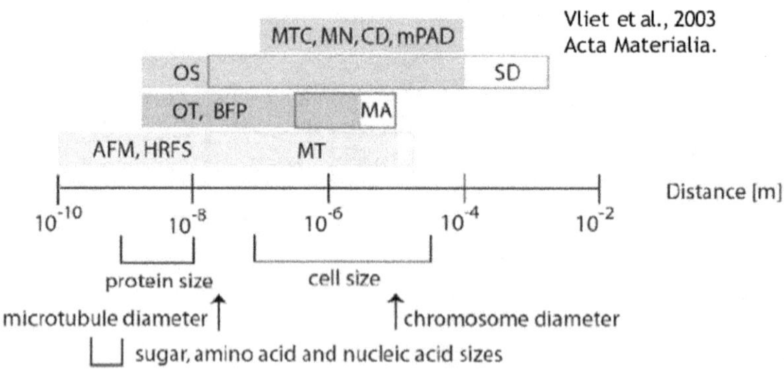

Figure 6.6 Biological Micro-actuators: displacements.

Biological actuators: forces

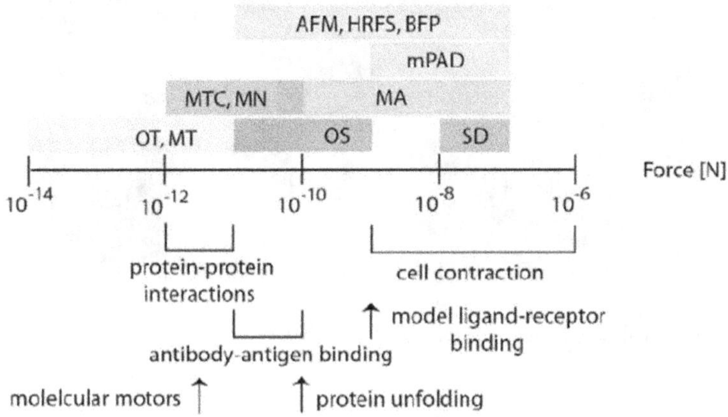

Figure 6.7 Biological actuators: forces.

Mechanical micro-actuators with a variety of input energies are discussed, as well as how they are utilised.

- Electrostatic
- Electro-magnet based
- Thermal
- Chemical
- Piezo-electric
- Shape memory alloy (SMA)
- Smart material-based
- Light-induced
- Biological [1-3].

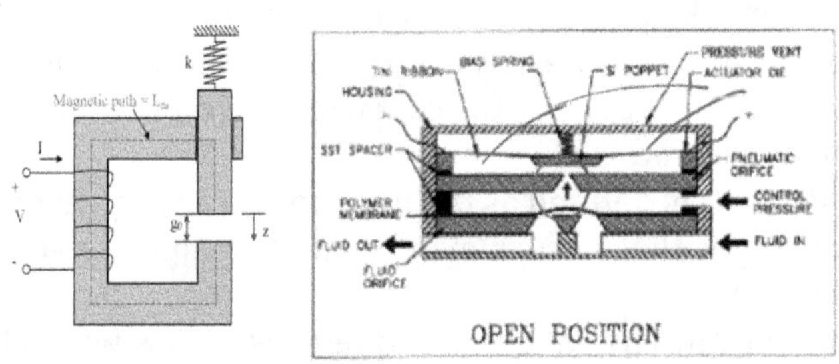

Figure 6.8 Mechanical actuators.

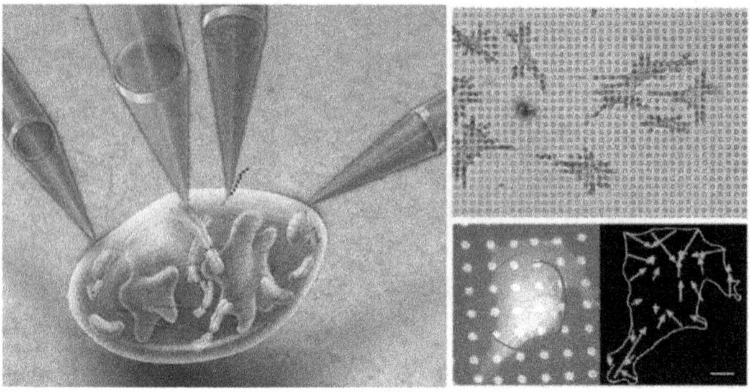

Figure 6.9 Biological actuators.

Characteristics of a mechanical (micro) actuator

- Stroke: The greatest amount of displacement that an actuator is capable of producing.
- Force/torque: An actuator's maximum force or torque is defined as the force or torque that the actuator can produce.
- Stiffness: Stiffness is defined as the rate at which the produced force/torque diminishes with the length of the stroke.
- Input energy: Input energy is the amount of energy (for example, voltage and current) that is supplied to the actuator.
- Efficiency: Efficiency is defined as the ratio of released energy to the amount of input energy.
- Linearity: Linearity is the degree to which the relationship between force and stroke is linear. In an ideal world, we'd want continuous force.
- Hysteresis: Hysteresis is defined as the difference in displacement/force for the same values of input energy in up and down strokes for the same values of input energy.
- Response time: Responding time is the amount of time it takes for the actuator to react from the moment the input signal is received.
- Drift: Drift is defined as the unexpected change in force or displacement that occurs even while the input energy remains constant.
- Bandwidth: The range of frequencies across which the actuator can consistently provide the specified displacement and force is referred to as its bandwidth [2, 3].

Some micro-actuators

- Electrostatic micro-motors
- Electrostatic comb-drive
- Magnetic actuators
- Thermal micro-actuators
- Pneumatic actuators
- Piezoelectric actuators
- Surface-tension driven fluidic actuators

Electrostatic comb-drive

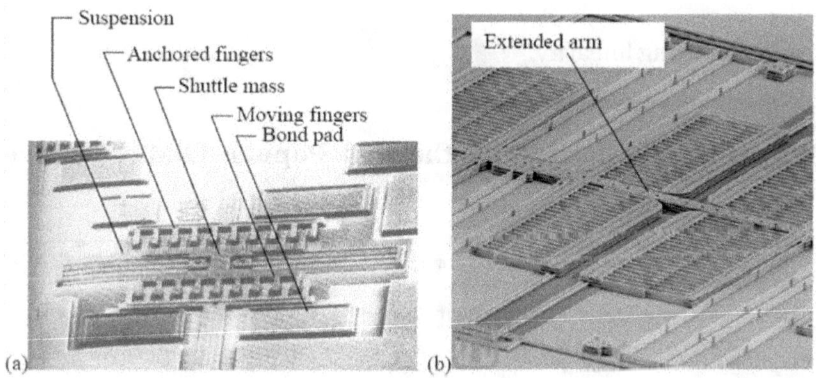

Figure 6.10 Electrostatic comb-drive.

Comb-drive: a schematic

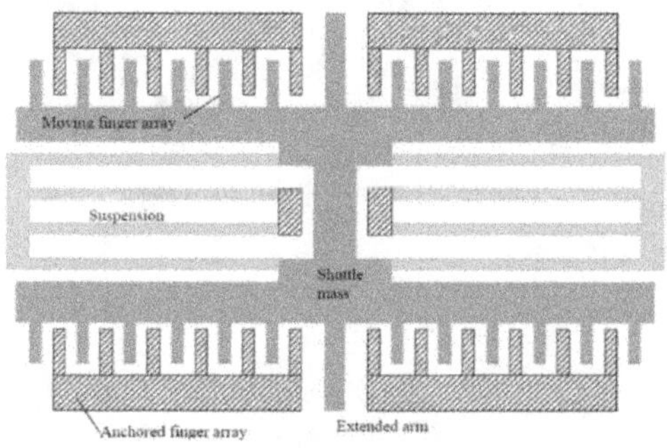

Figure 6.11 Comb-drive.

Getting linearity

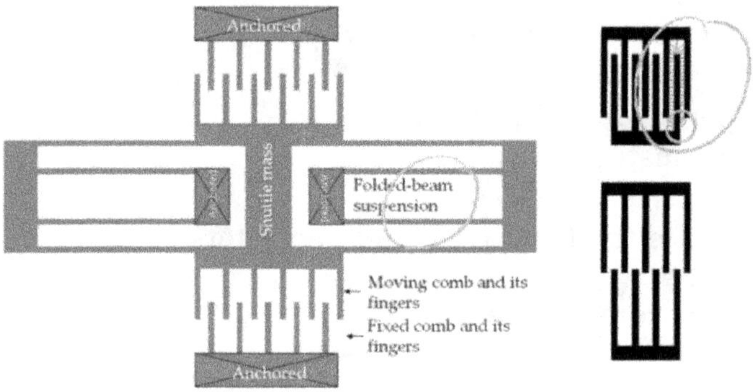

Figure 6.12 Getting linearity.

6.3 Why is Electrostatic Force the Most Popular in Micro-actuators?

E	Young's modulus	w	Thickness of beams	g	Gap between comb-fingers
t	Width of the suspension beams	l	Length of the suspension beams	N	Number of comb-pairs
				V	Voltage

$$k = \frac{2Etw^3}{l^3}$$ Lumped mechanical stiffness of the suspension

$$F_e = \frac{N\varepsilon tV^2}{2g}$$ Electrostatic force

$$\delta = \frac{F_e}{k} = \left(\frac{N\varepsilon V^2}{4E}\right)\frac{l^3}{gw^3}$$ Deflection of the shuttle

$$\frac{\delta}{l} = \left(\frac{N\varepsilon V^2}{4E}\right)\frac{l^2}{gw^3} \propto L^{-2}$$ Scaling of relative deflection with size for fixed voltage

$$V = \sqrt{\left(\frac{\delta}{l}\right)\left(\frac{4E}{N\varepsilon_0}\right)\left(\frac{gw^3}{l^2}\right)} \propto L$$ Scaling of voltage for fixed relative deflection

Electrostatic micro-motors

Figure 6.13 Electrostatic micro-motors.

There was a race between U.C. Berkeley and MIT to make the first rotary motor in silicon.

Harmonic micro-motor (MIT)

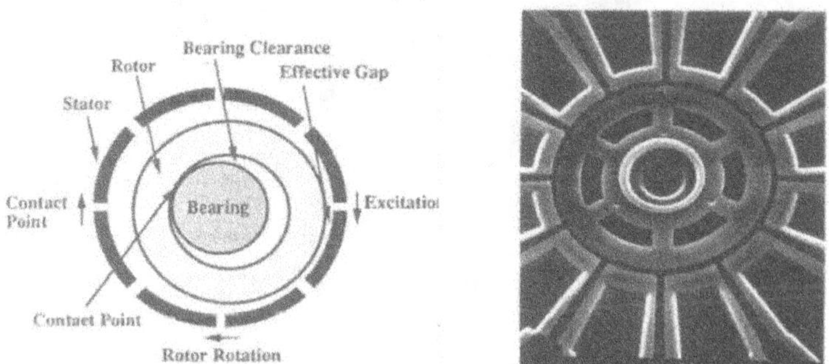

Figure 6.14 Harmonic micro-motor (MIT).

Micro-gear train (Bell Labs)

Figure 6.15 Micro-gear train (Bell Labs).

Micromechanical lock (Sandia)

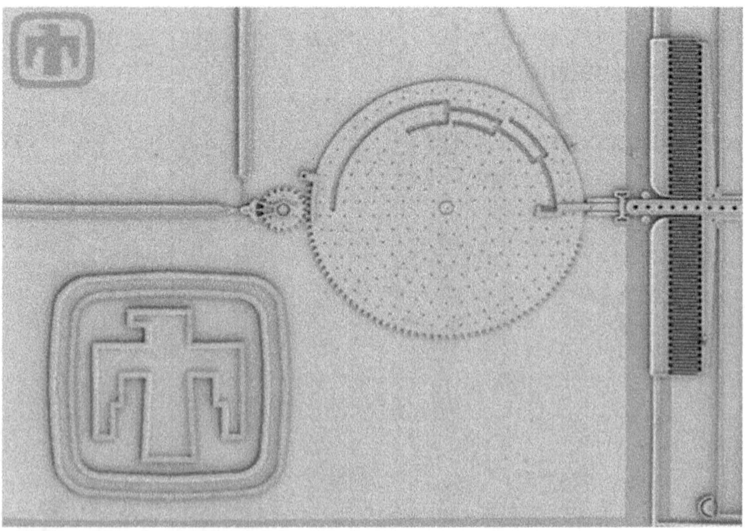

Figure 6.16 Micromechanical lock (Sandia).

SUMMiT of Sandia

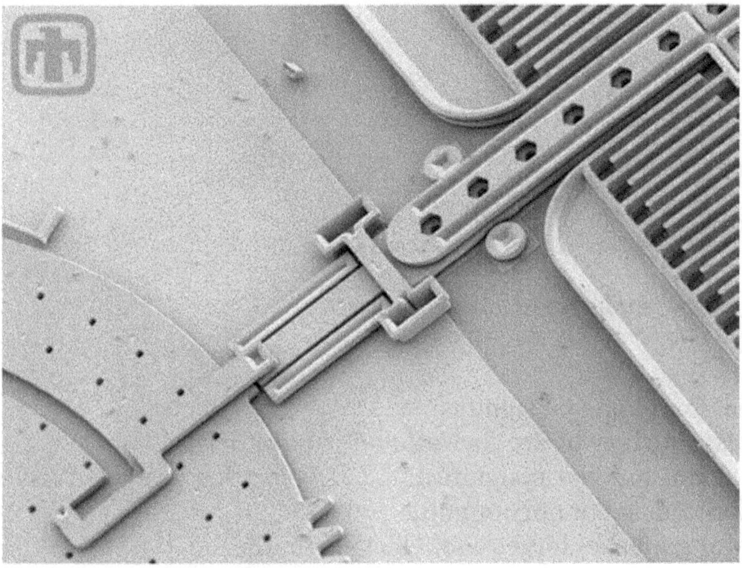

Figure 6.17 SUMMiT of Sandia.

Sandia's micro-engine gear train

Figure 6.18 Sandia's micro-engine gear train.

Sandia National Laboratory

Figure 6.19 Sandia National Laboratory.

Inertial forces can be significant.
Figures: courtesy of Sandia National Laboratory
The inertia may be insignificant, but velocities are relatively huge (million rpm is not uncommon).
Therefore, inertial forces could be substantial.

Wedge motor of Sandia

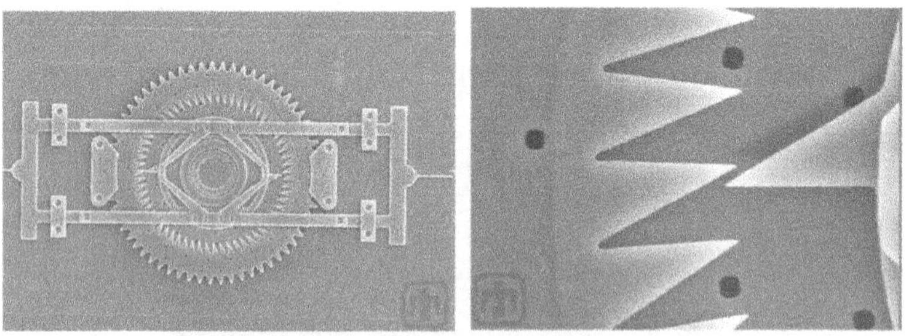

Figure 6.20 Wedge motor of Sandia.

Prime mover linkage c/o micro

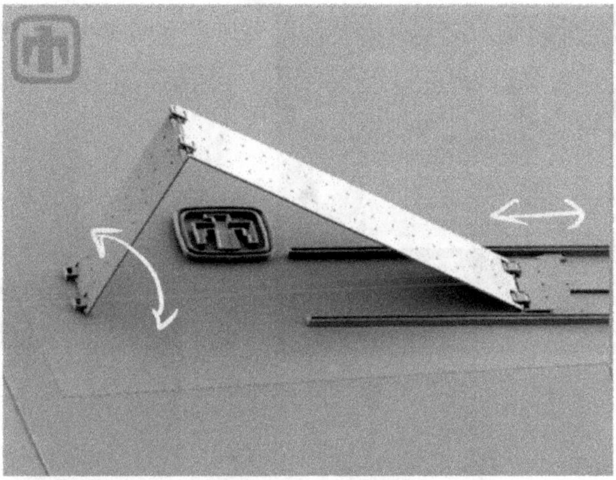

Figure 6.21 Prime mover linkage c/o micro.

6.4 Is electro-magnet Force Not Scalable to Micro?

- Electromagnetic force between two coils

$$F = \frac{\mu_0}{2\pi} I_1 I_2 \frac{l}{d}$$

Case 1: Constant current density

$$J = \frac{I}{A_{cs}} = \text{constant} = \Rightarrow I \propto L^2 \Rightarrow F \propto L^4$$ Very, very bad

Case 2: Constant temperature rise

$$\rho J^2 A_s \propto k\square T \Rightarrow J \propto \frac{1}{L} \Rightarrow I \propto L \Rightarrow F \propto L^2$$ Bad

Magnetic actuation-based pump

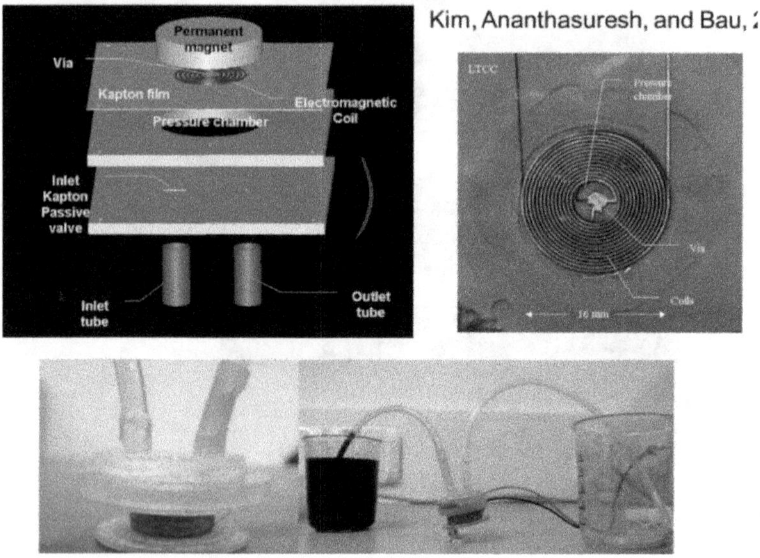

Figure 6.22 Magnetic actuation-based pump.

6.5 Coils on the Chip

Although the fabrication of micro coils is feasible, the process complexity rises as a result.

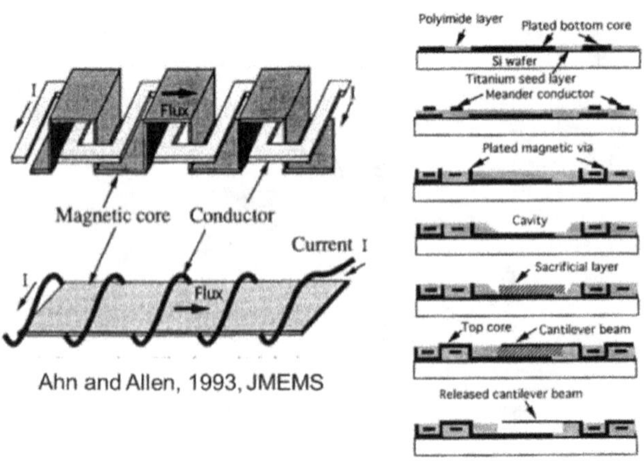

Figure 6.23 Coils on the chip.

6.6 Electro-thermal Actuation

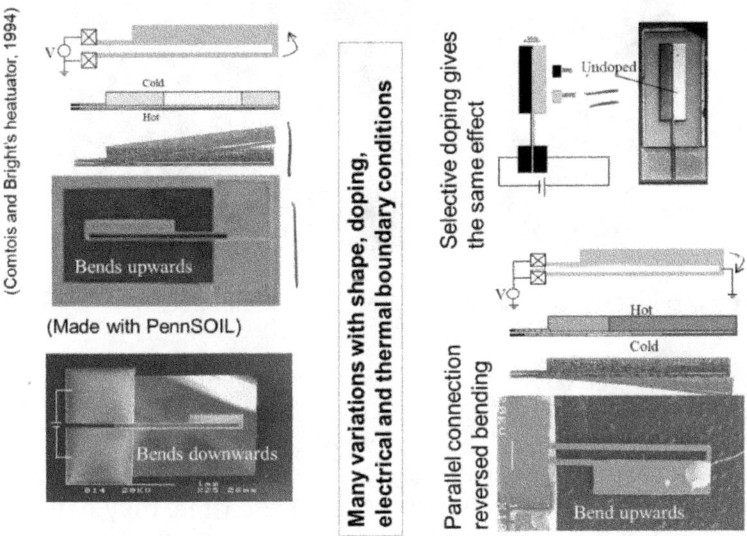

Figure 6.24 Electro-thermal actuation.

6.7 Enhanced Thermal Expansion

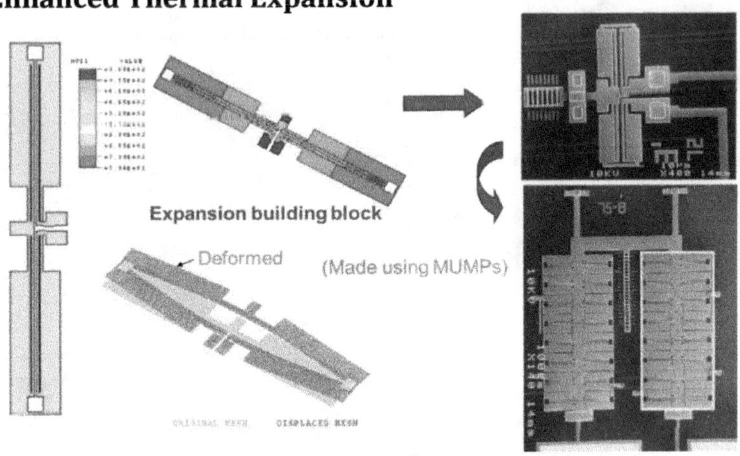

Figure 6.25 Enhanced thermal expansion.

6.8 Electro-thermally Actuated Parallel Linkages

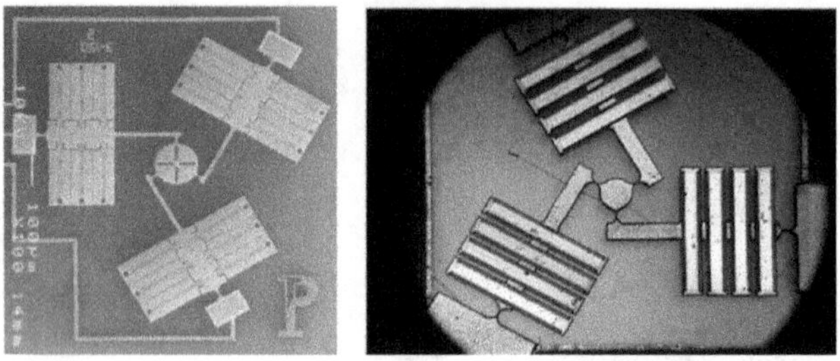

Figure 6.26 Electro-thermally actuated parallel linkages.

6.9 Piezoelectric Actuators

Electrical charge is generated by active materials when mechanical strain is applied, and vice versa. Most of the time, they have significant forces but just modest displacements. Typically, bimorph mode is used. Make sure your movements are accurate. Large voltages are required for operation.

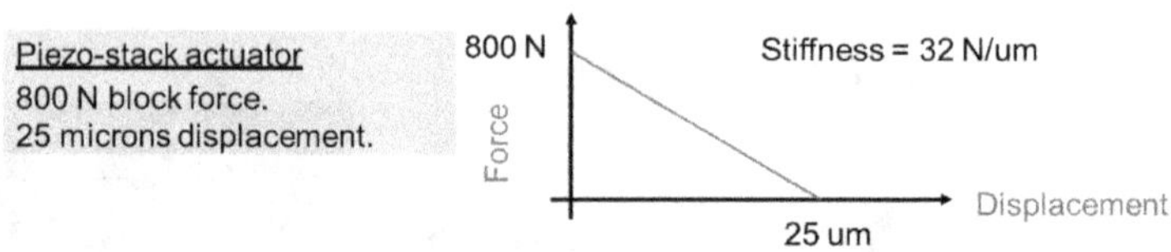

Figure 6.27 Thermo-pneumatic actuation.

6.10 Thermo-pneumatic actuation

A normally closed fluidic valve

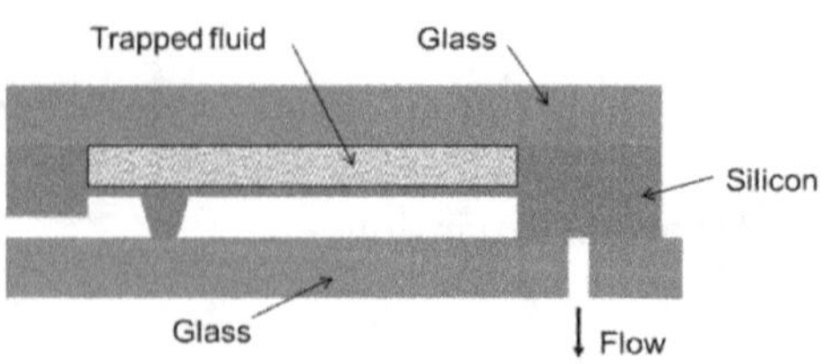

Figure 6.28 Closed fluidic valve.

6.11 Thermo-pneumatic Actuation

Redwood Microsystems's thermo-pneumatic normally closed valve

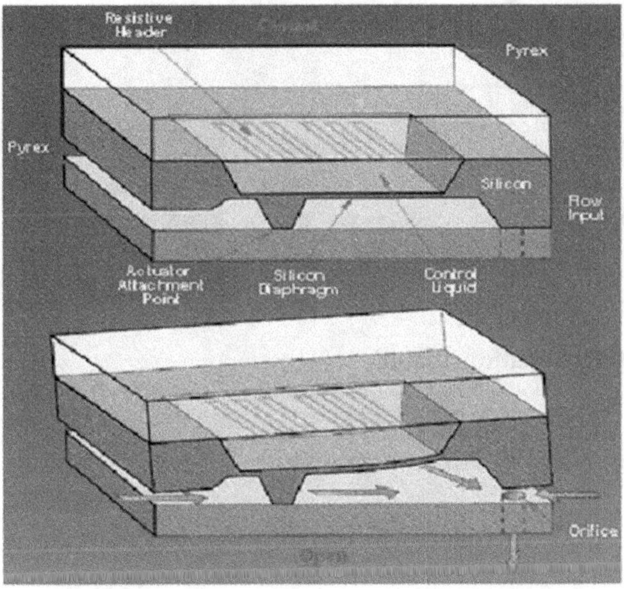

Figure 6.29 Redwood Microsystems's thermo-pneumatic actuators.

6.12 Pneumatic Actuation

1. Bi-metal cantilevers curled due to residual stress.
2. Opened with actuating the bottom membrane.

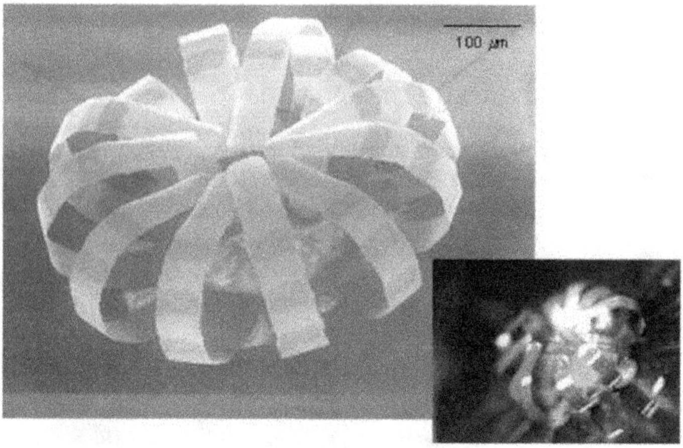

Figure 6.30 Pneumatic actuation.

6.13 Other Ways of Micro-actuation

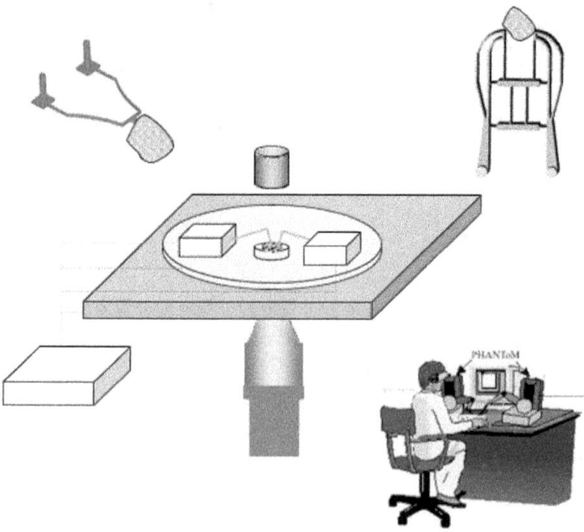

Figure 6.31 Other ways of micro-actuation.

6.14 Micro-actuation with Precision Stages

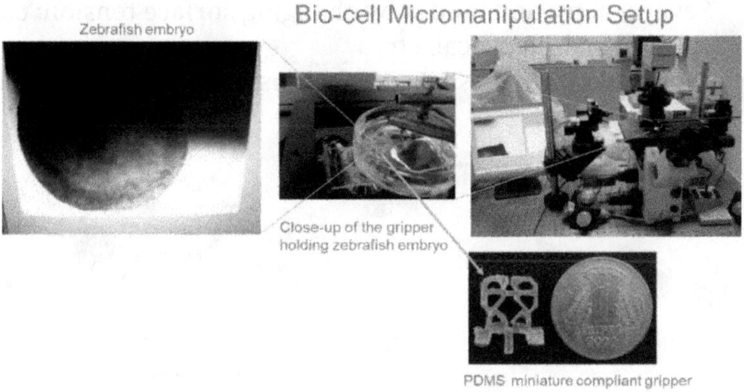

Figure 6.32 Micro-actuation with precision stages.

6.15 Novel Digital Speakers

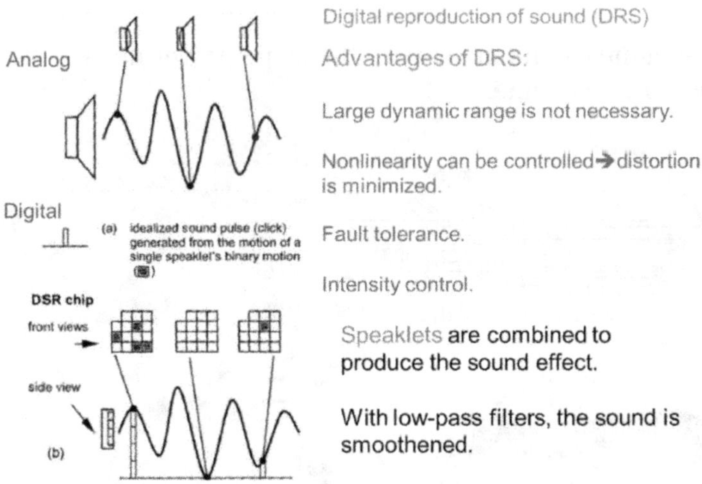

Figure 6.33 Novel digital speakers.

6.16 Surface-driven Flows

- Several actuation modes by changing surface tension: chemical, thermal, electrical, etc.
- Digital Microfluidics by surface tension driven droplets.

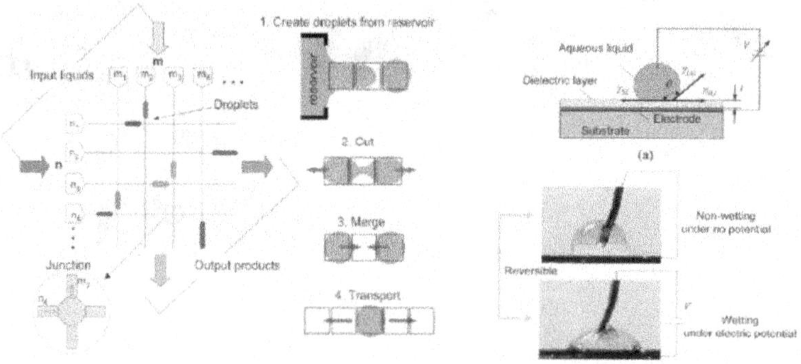

Figure 6.34 Surface-driven flows.

6.17 Power on a Chip: MIT Micro-engine

A micro-engine can be made with the same power density as that of a big gas turbine engine.

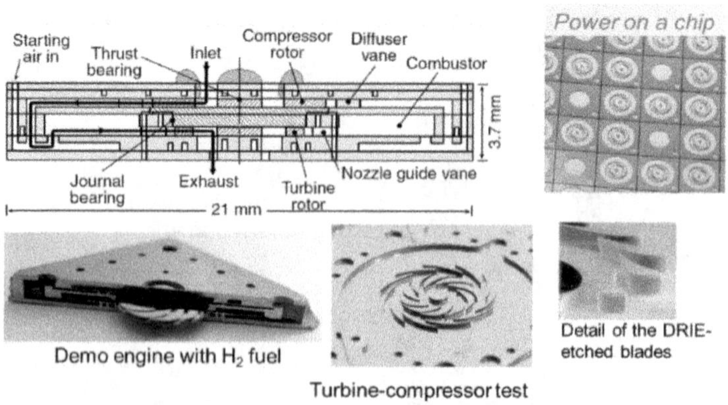

Figure 6.35 MIT micro-engine.

6.28 Example of a Smart Actuator

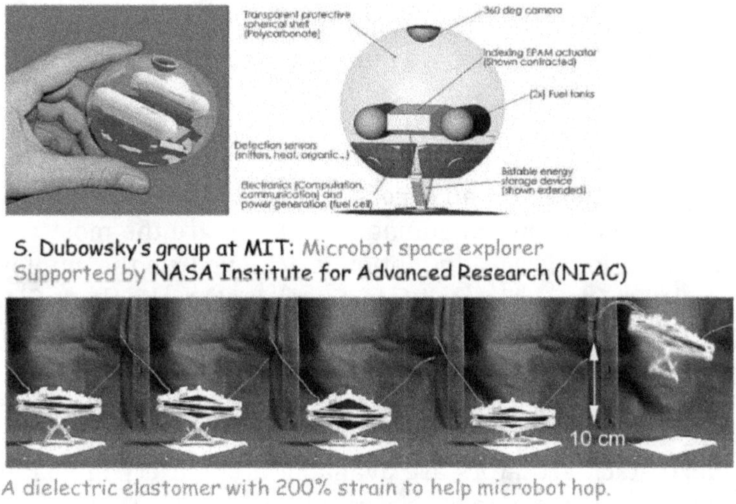

Figure 6.36 Smart actuator.

6.19 Micro Actuation – Macro Effect

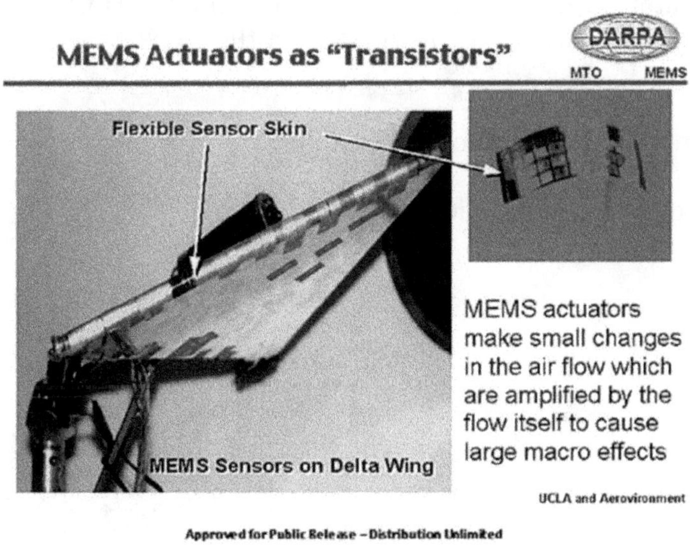

Figure 6.37 Micro actuation – macro effect.

6.20 Characteristics of Actuators

It is essential to reduce the size of things. Because we need to move solids and make fluids flow, we must first move the solids.

Status

Electrostatic actuation is an excellent choice for sensor applications. Thermal, magnetic, and pneumatic actuation are the most common types of actuation. Precision motion is achieved via the employment of piezoelectric and voice coil (electromagnet) technology. SMA and other active materials are intended for use in specialised applications. Surface tension, electrophoresis, electro osmosis, and other techniques are being used in micro fluidics. Commercial micro-actuators are not available as stand-alone devices at the time of publication. Piezo stack actuators are available in a variety of shapes and sizes. Actuators need a great deal more investigation [3].

6.21 Trajectory Planning for Robot Manipulators

Trajectory Planning

Manipulators with many degrees of freedom for performing a wide range of sophisticated manipulations in the work environment [4-20].

Path: only geometric description

Trajectory: timing included

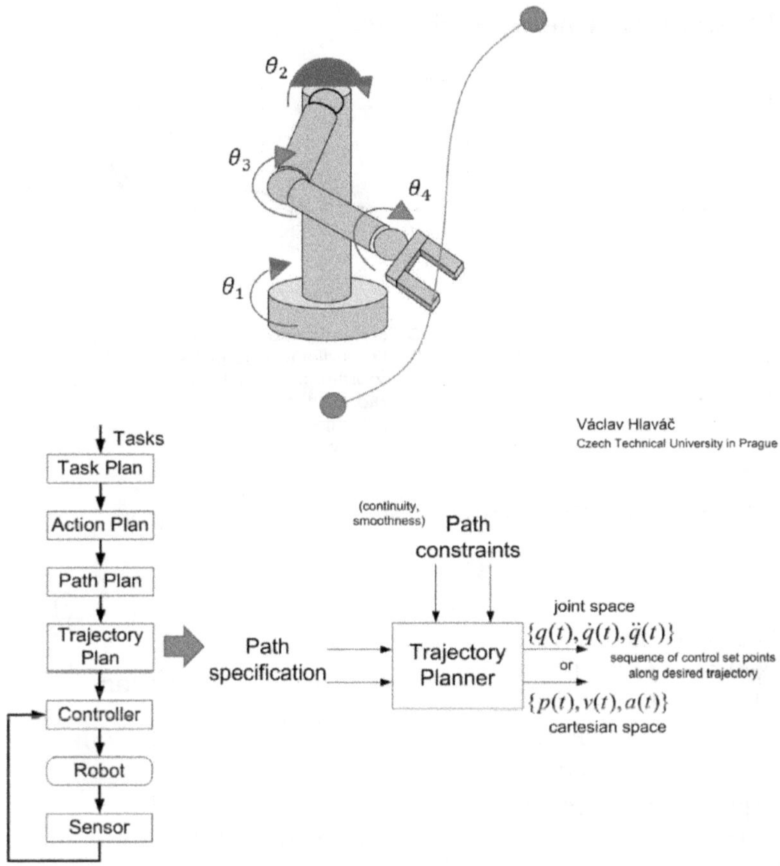

Figure 6.38 Trajectory planning for a robot.

6.22 Joint Space Vs Operational Space

- Joint-space description:
  - The description of the motion to be made by the robot by its joint values.
  - The motion between the two points is unpredictable.
- Operational space description:
  - In many cases operational space = Cartesian space.
  - The motion between the two points is known at all times and controllable.
  - It is easy to visualize the trajectory, but it is difficult to ensure that singularity does not occur.

6.23 Planning in Operational Space

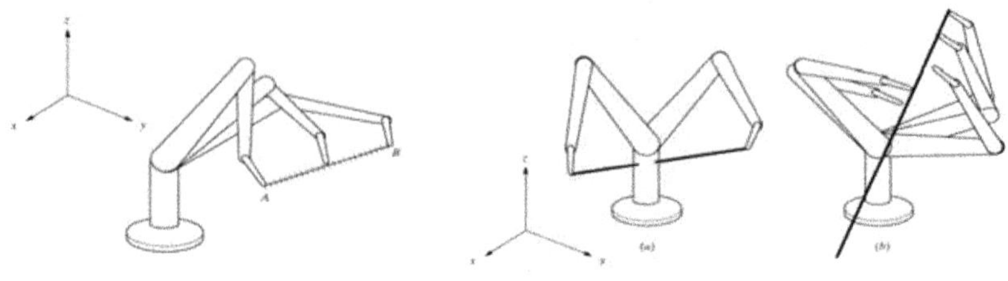

Sequential motions of a robot to follow a straight line.

Cartesian-space trajectory
(a) The trajectory specified in Cartesian coordinates may force the robot to run into itself, and
(b) the trajectory may requires a sudden change in the joint angles.

Figure 6.39 Planning in operational space.

- Calculate path from the initial point to the final point.
- Assign a total time T_{path} to traverse the path.
- Discretize the points in time and space.
- Blend a continuous time function between these points
- Solve inverse kinematics at each step.

- Advantages
  - Collision free path can be obtained.
- Disadvantages
  - Computationally expensive due to inverse kinematics.
  - It is unknown how to set the total time T_{path}.

6.24 Planning in Joint Space

- Calculate inverse kinematics solution from initial point to the final point.
- Assign total time T_{path} using maximal velocities in joints.
- Discretize the individual joint trajectories in time.
- Blend a continuous function between these point.

- Advantages
  - Inverse kinematics is computed only once.
  - Can easily take into account joint angle, velocity constraints.
- Disadvantages
  - Cannot deal with operational space obstacles.

Path Definition

"Expressing the desired positions of a manipulator in the space, as a parametric function of time"

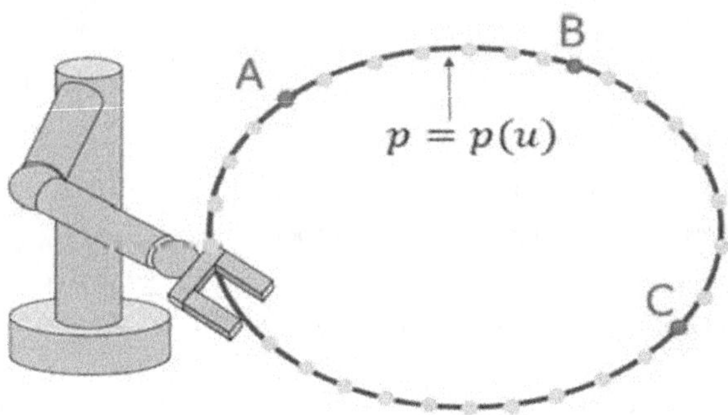

Figure 6.40 Path of robotic arm.

- B. Siciliano et al. Robotics (Modelling, planning and control), Springer, Berlin, 2009, chapter 4: Trajectory planning, pages 161-189.

Task to Trajectory

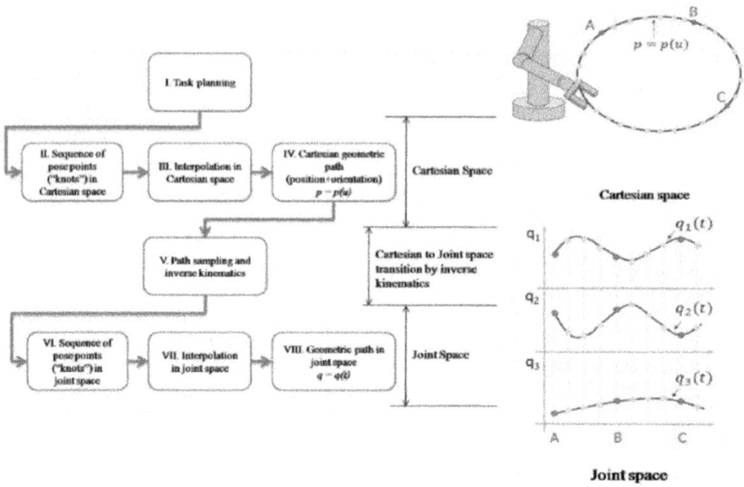

Figure 6.41 Task to trajectory.

Types of Motion

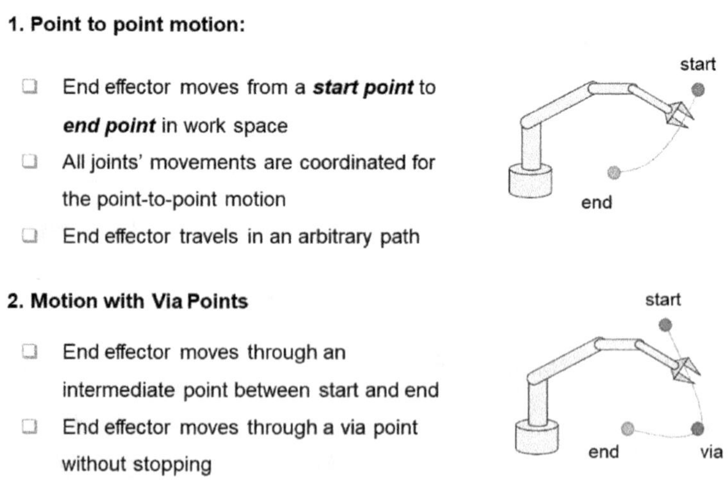

Figure 6.42 Types of motion.

Joint Space Planning

Point to point motion:

"Smooth functions are used to describe the movements of joints from beginning to end."

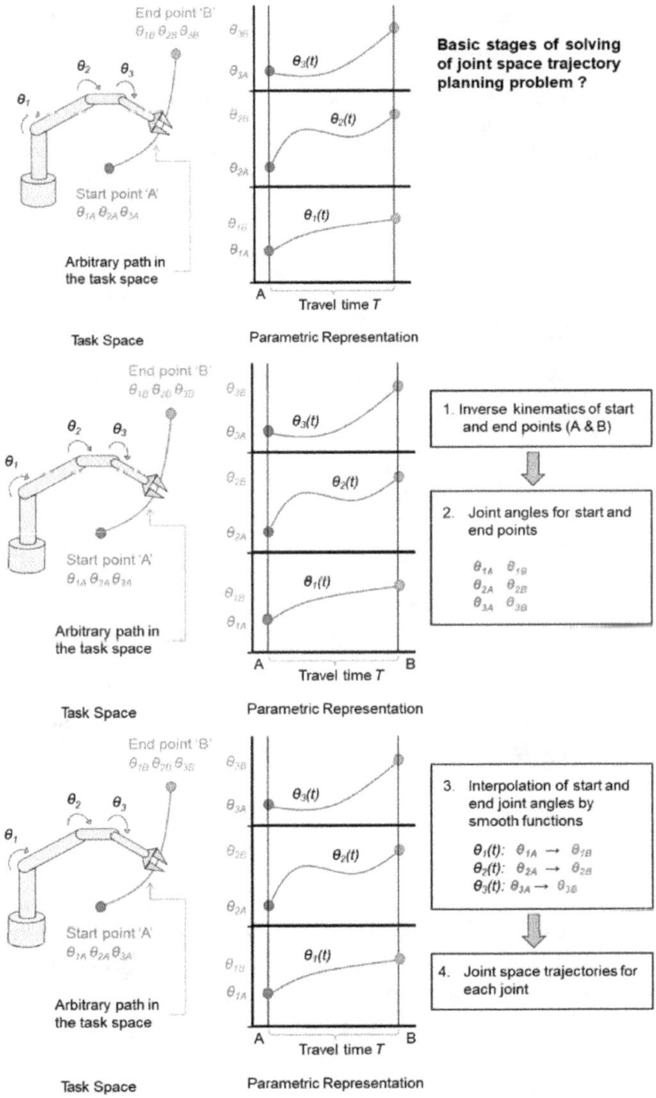

Figure 6.43 Point to point motion.

6.25 Smooth Motion, Quality of Work

Non-smooth trajectories result in poor quality in the manufacturing process.

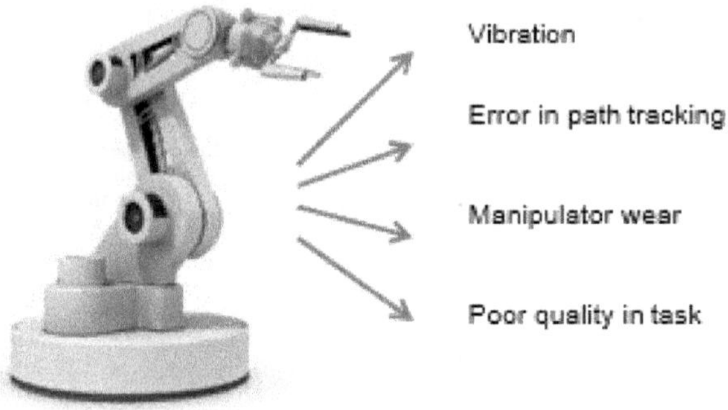

Figure 6.44 Features of planning of robot.

Linear Trajectory

$$q(t) = q^s + \frac{q_g - q_s}{t_g} t$$

$$\dot{q}(t) = \frac{q^g - q^s}{t_g}$$

$$\ddot{q}(t) = \begin{cases} \infty & t = 0, t_g \\ 0 & 0 < t < t_g \end{cases}$$

- ☐ Infinite accelerations at endpoints
- ☐ Discontinuous velocity when two trajectory segments are connected (at via points)

Triangular Velocity Trajectory

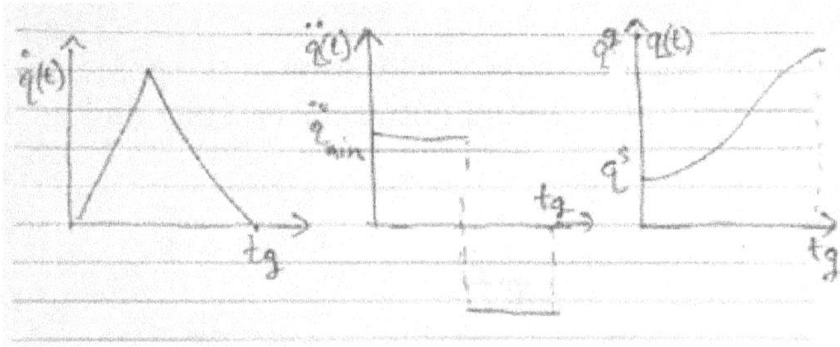

Figure 6.45 Triangular velocity trajectory.

Acceleration discontinuity at endpoints and at the midpoint of the trajectory

Linear Trajectory with Parabolic Blends

- Zero acceleration in middle segments.
- Constant acceleration at end segments.

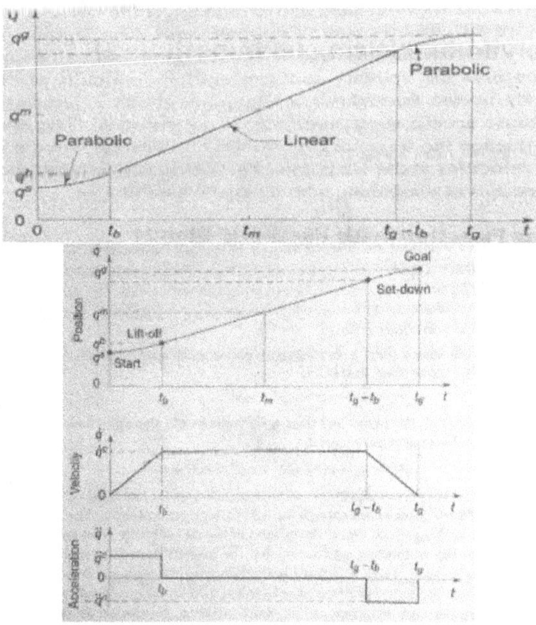

Figure 6.46 Linear trajectory with parabolic blends.

6.26 Inclusion of Via Points into a Linear Trajectory with Parabolic Blends

In order to achieve continuous acceleration at the point, it is possible to add via points (knot points), between the start and goal (qs, qg) locations [3].

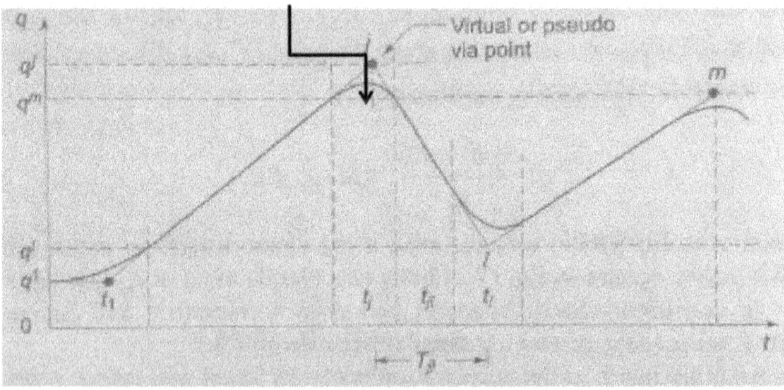

Multi-stage linear parabolic blend spline

Figure 6.47 Inclusion of via points into a linear trajectory with parabolic.

6.27 Cubic Polynomial (Bring in Smoothness)

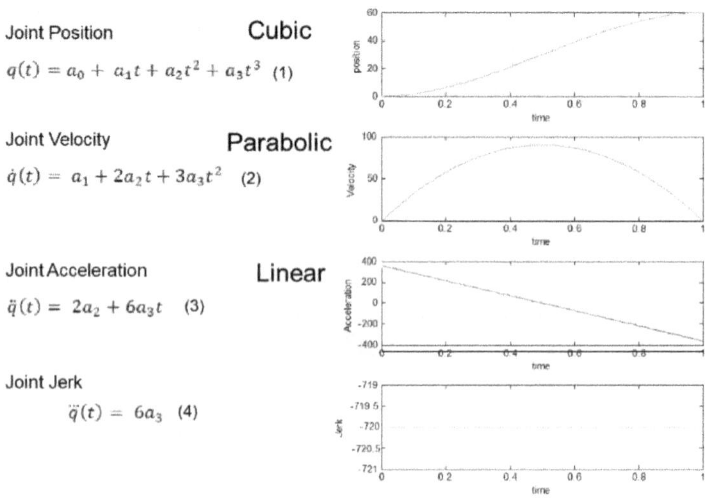

Figure 6.48 Cubic polynomial (bring in smoothness).

Cubic Polynomial (zero sped at end- points)

- ☐ Satisfies position and velocity at end-points
- ☐ E.g.: zero speed at end-points

$$q(0) = \qquad q(t\ ) =$$

$$\Rightarrow \begin{aligned} &(1)\text{ for } t=0, && q^s = a_0 \\ &(1)\text{ for } t=t_g, && q^g = a_0 + a_1 t_g + a_2 t_g^2 + a_3 t_g^3 \\ &(2)\text{ for } t=0, && q^s = {}_1 \\ &(2)\text{ for } t=t_g, && q^g = a_1 + 2a t_{2g} + 3a_2 t_g^2 \end{aligned}$$

$$q(t) = q^s + \frac{3}{t_g^2}(q^g - q^s)t^2 - \frac{2}{t_g^3}(q^g - q^s)t^3 \qquad 0 \le t \le t_g$$

$$a_0 = q^s \qquad a_1 = q^s(=0) \qquad a_2 = \frac{3(q^g - q^s)}{t_g^2} \qquad a_3 = \frac{-2(q^g - q^s)}{t_g^3}$$

- ☐ Acceleration is linear and uncontrollable

$$\ddot{q}(t) = 2a_2 + 6a_2 t \qquad 0 \le t \le t_g$$

Cubic Polynomial (zero sped at end- points)

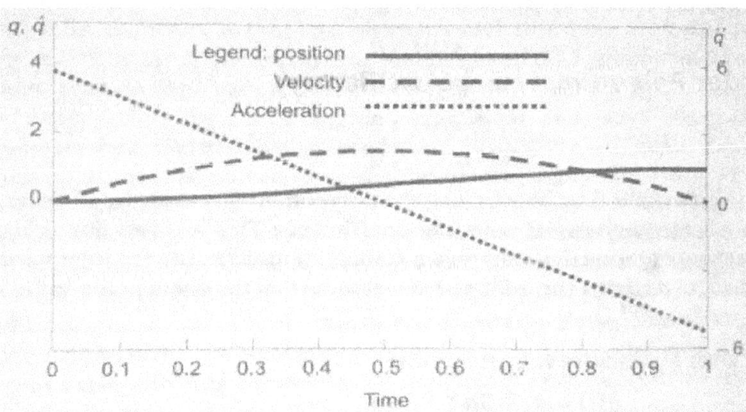

Figure 6.49 Cubic polynomial (zero sped at end- points).

Cubic Spline Trajectory

- Stitching cubic polynomials together [2].
- Acceleration is not continuous at via (stitching points)

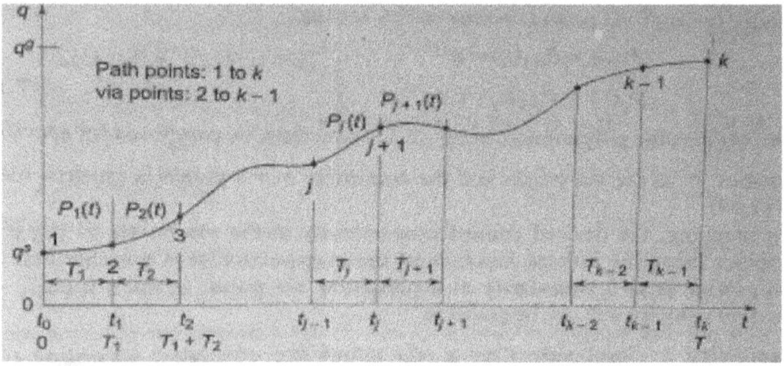

Figure 6.50 Cubic spline trajectory.

5th Order Polynomial (more oscillatory)

$$q(t) = a_0 + a_1 t + a_2 t^2 + a_3 t^3 + a_4 t^4 + a_5 t^5$$

$$\dot{q}(t) = a_1 + 2a_2 t + 3a_3 t^2 + 4a_4 t^3 + 5a_5 t^4$$

$$\ddot{q}(t) = 2a_2 + 6a_3 t + 12a_4 t^2 + 20a_5 t^3$$

$$\dddot{q}(t) = 6a_3 + 24a_4 t + 60a_5 t^2$$

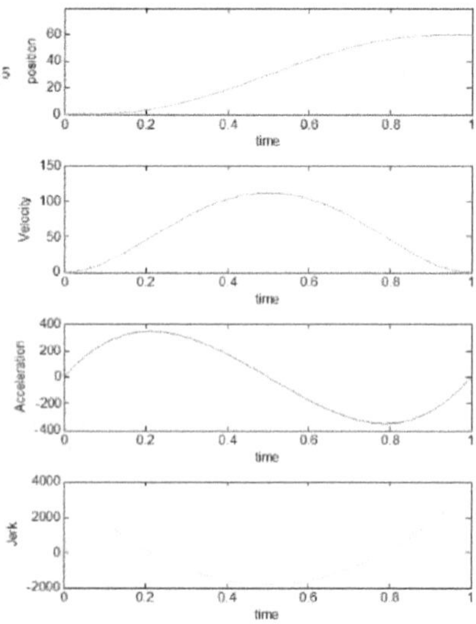

Figure 6.51 5th order polynomial (more oscillatory).

References

1. Borboni, A., Aggogeri, F., & Faglia, R. (2013). Design and analysis of a fibre-shaped micro-actuator for robotic gripping. International Journal of Advanced Robotic Systems, 10(3), 149.
2. Fearing, R. S. (1991, October). Control of a micro-organism as a prototype micro-robot. In Proc. 2nd Int. Symp. Micromachines and Human Sciences.
3. Koga, A., Suzumori, K., Miyagawa, T., & Sekimura, M. (1997). Electrostatic linear micro actuator with vibrating motion (application to the focusing mechanism of a miniature CCD camera). Advanced Robotics, 12(2), 103-113.
4. Katoch, R., & Ueda, J. (2015, July). Trajectory planning for antagonistic non-linearly quantized SMA actuator arrays based on evolutionary optimization. In 2015 American Control Conference (ACC) (pp. 2631-2636). IEEE.
5. Ikegami, T., Torii, A., Doki, K., & Ueda, A. (2006, November). Motion analysis of a micro-actuator using three piezoelectric actuators. In 2006 IEEE International Symposium on MicroNanoMechanical and Human Science (pp. 1-6). IEEE.
6. Towfighian, S., Seleim, A., Abdel-Rahman, E. M., & Heppler, G. R. (2011). A large-stroke electrostatic micro-actuator. Journal of Micromechanics and Microengineering, 21(7), 075023.
7. Lee, D. Y., & Cho, H. (2002, September). Precision force control via macro/micro actuator for surface mounting system. In IEEE/RSJ International Conference on Intelligent Robots and Systems (Vol. 3, pp. 2227-2232). IEEE.
8. Kandula, P., & Dong, L. (2018). Robust voltage control for an electrostatic micro-actuator. Journal of Dynamic Systems, Measurement, and Control, 140(6), 061012.
9. Pawinanto, R. E., Yunas, J., Majlis, B. Y., & Hamzah, A. A. (2016). Design and fabrication of compact MEMS electromagnetic micro-actuator with planar micro-coil based on PCB. Telkomnika, 14(3), 856-866.
10. Kreth, P., Gustavsson, J., Alvi, F., & Oates, W. (2012). Studies on the Resonance-Enhanced Micro-Actuator with Active Structures. In 6th AIAA Flow Control Conference (p. 3240).
11. Gilbertson, R. G., & Busch, J. D. (1996). A survey of micro-actuator technologies for future spacecraft missions. Journal of the British Interplanetary Society, 49(4), 129-138.
12. Hsieh, J. C., Lin, D. T., & Suen, M. S. (2020). The design of high strength electro-thermal micro-actuator based on the genetic algorithm. Microsystem Technologies, 26(4), 1113-1119.
13. Gasparetto, A., Boscariol, P., Lanzutti, A., & Vidoni, R. (2012). Trajectory planning in robotics. Mathematics in Computer Science, 6(3),

269-279.

14. Shin, K., & McKay, N. (1986). A dynamic programming approach to trajectory planning of robotic manipulators. IEEE Transactions on Automatic Control, 31(6), 491-500.
15. Guan, Y., Yokoi, K., Stasse, O., & Kheddar, A. (2005, July). On robotic trajectory planning using polynomial interpolations. In 2005 IEEE International Conference on Robotics and Biomimetics-ROBIO (pp. 111-116). IEEE.
16. Field, G., & Stepanenko, Y. (1996, April). Iterative dynamic programming: an approach to minimum energy trajectory planning for robotic manipulators. In Proceedings of IEEE international conference on robotics and automation (Vol. 3, pp. 2755-2760). IEEE.
17. Almasri, M. M., Alajlan, A. M., & Elleithy, K. M. (2016). Trajectory planning and collision avoidance algorithm for mobile robotics system. IEEE Sensors journal, 16(12), 5021-5028.
18. Gasparetto, A., Boscariol, P., Lanzutti, A., & Vidoni, R. (2015). Path planning and trajectory planning algorithms: A general overview. Motion and operation planning of robotic systems, 3-27.
19. Angeles, J., Rojas, A., & Lopez-Cajun, C. S. (1988). Trajectory planning in robotic continuous-path applications. IEEE Journal on Robotics and Automation, 4(4), 380-385.
20. McCarragher, B. J., & Asada, H. (1995). The discrete event modeling and trajectory planning of robotic assembly tasks.

7

Study of Different Features of Robots in Diverse Applications of Industrial Engineering

Arati Rath[1], Dilip Kumar Bagal[2], Ritesh Barkey[2], Sriparna Paul[3], Abhishek Barua[4], and Siddharth Jeet[4]

[1]School of Computer Engineering, Kalinga Institute of Industrial Technology, Bhubaneswar, Odisha India
[2]Department of Mechanical Engineering, College of Engineering and Technology, Bhubaneswar, Odisha, India
[3]Council of Scientific and Industrial Research–Advanced Materials and Process Research, Bhopal, Madhya Pradesh, India
[4]Department of Mechanical Engineering, Centre for Advanced Post Graduate Studies, BPUT, Rourkela, Odisha, India

7.1 Material Handling Using Robots

This kind may be further divided into two categories: (1) loading and unloading of the machine and (2) Transfer of materials. Transferring components from one machine to another is accomplished via the employment of material handling robots. Material handling is a mix of art and science that includes the following tasks: moving, storing, protecting, and controlling the material.

Figure 7.1 Material handling.

Material handling is defined as delivering the appropriate quantity of the appropriate material in the appropriate condition at the appropriate location in the appropriate position in the appropriate sequence in the appropriate time via the appropriate technique.

Types of Material Handling Robots

1. **Palletizing Robots:** The use of an industrial robot arm to automate the stacking of bags, cases, boxes, and other containers improves production while reducing the risk of harm to humans.
2. **Pick and Place Robots:** Manual pick and place is a time-consuming procedure that is similar to manual palletizing.
3. **Part Transfer Robots:** A portion of the material handling process is completed by means of these robots [1].

Goals of Material Handling

The primary goal of material handling is to reduce the unit costs of manufacturing as much as possible.

- Product quality should be maintained or improved, damages should be minimised, and materials should be protected.
- Improve the safety of employees and the working environment.
- Increase your productivity.
- Encourage more people to utilise the amenities.
- Inventory control is essential.

7.2 Machine Loading and Unloading

During the machine loading and unloading operation, a robot will be utilised to transfer the work pieces to and from the manufacturing machine, respectively. In the field of material handling operations, this application falls under the umbrella term "material handling operations." It is necessary to do the following three procedures in order to properly load and unload the machine:

- ✓ Loading the machine
- ✓ Unloading the machine
- ✓ Loading and unloading machines [2]

Figure 7.2 Machine loading and unloading.

Machine Loading

In this process, the robot puts raw work parts into the machine, and additional systems are utilised to unload the completed work parts from the machine once they have been finished by the robot. When working with sheet metal, an example would be loading the sheet metal into a press using a robot and then removing the completed pieces of work using gravity to take them from the press once they have been done. Machine unloading involves the removal of completed work components from the machine with the assistance of a robot, while the loading of raw materials is accomplished without the assistance of a robot. Plastic modelling and die casting are two examples [3].

Machine Loading & Unloading

Work components are loaded and unloaded from and into the machine using a robot, which accomplishes both tasks in this procedure. For instance, consider the machining process.

Benefits of Precision Machine Loading

It is possible to calculate the profitability of precision machine loading machines in the following ways:

- The accuracy and safety of the system have been improved.
- Speed.
- Reliability.
- Flexibility.
- It is possible to utilise it in a hazardous work environment.
- Machine loading robots do their tasks effectively, relentlessly, and with pinpoint accuracy.

7.3 Assembly

The use of assembly line robots to assemble components falls between the extremes of human labour and dedicated or "hard" automation in the manufacturing industry. A robot assembly worker can move more quickly and precisely than a human assembly worker.

Figure 7.3 Assembly line in automobile industry.

Assembly jobs are usually those that include the insertion of a peg into a hole in a piece of machinery.

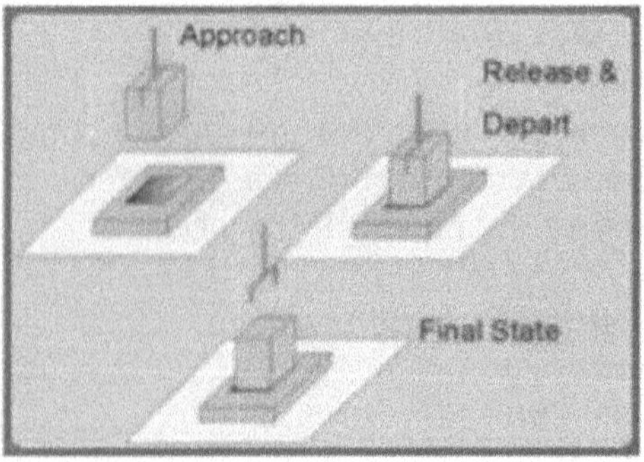

Figure 7.4 Assembly.

Application

Applications for robots may be found in assembly areas including the following tasks:

1. Screwing studs and screws into threaded holes.
2. The act of screwing and unscrewing nuts.
3. Shafts are inserted into holes in the frame.
4. Installation of electronic components in electric assemblies is the fourth step.
5. Assemblies of tiny electric motors, connectors, switches, and other similar components

7.4 Inspection

It is anticipated that certain sensors will be used in the robotic inspection process to determine the value of a produced component during the inspection process. It inspects the completed components using mechanical probe tests, which are performed on the parts. It is also possible to employ a robot arm equipped with a vision camera for non-destructive testing and 3D measurements. It is capable of objectively identifying and pinpointing faults or defective components prior to the packaging or shipping process.

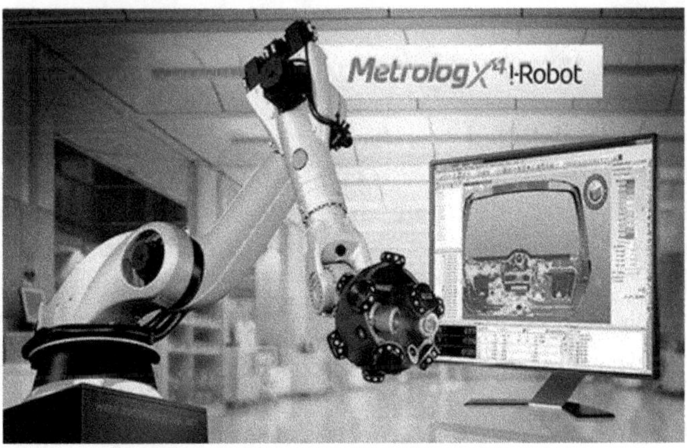

Figure 7.5 Inspection.

7.5 Welding

Robot welding is a technique that joins various types of materials together. Metals and their alloys constitute the vast majority of the materials that are welded, but welding is also used to connect other types of materials, such as thermoplastics, to form joints. Different metals or alloys are joined together via the use of a variety of welding techniques in which heat is provided either electrically or through the use of a gas torch.

Figure 7.6 Welding.

Why Robot Arc Welding?

An arc welder is a skilled worker who is helped by a fitter, who is a person who knows how to weld. The fitter's role is to arrange the job and secure the components of the welder in their proper positions. The working environment for a welder is usually uncomfortable and sometimes dangerous. The arc created by the welding process produces ultra-violet radiations that are harmful to human eyesight. As a consequence, welders must wear eye protection in the form of a welding helmet with a dark glass to shield their eyes from the heat. While the operation is in progress, sparks and smoke are produced, which may pose a danger to the operators. In light of the potential dangers for human employees involved in continuous arc welding, it makes sense to explore the employment of industrial robots for a variety of operations [1].

Benefits of Robot Arc Welding

Productivity is increased as a result of improved product quality, improved safety, and improved quality-of-work life.

Features of Arc Welding Robots

Precision of motion programming in a motion control system interface with other systems work volume and degrees of freedom work volume and degrees of freedom which gives precision of motion programming of arc welding process.

Arc Welding Robot

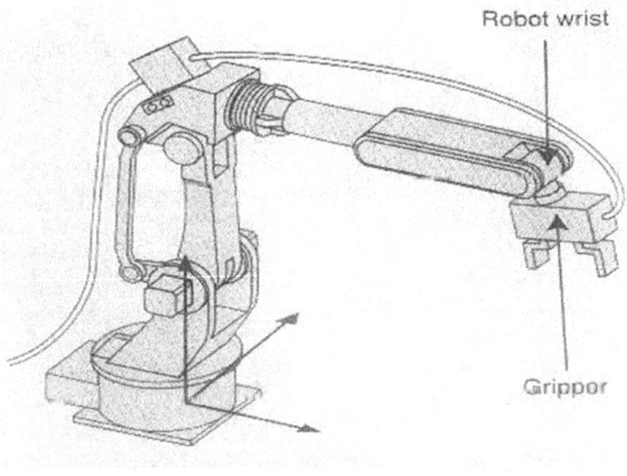

Figure 7.7 Arc welding robot.

Problems for Robots in Arc Welding

The fact that arc welding is often conducted in restricted spaces that are difficult to reach, such as the inside of tanks, pressure vessels, and ship hulls, creates a similar issue. Humans are more adept at positioning themselves in certain locations than robots. It is one of the most challenging technical issues to deal with while working on a batch manufacturing job because of the variance in the size of the components. Having dimensional variations of this nature means that the arc-welding path that needs to be followed will differ slightly from one part to another. Another technical challenge is the diversity in the edges and surfaces that must be welded together in a seamless manner. Rather from being straight and uniform, the edges are frequently wavy and uneven. This results in variances in the distance between the components, as well as other issues with the way the pieces join together prior to the welding process being performed.

Arc Welding Robots Performing in a Workshop

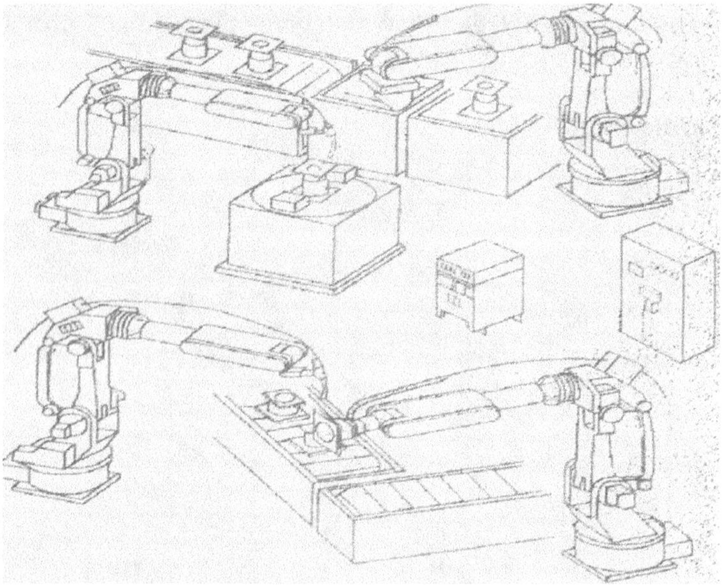

Figure 7.8 Arc welding robots performing in a workshop.

Why Robot Spot Welding?

When doing bigger spot-welding jobs, the welding guns with wires connected are very heavy, and may easily weigh more than 100 pounds in total. The equipment is hung from an overhead hoist system in order to aid the operator in moving the cannon. Even with this help, the spot-welding gun is a large and heavy object that is difficult to operate by a human worker at the high rates of production required on a vehicle body manufacturing line. As a result of this difficulties, there are often issues with the uniformity of the welded items produced on such a manual line. A consequence of these challenges, robots have been successfully used on this kind of manufacturing line to execute tasks with high levels of accuracy and efficiency [1, 4].

Benefits of Robot Spot Welding

The advantages of robot spot welding include higher product quality, increased operator safety, and more control over the manufacturing process.

Features of Spot-Welding Robots

Robots must be of a substantial size. It must have a large enough payload capacity to allow for easy manipulation of the welding gun for the intended application. The work volume must be sufficient in relation to the size of the finished product. The robot must be able to position and orient the welding gun in areas of the product that may be difficult to reach with a human operator present. It is possible that an increase in the number of freedoms may be required as a consequence of this. To complete the many positioning steps needed for the spot-welding cycle, the controller memory must have sufficient capacity.

A Typical Spot-Welding Robot

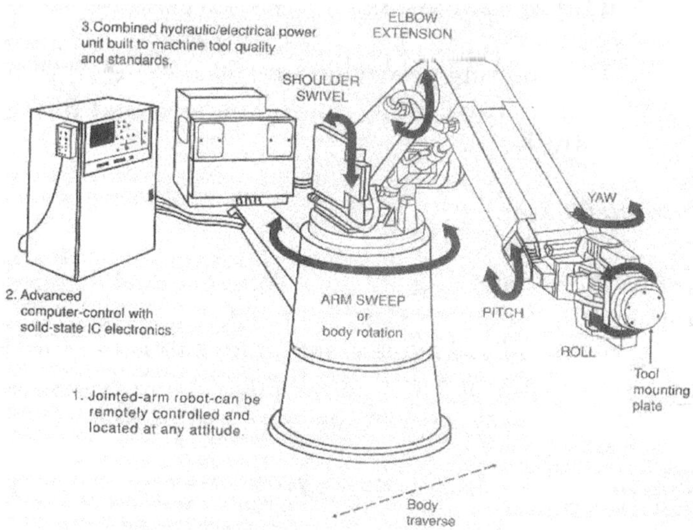

Figure 7.9 A typical spot-welding robot.

7.6 Spray Painting

In order to prevent metallic materials from corrosion, they will all be coated at the end of the manufacturing process. To attract consumers in the automotive business, the completed metals are painted with a variety of colours to make them stand out.

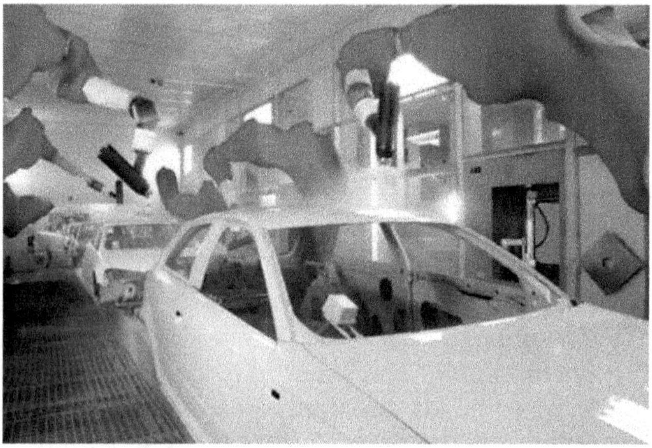

Figure 7.10 Spray painting.

Methods used for Spray Painting

Traditionally, this painting procedure was accomplished via the use of two techniques: spray coating and immersion and flow coating methods, respectively.

Spray Coating Method:

Spray coating is a technique of applying paint to metal that makes use of a spray gun to cover the metal with paint. All of the work is done manually by highly trained human labourers. This procedure may be carried out in three distinct methods, namely, via airless spray technique, electrostatic spray method, and atomization method.

Immersion and Flow Coating Methods:

Operationally, the immersion coating technique and the flow coating method are quite similar to one another. Both of these approaches are extremely straightforward techniques to use throughout the painting process. The immersion technique involves dropping a metal into a paint tank and then pulling it out. The paint that has been left over is returned to the tank. In the flow coating technique, the metals to be coated are put on top of the paint tank, which allows the paint to flow through them. The paint is created to flow on the canvas during this procedure.

Why Robot for Spray Painting

During the painting operation, the spray coating technique is the most hazardous procedure. It poses a number of risks to human workers, including the following:

- The production of fire when combustible paint and air are combined.
- Toxic gases and mist are emitted into the surrounding environment.
- Possibility of developing a cancerous condition.
- Hearing difficulties will be caused by the noise produced by the spray gun nozzle.

This robot connects a spray cannon to its wrist and uses it as an end effector, similar to a human. A person is responsible for programming the operation with the assistance of teach via techniques. In order to complete this procedure, a robot must be equipped with a number of features such as two or more programme storage, continuous route control, manual lead through programming technique, and hydraulic drive system, among others. This kind of robot is most often seen in the automotive industry, where it is employed to paint the outside and inside of a vehicle.

Advantages of Spray Painting Robot

The following are some of the benefits of using a spray painting robot: high consistency, increased production, increased worker safety, reduced power consumption, and reduced paint usage.

7.7 Mobile Robots

Robots that move from one location to another on legs, rails, or wheels are known as mobile robots. For example, in 1979, a nuclear accident in the United States produced a leak of radioactive material, which resulted in the development of a specific robot capable of handling hazardous materials.

Figure 7.11 Mobile robots.

Why Go from Place to Place?

- Transport products and supplies;
- Deliver communications;
- Get to your destination quicker;
- Complete a job while you're there;
- Gather information about what's there;
- Get away from something or someone.

Most Robots Get Around by Rolling

Walking is challenging because it requires the maintenance of balance. Swimming is only beneficial when done in a body of water. Flying requires tremendous amounts of speed and energy. The use of wheels and threads to move over uneven terrain makes moving more convenient and efficient. Asthey have a large number of points of contact with the ground, they are very stable [1].

How do Mobile Robots Work?

Robotic mobile devices consist of the following components: sensors, motors, wheels, and software for controlling them.

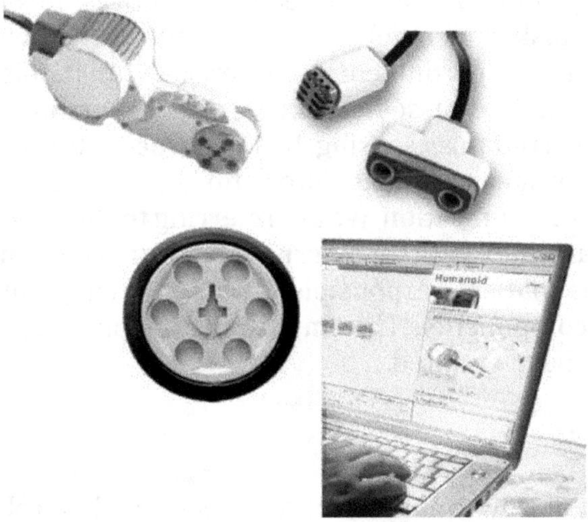

Figure 7.12 Components of a mobile robot (an example).

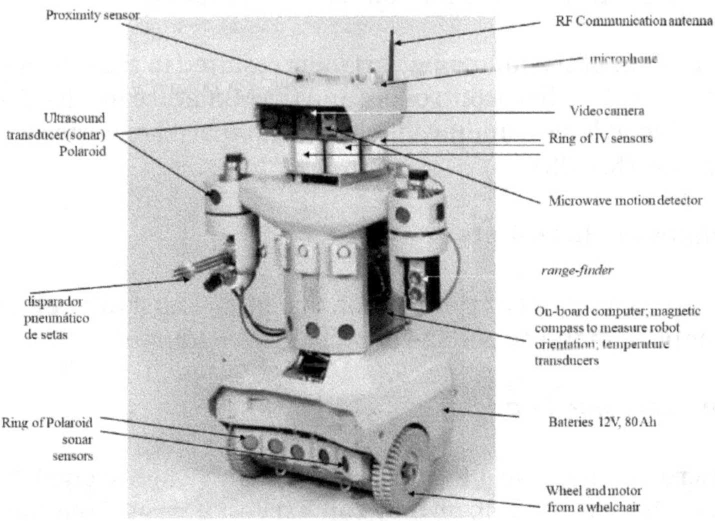

Figure 7.13 Components of a mobile robot.

Components of a Mobile Robot

In order to function, a mobile robot must be composed of a range of physical components (hardware) and computing components (software). If we think of a mobile robot in terms of hardware components, we may think of it as a collection of subsystems that execute the following tasks: Locomotion refers to the way in which a robot moves through its environment and interacts with it. It is the process through which a robot assesses its own and its surroundings' attributes that is referred to as sensing. It is the process through which the robot translates measurements into actions that is known as reasoning. In robot communication, we are referring to the way in which the robot communicates with an external operator. There are a number of subsystems that are responsible for the following duties in terms of software components. The importance of planning in all of its forms cannot be overstated.

7.8 Micro-bots

It is a branch of tiny robotics that focuses on mobile robots with dimensions less than one millimetre in size. Micro-bots (also known as micro-robots) with dimensions less than one millimetre in size. Robots that are capable of handling components as tiny as microns in size are also referred to as micron-handling robots.

A combination of the following variables resulted in the development of micro-bots: (a) Microcontrollers and (b) Miniature mechanical systems on silicon have a unique look when compared to larger mechanical systems (MEMS).

Challenges of Micro-Bots

In the development of a micro-robot, one of the most difficult tasks is to accomplish mobility with a very low power source.

How to Overcome Challenges

The micro-robots may be powered by a tiny, lightweight battery source such as a coin cell, or they can harvest energy from their surroundings in the form of vibration or light. A new generation of micro-robots is using biological motors as power sources, which extract chemical energy from the surrounding fluid in order to operate the robotic device.

Medical Micro-bots

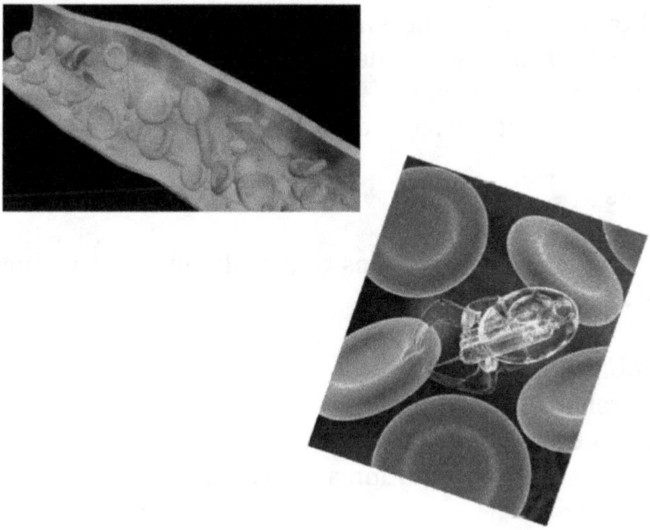

Figure 7.14 Medical micro-bots.

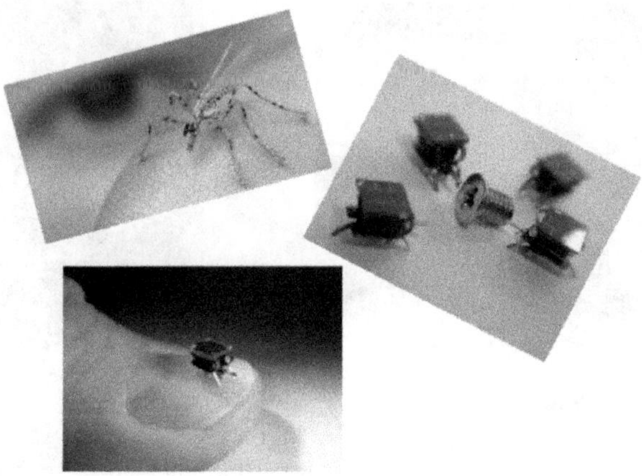

Figure 7.15 Medical micro-bots.

Tasks

A Micro-bot is also equipped with a mobile communication module, which allows it to accept instructions from control centres and respond to them in real time. Listed below are the most important functions of a micro-bot:

1. Obtaining and quantifying information via sensing.
2. Collecting samples is the second step.
3. Image recording and
4. Image transfer to a Central Control Station are the next steps.

Subsystems

There are four basic subsystems that make up a Micro-bot system, which are as follows:

- Mobility
- Power
- Communication
- Control and computation are two important concepts
- Sensors.

7.9 Robots in Space

Figure 7.16 Robots in space.

Robots in Hazardous Environments

TROV in Antarctica operating under water

HAZBOT operating in atmospheres containing combustible gases

Figure 7.17 Robots in hazardous environments.

7.10 Medical Robots

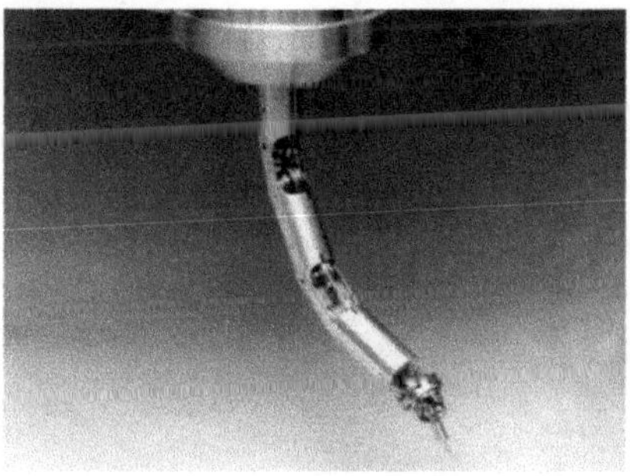

Robotic assistant for micro surgery

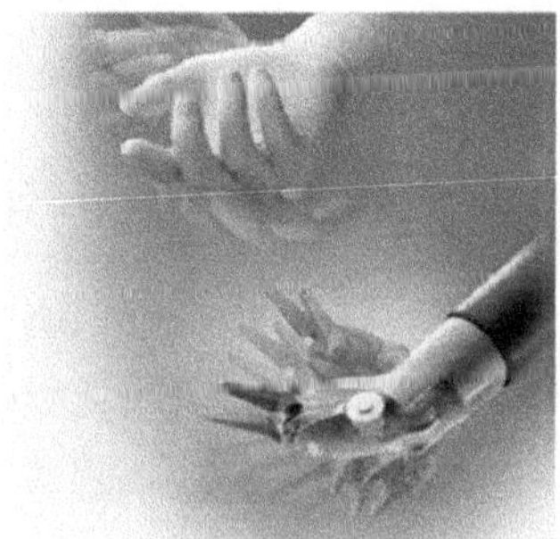

the operating system in robotic surgery

Figure 7.18 Medical robots.

7.11 Robots at Home

Sony SDR-3X Entertainment Robot

Sony Aido

Figure 7.19 Robots at home.

7.12 Future of Robots

Artificial Intelligence

Cog

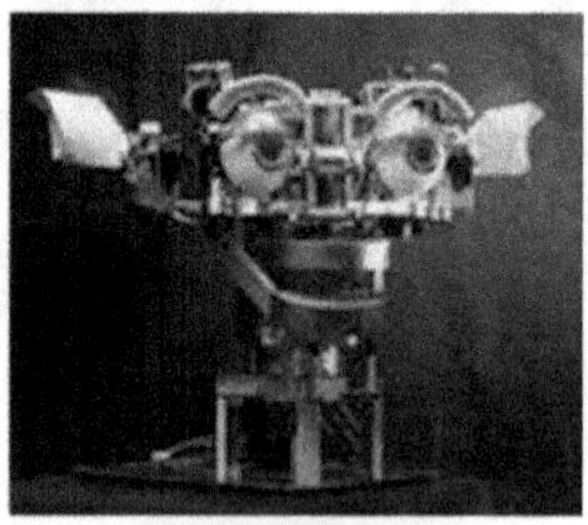

Kismet

Anatomy

Robot Work Crews

Garbage Collection Cart

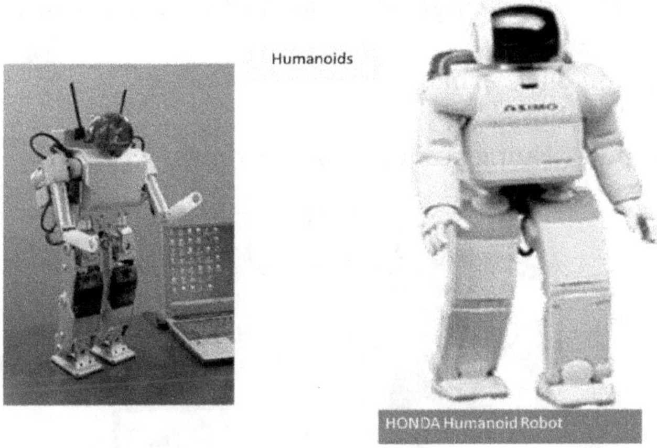

Figure 7.20 Future robots.

7.13 Ideal Tasks of a Robot

Tasks which are

- Dangerous
    - Space exploration
    - chemical spill clean-up
    - disarming bombs
    - disaster clean-up
- Boring and/or repetitive
    - Welding car frames
    - part pick and place
    - Manufacturing parts.
- High precision or high speed
    - Electronics testing
    - Surgery
    - Precision machining [2].

Automation – Machinery designed to carry out a specific task

- Bottling machine robots, they are always better than robots because they can be optimally designed for a particular task).
- Dishwasher
- Paint sprayer

Robots – machinery designed to carry out a variety of tasks

- Pick and place arms
- Mobile robots
- Computer Numerical Control machines

Figure 7.21 Automation – machinery.

7.14 Recent Developments in Robotics

Atlas, a new version of a humanoid catastrophe robot, can do half-turns in the air and even a back flip, according to the company. The astonishingly humanoid robot, named Atlas, stands 4.9 feet (1.5 metres) tall and weighs 165 pounds (75 kilogrammes). According to Boston Dynamics, which manufactures the robot, it navigates around its environment using Lidar and stereovision. Atlas is intended to be able to respond to emergency circumstances in which human life would usually be placed at danger, such as entering buildings that have collapsed after an earthquake or dealing with patients who have fatal, highly contagious illnesses, without putting human life at risk. This robot performs incredible acrobatic feats, such as backflips and half-turn leaps, on command. Besides having human-like skills, Atlas has other traits as well, such as a sense of balance, which allows it to avoid falling over when pushed and to get back up after a particularly hard shove.

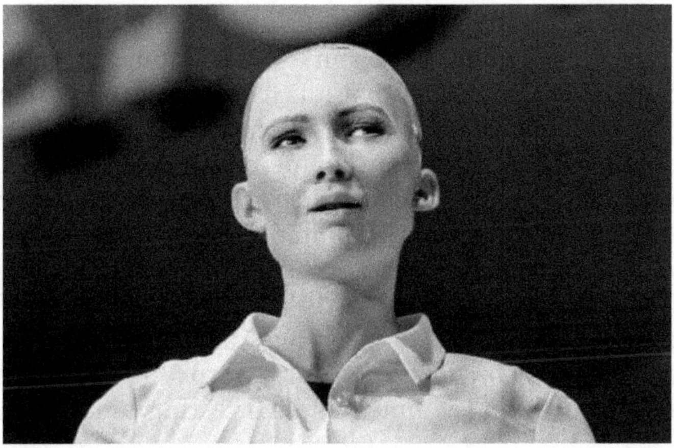

Figure 7.23 Hanson Robotics' "Sophia".

On July 12, 2017, Hanson Robotics' "Sophia" visits the RISE Conference at the Hong Kong Convention and Exhibition Centre, where she was developed by the company. A robot with an uncannily human-like look recently took another step toward human status when it was awarded citizenship by the Kingdom of Saudi Arabia during the Future Investment Initiative, a tech conference held in the country (FII). The robot, named "Sophia," was developed by Hanson Robotics (HR) and has a pale-skinned face with features that are capable of being extremely active and expressive, as well as exhibiting a wide variety of emotional responses. Following a statement on the company's HR website, the "latest and most advanced robot" addressed hundreds of people in Riyahd, Saudi Arabia, during the FII conference on Oct. 25. She also announced that she had been granted citizenship, the first time such a distinction has been bestowed upon a robotic being [1, 4].

Figure 7.24 Robots in agriculture.

It is operated by academics from Harper Adams University in the United Kingdom, who have set up an experimental farm called Hands Free Hectare. Despite the fact that it is harvest season in many areas of the globe, on one farm in the United Kingdom, robots are performing all of the hard lifting, rather than people. Hands Free Hectare, an experimental farm run by researchers from Harper Adams University in the village of Edgmond in the United Kingdom, has harvested approximately 5 tonnes (4.5 metric tonnes) of spring barley from the world's first robotically tended farm. The farm is the first of its kind in the world. According to the researchers, autonomous vehicles have completed every task on the farm from start to finish, including planting, fertilising, collecting samples, and harvesting, among others.

Figure 7.25 Exoskeletons.

Exoskeletons for robots are being developed in the same way that humans wear wet suits to work underwater or spacesuits to operate in space. This will allow the machines to wear a variety of costumes suited to various tasks, similar to how humans wear wet suits to work underwater. According to the latest research, self-folding, heat-activated origami suits developed for robots were shown to be effective in tests, allowing the machines to walk, roll, sail, and glide.

Figure 7.26 Robo-bee.

The Robo-bee is around the size of a centimetre in length. Its flapping wings let it to fly through the air at a speed 12 times faster than a paperclip, yet it is smaller and lighter than a quarter. That isn't even the most amazing aspect about it. The tiny bot is able to adhere to the bottom of any flat surface thanks to a technique using electrostatic charge. This includes anything from tree leaves to glass skylights to your plaster ceiling. This electric-powered perching is almost easy, using 1,000 times less energy than flying does to do the same task. The insectoid robot is known as Robo-Bee, and it was created by a team of roboticists headed by Moritz Graule, a mechanical engineer at the Massachusetts Institute of Technology. The wings of Robo-Bee beat nearly as fast as the wings of a real honeybee.

7.15 Robotics - Fields of Application

In the following disciplines, contemporary robots are making important contributions, and the following are some examples:

1. Agricultural production, automobile production, and construction production are the first three industries.
2. Having a good time.
3. Healthcare includes hospitalisation and patient-care services as well as surgery and research.
4. Law enforcement activities such as monitoring, patrolling, and so on.
5. Manufacturers and producers of goods and services
6. Defense activities such as demining, monitoring, and attacks are included in this category.
7. Mining, excavation, and exploration are among the ten transportation modes, which include air, ground, rail, space, and other modes.
8. Material handling and transport are covered in number eleven.
9. Spot welding and continuous arc welding are used in the steps of the machine loading and unloading procedure.
10. Painting using a spray gun
11. Assemblies
12. Inspection
13. Laboratories (science, engineering, and so forth.) [1-4].

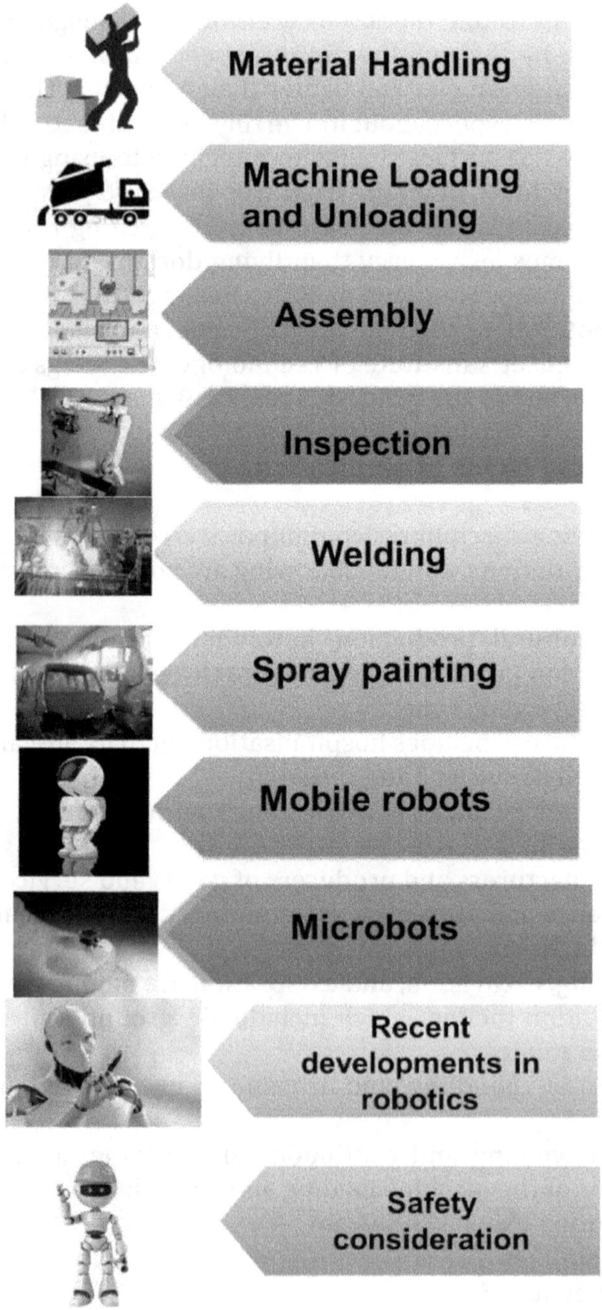

Figure 7.27 Applications of industrial robots.

7.16 The Purpose of Robots

Besides these duties, robots are also employed for the following [1-20]:

- Messy tasks.
- Tasks that must be repeated.
- Tasks that are potentially hazardous.
- Tasks that are impossible to do.
- Robots helping the physically challenged.
- Is capable of operating machinery with much more accuracy than humans.
- More cost-effective in the long run.

References

1. Irshat, K., Petr, R., & Irina, R. (2018, October). The selecting of artificial intelligence technology for control of mobile robots. In 2018 International Multi-Conference on Industrial Engineering and Modern Technologies (FarEastCon) (pp. 1-4). IEEE.
2. Wallén, J. (2008). The history of the industrial robot. Linköping University Electronic Press.
3. Nof, S. Y. (Ed.). (1999). Handbook of industrial robotics. John Wiley & Sons.
4. Athawale, V. M., Chatterjee, P., & Chakraborty, S. (2012). Selection of industrial robots using compromise ranking method. International Journal of Industrial and Systems Engineering 1, 11(1-2), 3-15.
5. Li, L., Fan, Y., Tse, M., & Lin, K. Y. (2020). A review of applications in federated learning. Computers & Industrial Engineering, 106854.
6. Chatterjee, P., Mondal, S., & Chakraborty, S. (2014). A comparative study of preference dominance-based approaches for selection of industrial robots. Advances in Production Engineering & Management, 9(1).
7. Kim, Y., Kim, S. H., & Kwak, Y. K. (2005). Dynamic analysis of a nonholonomic two-wheeled inverted pendulum robot. Journal of Intelligent and Robotic Systems, 44(1), 25-46.
8. Zhang, B., Xie, Y., Zhou, J., Wang, K., & Zhang, Z. (2020). State-of-the-art robotic grippers, grasping and control strategies, as well as their applications in agricultural robots: A review. Computers and Electronics in Agriculture, 177, 105694.
9. Liu, J., Zhang, X., & Hao, G. (2016). Survey on research and development of reconfigurable modular robots. Advances in Mechanical En-

gineering, 8(8), 1687814016659597.
10. Nayak, S., Kumar, N., & Choudhury, B. B. A Case-Based Approach for the Selection of Industrial Robots in Manufacturing Sectors.
11. Fu, Y., Li, M., Luo, H., & Huang, G. Q. (2019). Industrial robot selection using stochastic multicriteria acceptability analysis for group decision making. Robotics and Autonomous Systems, 122, 103304.
12. Whitney, D. E., Lozinski, C. A., & Rourke, J. M. (1986). Industrial robot forward calibration method and results.
13. Slamani, M., Nubiola, A., & Bonev, I. (2012). Assessment of the positioning performance of an industrial robot. Industrial Robot: An International Journal.
14. Quarta, D., Pogliani, M., Polino, M., Maggi, F., Zanchettin, A. M., & Zanero, S. (2017, May). An experimental security analysis of an industrial robot controller. In 2017 IEEE Symposium on Security and Privacy (SP) (pp. 268-286). IEEE.
15. Norrlof, M. (2002). An adaptive iterative learning control algorithm with experiments on an industrial robot. IEEE Transactions on robotics and automation, 18(2), 245-251.
16. Massa, D., Callegari, M., & Cristalli, C. (2015). Manual guidance for industrial robot programming. Industrial Robot: An International Journal.
17. Švaco, M., Šekoranja, B., Šuligoj, F., & Jerbić, B. (2014). Calibration of an industrial robot using a stereo vision system. Procedia Engineering, 69, 459-463.
18. Summers, M. (2005). Robot capability test and development of industrial robot positioning system for the aerospace industry. SAE transactions, 1108-1118.
19. Bingol, M. C., & Aydogmus, O. (2020). Performing predefined tasks using the human–robot interaction on speech recognition for an industrial robot. Engineering Applications of Artificial Intelligence, 95, 103903.
20. Bottin, M., Cocuzza, S., Comand, N., & Doria, A. (2020). Modeling and identification of an industrial robot with a selective modal approach. Applied Sciences, 10(13), 4619.

8

A Review of Identification Technologies

Ram Krishan

Department of Computer Science, Mata Sundri University Girls College, Mansa, India

8.1 Introduction

This chapter provides the details of RFID tags and Barcodes [1] technologies for digital information gathering. The bar code is a square or rectangular image which is a combination of series of parallel black lines and white spaces of varying widths that can be read by the barcode reader. Barcode uses a sequence of vertical lines and spaces to represent numbers and other symbols used to identification item number, location, price, quantity in stock, etc. [2]. A barcode symbol [3] contains five parts: a quiet zone, a start character, data characters, a stop character, and another quiet zone. Barcodes can be scanned by a barcode reader by using a laser or camera-based system that is sensitive to the reflections from the line and space thickness and variation. The reader translates the reflected light into readable data that is transferred to a computer that is used to identify the item detail with more accuracy than manual data.

8.2 Related Work

From the last decade, various utilizations of RFID have been explored and discussed. The RFID passive tags were explored in a way that they could predict the lifetime of the inventory control system and efficiently manage them [4] [5]. In a manufacturing process, the RFID framework increases production efficiency and improves the quality level. RFID distinguished the items during the delivery stages and update the information of data in the database. With this, the manager will be able to check and monitor the real-time data of the production process [6]. The use of the RFID framework for process control of assembling systems has been introduced [7]. The work presented in [8] implemented RFID system to automatically collect the workplace data which is used to identify any shortcomings of the running process. [3] presented a technical review of the design frameworks, used technologies and process of implementation to track and trace the products. The joining of RFID and smart system framework has been examined [9] for monitoring of systems. A few methodologies are utilizing standardized barcode tag in the industry presented in [10]. [11] presented an algorithm to monitor health by integrating the RFID using sensor-tags to recognize the communication between the human and system. The survey [2] uses a barcode identification scanner for the packing process to guarantee the accuracy of labels. Barcode readers detect the labels and if any incorrect label is found, they split them from the correct labels. The use of a barcode reader for identification in storage systems and recovery frameworks (SS/RF) is presented [7] to find various items. The range of RFID readers for coverage to detect the items is discussed [12] [13]. This chapter examines the fundamental comparison of RFID and barcode technology framework. Statistical data analysis is also presented to show the scanning time difference of both technologies.

This chapter is organized into six sections. The first section presents the introduction about RFID and Barcode. The second section focuses on the previous work and the current scenario of the identification technologies. The third section discusses the RFID and barcode technologies in detail. Section four highlights the comparative study of the identification technologies. Section five presents the scan time analysis of both RFID and barcode-based on the data collected from an inventory control system. Section six concludes the chapter.

8.3 Identification Technologies

Presently makers use two types of identification technologies for computerized information gathering known as RFID and Barcode systems[14][15]. A one-dimensional and two-dimensional barcodes are shown below in Figures 8.1 and 8.2. Barcodes are mainly used in retail stores, industries etc. for inventory management. In barcodes, the information about the product or item is encoded in horizontal as well as in vertical lines. To encode more information in a barcode, the size of the tag will expand [16]. 2D Barcodes [3] are of square or rectangle shapes having the number of dots on them. These dots are used to store the information about the product on which these barcodes are printed. There are not any standard barcodes, the different type of barcodes are used as per the requirements of the workplace [17].

Figure 8.1 One-dimensional product barcode.

Figure 8.2 Two-dimensional product barcode.

Different types of barcodes are known as:

a) Uniform Product Code (UPC): Mostly used in retail stores to check stock, price etc.
b) Code-39: Used for inventory, check shipment detail and industrial applications etc.
c) POSTNET: Use to encoding postal service mail assist zip codes in the US etc.

On the other hand, RFID is an acronym for "Radio-Frequency Identification" [7][15]. Detail of object is stored in the form of a tag RFID tag or smart labels and captured by a reader via radio waves [14][8]. RFID tags are working as Automatic Identification and Data Capture (AIDC) [18]. AIDC automatically capture the identification of the object and convert it into readable data and transfer it into the computer in lesser time and with more accuracy. RFID is available in different shapes and size. There are two types of tags: Passive and Active tags [19].

- Passive tags are widely used due to less expensive, small in size and a very important point is these tags must be powered up by the reader before data transmitting without any external battery [20].
- Active tags powered up by using a battery which is affecting transmitting data all time.

A basic RFID system is depicted in Figure 8.3 [21]. RFID is used in warehousing management to maintain stock, used in entering gate passes at restricted areas etc. RFID is consisting of three parts are: RFID Tag contains an integrated circuit in which data is stored. RFID Reader [5] used to convert data from tag or labels to readable data by converting radio waves into valuable data [8]. RFID antenna [6] [22] used to transmit data from the reader to the host computer system. This data can be stored in the database of the computer and analyzed at a later time.

The system of RFID works on the number of frequencies [23] like (LF- Low Frequency), (HF- High frequency, (UHF - Ultra High-frequency) and Microwave as presented in Table 8.1. Table 8.1 presents the range of various frequencies in hertz, read range in meters and different application of the RFID system [6]. All these frequencies apply to the different regions of the globe.

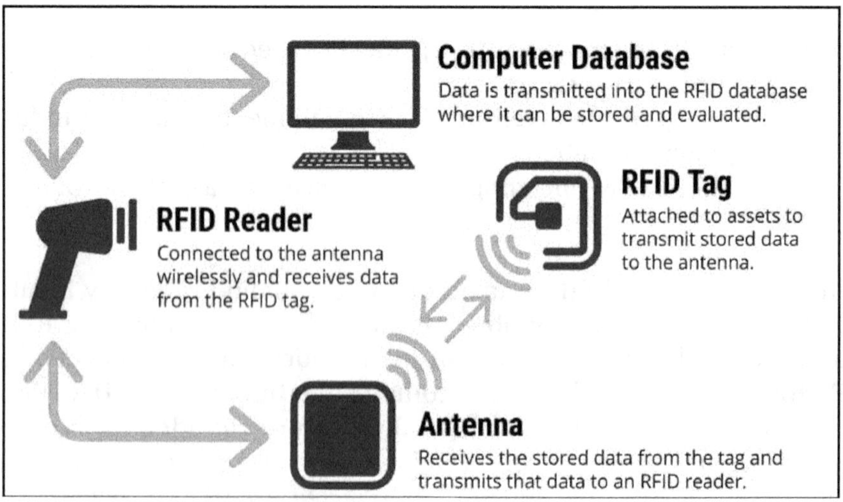

Figure 8.3 Basic RFID system.

Table 8.1 RFID characteristics

Frequency	Range	Read Range	Applications
LF	125 kHz 134 kHz	0.1 m	Smart cards, Animal tag, laundry, Ticket-ing.
HF	13.56 MHz	1 m	Item management of the small unit, Li-brary management, Supply chain, Theft prevention
UHF	900 MHz 902-928 MHz 868-871 MHz 950-956 MHz	2-20m	Vehicle Ids in trans-portation, Security, Item management on large scale, Supply chain.
Microwaves	2.4 GHz	10 m	Vehicle Ids in trans-portation, Toll tax collection, Security, Item management on large scale, Supply chain.

8.4 Comparison of Barcode and RFID Technology

Barcodes and RFID both are used to store data in different ways, in barcodes data is stored in the form of digital encodes and in RFID data is stored in the form of tag or labels. Both are used for tracking, inventory control management etc. The uses and implementation of both technologies are different as per Table 8.2.

Table 8.2 Comparison of barcode and RFID technology

Sr. No.	Barcode Technology	RFID Technology
1	Optical Scanner is used to read digital encodes.	Radio waves are used to read tags or labels.
2	Line of sight is required to read bar codes.	The line of the sign is not required to read the RFID tag. It is a primary advantage of RFID tags.
3	Barcodes easily scan from 0-10 inches distance.	RFID scan at distance from 0-50 feet with more accuracy than barcodes.
4	Scans one barcode at a time by scanning each item using individually positions.	RFID scan 40 to 100 tags in single scanning without using the line of sight.
5	Barcodes codes are commonly used on paper. Due to the use of paper barcodes are less durable.	RFID tags form tags or labels with an integrated circuit, microchip made from a durable material.
6	Due to the use of the paper cost of barcodes are less than RFID.	Due to integrated circuit and microchip, RFID tags are more expensive than barcodes.
7	Barcodes provide lesser accuracy than RFID	RFID tags give higher accuracy and speed compared to barcodes.

8.5 Scan Time Analysis of Barcode and RFID

To compare the scanning time for both RFID and barcode technologies the data of an inventory control system is analyzed. Table 8.3 presents the statistical data of items used in the invetory control system oraganized in 10 different groups. The data shows the number of items scanned using the barcode scanner and the same items scanned using RFID scanners. The data collected for 611 items and barcode and RFID scan time is computed in minutes as presented in Table 8.3.

Table 8.3 Statistical data of items scanned by barcode and RFID (*Source* – GS1 US [24])

Item Group	Number of Items	RFID Scan Time (in minutes)	Barcode Scan Time (in minutes)
1	62	1.4	3.64
2	48	1.24	3.10
3	54	1.3	3.44
4	71	1.52	4.0
5	50	1.25	3.28
6	66	1.47	3.81
7	38	1.10	2.95
8	81	1.78	4.15
9	69	1.49	3.90
10	72	1.54	4.01

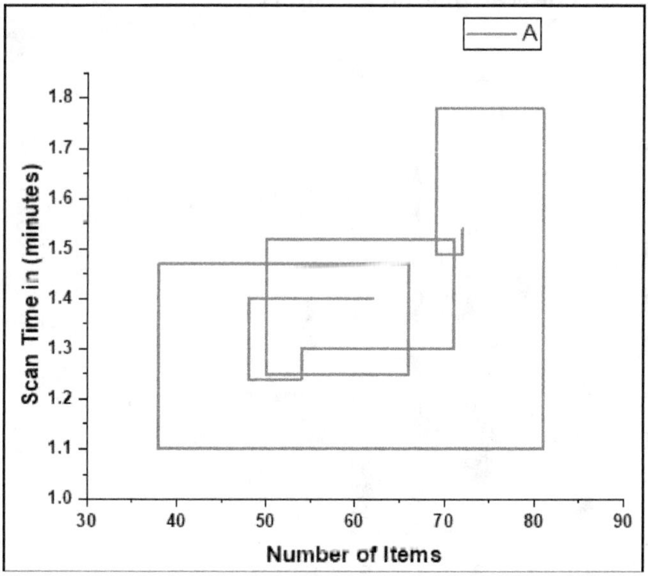

Figure 8.4 RFID scan time data analysis.

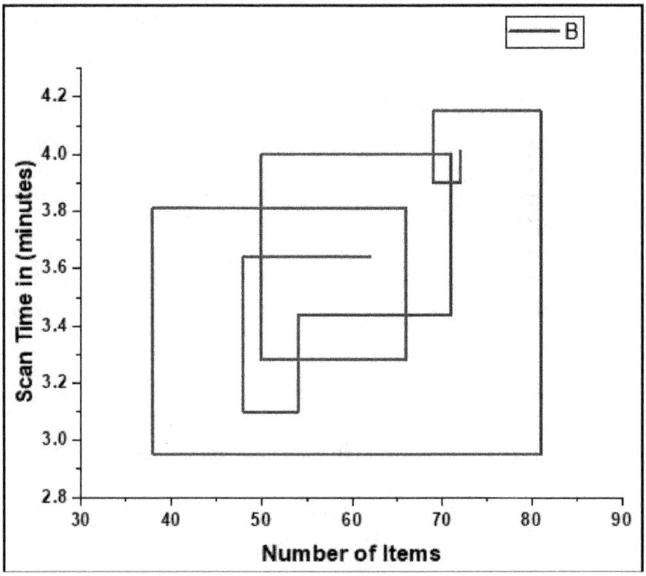

Figure 8.5 Barcode scan time data analysis.

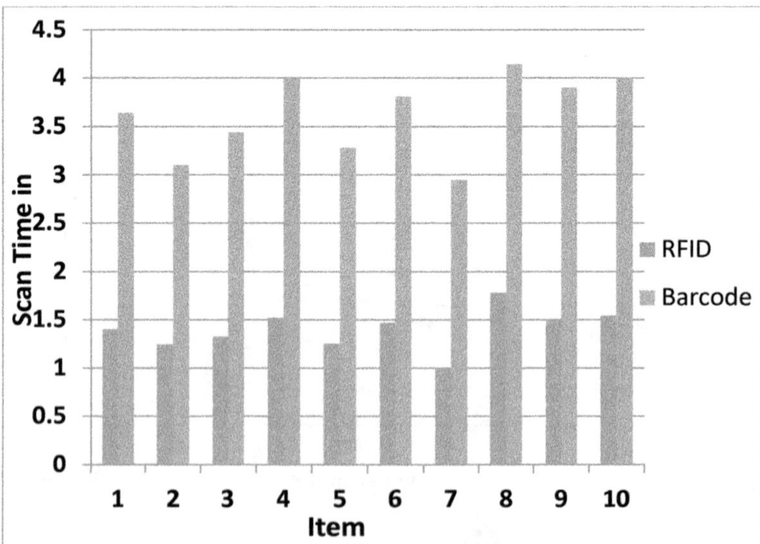

Figure 8.6 Graphical representation of comparative scan time analysis of barcode and RFID.

Figures 8.4 and 8.5 depict the data analysis of barcode and RFID scan time presented in Table 8.3. The X-axis of the figure presents the number of items scanned and Y-axis shows the scan time of items in minutes. The average time taken by barcode to scan the items is 3.628 but in case of RFID it reduces to 1.409. The RFID reduces the scan time 61.16% than the barcode thus increases the efficiency and productivity of the process.

Figure 8.6 shows a comparative data analysis of both Barcode and RFID graphically. It has been observed from the analysis that RFID scan time is very fast as compared to the barcode scanning of items. RFID is a much better option in terms of speed, accuracy and durability and the use of present and future technology.

8.6 Conclusion

This chapter reviews the identification technologies RFID, barcode and also presents the comparative analysis between them. From the results, it has been observed that RFID gives accurate and fast scanning time as compared to barcodes. The literature review also supports the benefit of RFID's scanning without the need for line of sight. This makes it promptly ideal over barcode technology in most operational conditions where quick throughput of an item is of vital significance. The various developments in two-dimensional barcode identifications and laser examining hardware could help to improve the time it takes for scanning barcodes. It is concluded that RFID is a simpler technology to use and it helps to improve the functioning of processes effectively.

References

1. Moselhi, O., Bardareh, H., & Zhu, Z. (2020). Automated data acquisition in construction with remote sensing technologies. Applied Sciences, 10(8), 2846.
2. Gratton, M. S., & Atkinson, D. S. (2013). U.S. Patent No. 8,534,540. Washington, DC: U.S. Patent and Trademark Office.
3. Schuitemaker, R., & Xu, X. (2020). Product traceability in manufacturing: A technical review. Procedia CIRP, 93, 700-705.
4. (OECD), "Radio-Frequency Identification (RFID): A Focus on Information Security and Privacy," *IDEAS Work. Pap. Ser. from RePEc*, no. 138, 2008, [Online]. Available:

http://search.proquest.com/docview/1698932001/.

5. Chanchaichujit, J., Balasubramanian, S., & Charmaine, N. S. M. (2020). A systematic literature review on the benefit-drivers of RFID implementation in supply chains and its impact on organizational competitive advantage. Cogent Business & Management, 7(1), 1818408.
6. Zamora González, G. (2014). Radio frequency identification (RFID) tags and reader antennas based on conjugate matching and metamaterial concepts. Universitat Autònoma de Barcelona.
7. Haddud, A., & Lee, H. (2013). Use of RFID technology in lean manufacturing: A survey on inventory management. International Journal of Management Theory and Practices, 14(1), 55-72.
8. Vagaš, M., Galajdová, A., Šimšík, D., & Onofrejová, D. (2019). Wireless data acquisition from automated workplaces based on RFID technology. IFAC-PapersOnLine, 52(27), 299-304.
9. Akbari, A., Mirshahi, S., & Hashemipour, M. (2015, May). Application of rfid system for the process control of distributed manufacturing system. In 2015 IEEE 28th Canadian Conference on Electrical and Computer Engineering (CCECE) (pp. 497-501). IEEE.
10. Golparvar-Fard, M., Bohn, J., Teizer, J., Savarese, S., & Peña-Mora, F. (2011). Evaluation of image-based modeling and laser scanning accuracy for emerging automated performance monitoring techniques. Automation in construction, 20(8), 1143-1155.
11. Mirshahi, S., Akbari, A., & Uysal, S. (2015, May). Implementation of structural health monitoring based on RFID and WSN. In 2015 IEEE 28th Canadian Conference on Electrical and Computer Engineering (CCECE) (pp. 1318-1323). IEEE.
12. Duroc, Y., & Tedjini, S. (2018). RFID: A key technology for Humanity. Comptes Rendus Physique, 19(1-2), 64-71.
13. Jituri, V. V. (2020). RAdio Frequency IdentificatioN (RFID) and ease of life. International Journal of Research-GRANTHAALAYAH, 8(8), 49-55.
14. Tucker, S., Darcy, P., & Stantic, B. (2014, January). A comparative study of RFID technology measuring efficiency and acceptance when capturing attendance. In Proceedings of the Thirty-Seventh Australasian Computer Science Conference-Volume 147 (pp. 75-79).
15. Wamba, S. F., Anand, A., & Carter, L. (2013). RFID applications, issues, methods and theory: A review of the AIS basket of TOP journals. Procedia Technology, 9, 421-430.
16. Wang, Y., Jin, M., Chen, G., Cui, X., Zhang, Y., Li, M., & Wang, J. (2019). Bio-barcode detection technology and its research applications: A review. Journal of advanced research, 20, 23-32.
17. Elaskari, S., Imran, M., Elaskri, A., & Almasoudi, A. (2021). Using Barcode to Track Student Attendance and Assets in Higher Educa-

tion Institutions. Procedia Computer Science, 184, 226-233.

18. Oluyisola, O. E., Strandhagen, J. W., & Buer, S. V. (2018). RFId technology in the manufacture of customized drainage and piping systems: A case study. IFAC-PapersOnLine, 51(11), 364-369.
19. Juels, A. (2006). RFID security and privacy: A research survey. IEEE journal on selected areas in communications, 24(2), 381-394.
20. Sardroud, J. M. (2012). Influence of RFID technology on automated management of construction materials and components. Scientia Iranica, 19(3), 381-392.
21. Sun, C. (2012). Application of RFID technology for logistics on internet of things. AASRI Procedia, 1, 106-111.
22. Ting, S. L., Tsang, A. H., & Tse, Y. K. (2013). A framework for the implementation of RFID systems. International journal of engineering business management, 5(Godište 2013), 5-9.
23. Louw, L., & Walker, M. (2018). Design and implementation of a low cost RFID track and trace system in a learning factory. Procedia Manufacturing, 23, 255-260.
24. GS, U. (2015). GS1 US survey shows manufacturers and retailers embrace RFID to enhance inventory visibility.

9

Challenges in Multi-Robot Systems: Location Tracking, Sensing Traffic and Dynamic Coalitions

Varun Yogi S[1], B S Parinitha[1], Ajay Sudhir Bale[1], Naveen Ghorpade[2], Baby Chithra R[1], and Kaushik M[3]

[1]*Department of ECE, School of Engineering and Technology, CMR University, Bengaluru, Karnataka, India*
[2]*Department of CSE, School of Engineering and Technology, CMR University, Bengaluru, Karnataka, India*
[3]*Department of ECE, Bannari Amman Institute of technology, Sathyamangalam, Tamil Nadu, India*

9.1 Introduction

Tremendous early research activities and implementations concentrated on the basis of single robot system or distributed problem-solving systems deprived of the mechanisms of robotics. A rapid evolution has been observed in the study of multiple robotics.

Multi Robot Task Allocation (MRTA) is a significant research area whereas the task can be discrete or continuous in terms of specificity, timescale and complexity [1]. Reconfigurable robots, communication, biological inspirations, architectures, object transport, manipulation, mapping and motion coordination are the major research areas with reference to multi robot system [2]. Robot Coordinate based approaches requires low computational power [3]. A single task is allocated to a single robot whereas other robots intricate their tasks by themselves. The swarm targets a single task of navigation flanked by two targets that exhibits a distinctive situation. Thus, the swarm demonstrates a dynamic structure of robust in nature which is also self-organized [4].

It's a complex virtue to make each robot to allocate the task by itself since the entire sphere of the problem is not viewed by all the robots. This gives a noisy and unfinished data to the robots, that leads to a situation in such a way that the robot is unaware of the periodical actions done by other robots at various occurrences in the past and future. Intentional coordination approach guides the robots to accomplish their individual roles done by thoughtful communications and consultations. The following special feature makes the intentional coordination as a suitable tactic: serene design and implementation, acquiescent to the official analysis [5-6].

The properties of interference affects the individual robot's performance. Simulations of sensor-based with respect to multi-robot system and theoretical approaches have a very good sync [7]. Now a days the Multi robot System (MRS) comprises of Unmanned Ground Vehicles (UGVs) and Unmanned Aerial Vehicles (UAVs) hold good adaptability and robustness. A group of Heterogeneous robots permits a huge assortment of responsibilities in same environment with diverse features. Thus, autonomous and benign navigation of dissimilar kinds of vehicles are possible with a proper trajectory planning [8]. Set based task priority inverse kinematic system holds good for tasks involved in physical interaction in the same and external environments [9-10].

9.2 Location Tracking

Location Tracking can be termed as a procedure in knowing an object's location in an environment [11, 12]. A localization technique is necessary to determine the location and direction of the mobile robot based on one of the localization algorithms and earlier knowledge represents by the environment map or known beacons. It is difficult to perform any task by any mobile robot without knowing its position in the environment. Some applications based on IR systems using passive Landmark for localization of indoor mobile robots was developed [12]. Multi robot systems can be utilized in numerous tasks like inspection, formation and police work. In alternative situations, robots want to navigate info or to boost their path designing [13-14].

9.2.1 Approaches in Location Tracking

In [13], an indoor robot algorithm is being proposed where the location of a robot is determined using minimum bounded circle algorithm. The problem of localization can be resolved by using LED arrays that are uniformly distributed across the environment. The location of the robot is determined by the LDR sensors that are equipped in the robot which sends data to the central controlling unit (CCU).

Binary search algorithm is used to control the lighting of the LED array. By applying minimum bounded circle algorithm to the data received is analysed by the system at the CCU and thus determining the robot's location. The indoor environment which is referred in this paper comprises of uniformly distributed 2D holes which are later sued to fit the LEDs in an array. These LEDs behave like transmitters and the LDR groups installed in the robot act as receivers in determining the robot location.

The procedure used in [14] is similar to the procedure which is used to determine robot location in [13], here the identification for communication and transmitting the data to the CCU for each robot is different. But here a problem arises when the CCU does not know the origin of the data received by the CCU in the indoor environment. K-Means Clustering Algorithm [15] is employed to isolate the acquired data into clusters resolving this issue. The K-Means clustering algorithm is a simple yet perfect to fragment large group of data. In this paper the group of data represents labelled LEDs which is represented by the LEDs sensed by the LDR group equipped in the robot. Here each subgroup of labelled LEDs belongs to one robot which helps the CCU in determining the origin of the data and intern helps in determining the location of the robot in the environment.

In [16], a Lidar based robot location tracking system is being put forward where a mapper system which could be a dynamic robot or a handheld camera which explores the target environment. Firstly, from the incoming image data, using simultaneous localization and mapping (SLAM) a 3D cloud map is generated and the 3D map is then converted to a 2D representation. After this process, a wheeled robot called the follower with Laser Range Finder (LRF) is used in the mapped environment to carryout localization process. Here the LRF records, encoder records and for the localization system the map are the inputs and the follower robot's pose and scale factor are the outputs. The scalar drift in monocular SLAM is a tedious process which is why the local deformation of the built map is always equal and insignificant in all directions. In a visual SLAM map, a set of points is maintained in order to represent the landmarks in real world. The absolute distance between the camera centre and the landmark is unobservable for a mapper agent. The follower robot is directed to a specific local map, where the same landmark is measured from various angles using the LRF. Figure 9.1 shows the proposed system in

[16] which consists of the mapper agent and the follower agent or follower robot.

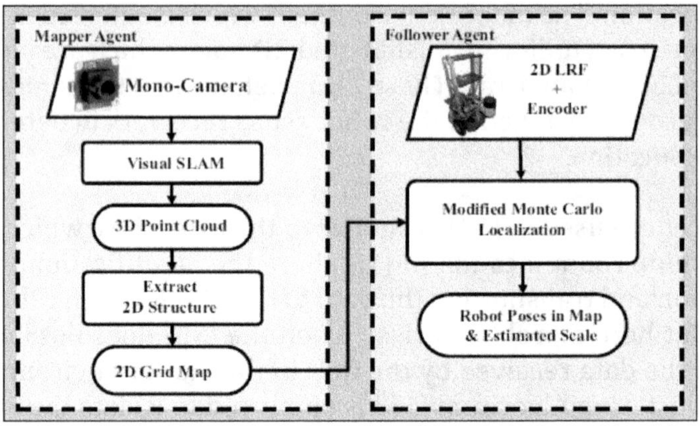

Figure 9.1 A system of mapper and follower robot [16]. Reproduced from an open-source article.

Based on possible roles, there are several navigational methods for multi-robot systems. Artificial forces are used to draw robots from a certain direction while repelling them from obstacles and other robots [17]. Rejection of a local minimum by reconstructing future roles by additional powers is one of the sensor oriented methods for mutual motion planning [18] and get away from local minima, thus disrupting the equilibrium of artificial powers [19]. Control granularity is an important feature of these interfaces [20-21] and it encompasses control that is robot-oriented, team-oriented, and environment-oriented. In robot-oriented control, the operator interfaces with a single robot, which may be statically or dynamically chosen [22-26]. An operator will delegate a goal to a group of robots as if the group were a single individual using team-oriented management [27-29]. Environment based regulation happens as the controllers alter the environment, usually through the use of virtual or mixed reality, to affect the actions of the robots or to signify targets [30].

9.2.2 Traffic Sensing

Smart urban spaces are on the rise nowadays, and the demand for sensing traffic in urban spaces is growing rapidly. IoT based sensor networks which collect traffic and other data such as accidents, weather information, pedestrians, pollution levels etc. and send these data to a cloud centre. These IoT sensor networks are a common sight in traffic sensing in the past years and have seen tremendous growth. Now its time to upgrade to a new traffic sensing method using multiple robots. Multi Robot Systems (MRS) based traffic sensing is getting popularized nowadays as a means of a more efficient way to manage traffic in urban scenarios. In this section we will see the different ways to apply MRS in traffic sensing. Here the main purpose of using a MRS based traffic sensing system is to allow the governing bodies or any citizen to have information about what's happening the city or to predict the future scenarios that will arise in the city. UAVs are commonly used as sensor-carrying airborne models in a range of uses, including traffic surveillance, urban management, mail delivery, and aerial cinematography. In order to collect high-level knowledge about the environment, machine vision algorithms are applied to aerial visual information. Figure 9.2 shows traffic sensing by an aerial robot which is sensing traffic in a junction.

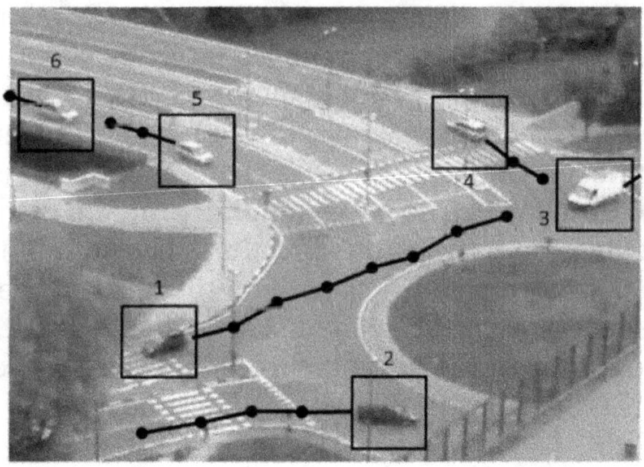

Figure 9.2 Traffic sensing by an aerial robot [31]. Reproduced from an open-source article.

Approaches in traffic sensing: In [32], a swarm city traffic sensing system is presented where a swarm city is developed using Unity game engine and using MATLAB a swarm intelligence systems is developed and using python some algorithms that performs data fusions are developed. The swarm city developed in this paper replicates a small European city. In addition to this, the simulator includes traffic, pollution, population models that replicates real life cities. The theoretical and empirical techniques are combined in this model and the density, speed and flow are usually managed by these models [33]. A swarm intelligence is developed where swarm drones are controlled by an algorithm which is used to monitor traffic in the swarm city. Since battery power on drones are constrained, there are charging hubs spread around the city so that the drones can land there charge up the batteries and get back to its respective task. Here a behaviour-based algorithm is implemented so that each drone share information with the fellow drones and decide the next cell to be visited.

In [34], a traffic management system for multiple Automatic Guided Vehicles (AGV) in the industrial transportation is presented. The focus is to get traffic management system that boosts the traffic in multi AGV systems in industries. To meet this requirement, a decentralized controlled algorithm is designed to make the vehicle fully autonomous and decide its own path to the destination and to synchronize its actions with other autonomous vehicles in the work environment to prevent collisions with other vehicles intern preventing traffic in the space making the work process smoother. Here both the models are two layered architecture where the topological primitives operate on the upmost layer while the underlying state lattice space operates on the bottom layer. The path planning problem is solved in two steps by utilising the path planner module where in initial stage of algorithm calculates the vehicle's path on the basis of the geography of the workspace and in the second step, the shortest feasible path along the geography of the work environment thus preventing collisions with other vehicles in the workspace is decided in the bottom layer algorithm.

In [35], Traditional traffic surveillance systems are used for traffic flow analysis and traffic monitoring, which is limited to the immediate vicinity of readily identifiable objects. Traditional traffic surveillance devices, on the other hand, are used for traffic movement analysis and tracking, which are restricted to the immediate vicinity of easily recognizable objects. At the same time, the robust tracking potential is much more discussed as a safety warning device under practical adverse circumstances, such as traditional occlusion and night-time backlights. In traffic surveillance devices, a number of deep-learning-based techniques have been used to close these holes. By merely retraining fantastical object detection algorithms, this data-driven approach produces reasonably decent performance. Large-scale structures gain more supervised knowledge in visual feature maps than small-scale vehicles. As a result, the gradient of small scales is quickly overwhelmed by large-scale gradients. As a result, the accuracy of small-scale vehicles is often inferior to that of large scale vehicles.

The existing autonomous car's crash avoidance system is proposed to be strengthened. The proposed solution combines the present autonomous car collision avoidance system with the DSRC system, allowing velocity and position data interchange between vehicles and enhancing the present collision avoidance system [36]. Collision avoidance algorithm creation for a bunch of mobile robots, as well as testing and validating the collision avoidance system algorithm for TurtleBot in Gazebo software simulations and practical experiments. To increase the precision of the collision avoidance device, a system that can determine the exact position of TurtleBot in a pre-stored chart, such as the beacon system, may be preferable to using an approximate position via AMCL. Adding a Wi-Fi modem to the TurtleBot-2 in order to derive a smoother communication between the TurtleBot and the workspace, as well as upgrading the new navigation stack algorithm enhancing the stability, versatility, and the sensitivity of the TurtleBot movement.

The AU-AIR dataset, a multi-modal UAV dataset compiled in a natural setting. With a various set of documented data types for UAVs, we intend to close the difference between computer vision and robotics. It can be implemented for various data fusion-related research areas, including object annotations, visual data and flight data. In [37] they use the AU-AIR dataset to provide a foundation for training and analyzing mobile object detectors, taking into account their real-time applicability. The distinctions in aerial and natural picture object identification are highlighted. Natural photographs of a single or multiple object in their natural surroundings are used in datasets. The vast majority of the photographs in the datasets were captured by humans using a point and shoot camera, but they are all side view.

9.2.3 Dynamic Coalitions

In today's world, robots are advanced to a point where they can work together to accomplish complicated tasks. Beginning with the most basic forms of cooperation motivated by nature, such as swarms or flocks, recent evolution has progressed to a more dynamic socio-inspired model of cooperation: alliance problem solving. Performing complex tasks like search and rescue, data collection is difficult due to the inherent complexity and the limited abilities of the available robot to perform these tasks and also performing all these tasks is impossible for one robot. Furthermore, each robot's control algorithms are unique, allowing it to make decisions depending on the current situation. When working together on a mission, robots must work together to obtain access to limited resources and optimize their gain. While carrying out such complex tasks, robot cooperation is fundamental. Addressing this issue, robot coalitions are formed where team formation of the existing robots takes place that is deployed for a specific task ensuring the task is completed on time and efficiently. Coalition formation by robots is a computationally hard task. Effective coalition formation is being studied from many years and an optimal and a near optimal coalition formation have been proposed in [38-40] respectively. In spite of developing many coalition algorithms, very few are targeted to be deployed in multi robot systems [41], despite the fact that economists, game theorists, and more recently computer scientists have researched alliance forming thoroughly, there are still a number of obstacles to overcome [42]. Figure 9.3 shows the block diagram of a robot coalition in a dynamic environment where the tasks are distributed among the robot team which makes the execution of the task more efficient.

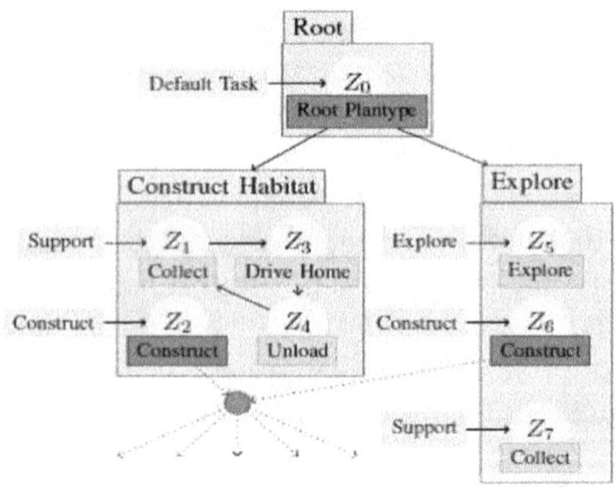

Figure 9.3 Robot coalition in dynamic environment [43]. Reproduced from an open-source article.

Approaches in dynamic coalitions: In [42], a couple of matching edges that do not share a vertex in classical bipartite matching equates to a task being assigned to only one robot and that robot being assigned to only one task. This problem of single robot, single job allocation can be reduced to a bipartite matching problem, according to the literature [44]. This aligns with the fact that traditional bipartite matching algorithms do not allow several edges incident on the same graph vertex in the final match. This allows several robots to be members of a single coalition assigned to a specific mission, but each robot may only be a member of one coalition. Classic bipartite matching techniques' polynomial worst-case time complexity aided us in developing a heuristic that seeks a successful solution with minimal computational effort.

In [44], a coalition algorithm is proposed for a group of mobile robots for task allocation where the algorithm is stored and runs in the robots on board computer. There is a novel value function which is used to assign the required number of robots for that task and also a distance-based cost function is presented to minimize the distance travelled by the robot. A clustering-based coalition method is implemented to perform coalitions of robot teams. In this coalition system, initially the robots are abruptly positioned in an environment and when a robot is given a job to do, it should travel there and execute the work. It is made sure that no two similar tasks are of same coalitions. The structure should be possessing maximum value.

In [45], a robot coalition interaction between each robots is presented for an agricultural scenario where robot coalitions are used to solve problems like selection of crops, field exploration, watering and harvesting of the crops. In order to perform the assigned task, the coalition of the robots should be in such a way that there is maximum efficiency. Efficiency can be computed for various subject areas depending on the gain that we receive by solving that issues such as the number of people retrieved for disaster medicine, maximum harvest for precise agriculture.

In [46] the robots are capable of multi- tasking. At the time of sowing and planting, they can even perform the tasks such as fertilizing and evaluating the soil structure in precision farming, and they can all work together to get the best results. Precision cultivation has the advantage of yielding the highest possible yield by using fewest capital possible. This requires determining the optimal growing conditions for each crop and planning the treatment and harvesting series. One of the coalition's defining characteristics is that all of its members have access to full details. They have access to information about the coalition's present situation, problem-solving progress, accessible and used tools, and incentive allocation. For coalition partners to communicate, operations related to mission output, supply, reception, and processing must be clear.

9.3 Conclusion

Some of the challenges and related solutions in relation with multi robot system are reviewed in this paper. Location tracking is one of the major challenges in multi robot which can be resolved by a localization technique accomplished with the help of IR systems by passive landmark and LDR integrated with K-means clustering algorithm. Lidar based robot system, LRF and SLAM also can exhibit location tracking of robots in both internal and external environment. SWARM intelligence system is helpful in the determination of position and direction of robots and helpful in successful implementation of smart city. Car collision avoidance system algorithm with Dedicated Short-Range Communication (DSRC) system and Adaptive Monte Carlo Localization (AMCL) can find out the velocity and position data interchange among the robots. The position, direction and velocity determination can enhance the traffic sensing system. Coalition in dynamic environment can be more efficient if the interaction between robots is through DLT, smart contracts by hyper ledger fabric program and visualized through Gazebo software. Thus, several works carried out to address the challenges like location tracking, traffic sensing and dynamic coalitions in multi robot systems are reviewed.

Table 9.1 Summary of the entire survey

Sl. No	Topics Presented	Challenges Addressed	Ref. No
1	An indoor robot algorithm is being proposed where the location of a robot is determined using LED arrays and minimum bounded circle algorithm.	Location Tracking	**13**
2	When the CCU does not know the origin of the data in the indoor environment, so they have a proposed a K-Means clustering algorithm based system where the received data is isolated and clustered assisting the CCU in knowing where the data is coming from and determining the location robot.	Location Tracking	**14**
3	A Lidar based robot location tracking system is being put forward where a mapper system which could be a dynamic robot or a handheld camera which explores the target environment.	Location Tracking	**16**
4	A swarm city traffic sensing system is presented where a swarm city is developed using Unity game engine and using MATLAB a swarm intelligence systems is developed and using python some algorithms that performs data fusions are developed.	Traffic Sensing	**32**
5	The focus is to get traffic management system that boosts the traffic in multi AGV systems in industries. To meet this requirement, a decentralized controlled algorithm is designed to make the vehicle fully autonomous and decide its own path to the destination and to synchronize its actions with other autonomous vehicles in the work environment	Traffic Sensing	**34**

	to prevent collisions with other vehicles intern preventing traffic in the space making the work process smoother.		
6	Traditional traffic surveillance systems are used for traffic flow analysis and traffic monitoring, which is limited to the immediate vicinity of readily identifiable objects. . In traffic surveillance devices, a number of deep-learning-based techniques have been used to close these holes	Traffic Sensing	**35**
7	A coalition algorithm is proposed for a group of mobile robots for task allocation where the algorithm is stored and runs in the robot's on board computer. A clustering based coalition method is implemented to perform coalitions of robot teams.	Dynamic Coalitions	**44**
8	A robot coalition interaction between each robots is presented for an agricultural scenario where robot coalitions are used to solve problems like field exploration, selection of crops, watering and harvesting of the crops. This paper speaks about secured interaction between coalition members through Distributed Ledger Technology (DLT) and smart contracts using hyper ledger fabric platform.	Dynamic Coalitions	**45**

References

1. Gerkey, B. P., & Matarić, M. J. (2004). A formal analysis and taxonomy of task allocation in multi-robot systems. *The International journal of robotics research, 23*(9), 939-954.
2. Arai, T., Pagello, E., & Parker, L. E. (2002). Advances in multi-robot systems. *IEEE Transactions on robotics and automation, 18*(5), 655-661.

3. Baglietto, M., Cannata, G., Capezio, F., Grosso, A., & Sgorbissa, A. (2009, June). A multi-robot coordination system based on RFID technology. In *2009 international conference on advanced robotics* (pp. 1-6). IEEE.
4. Ducatelle, F., Di Caro, G. A., Förster, A., Bonani, M., Dorigo, M., Magnenat, S., ... & Gambardella, L. M. (2014). Cooperative navigation in robotic swarms. *Swarm Intelligence, 8*(1), 1-33.
5. Lerman, K., Jones, C., Galstyan, A., & Matarić, M. J. (2006). Analysis of dynamic task allocation in multi-robot systems. *The International Journal of Robotics Research, 25*(3), 225-241.
6. Gerkey, B. P., & Matarić, M. J. (2004). A formal analysis and taxonomy of task allocation in multi-robot systems. *The International journal of robotics research, 23*(9), 939-954.
7. Lerman, K., & Galstyan, A. (2002). Mathematical model of foraging in a group of robots: Effect of interference. *Autonomous robots, 13*(2), 127-141.
8. Madridano, Á., Al-Kaff, A., & Martín, D. (2021). Trajectory planning for multi-robot systems: Methods and applications. *Expert Systems with Applications*, 114660.
9. Di Lillo, P., Pierri, F., Antonelli, G., Caccavale, F., & Ollero, A. (2021). A framework for set-based kinematic control of multi-robot systems. *Control Engineering Practice, 106*, 104669.
10. Di Lillo, P., Arrichiello, F., Antonelli, G., & Chiaverini, S. (2018, October). Safety-related tasks within the set-based task-priority inverse kinematics framework. In *2018 IEEE/RSJ International Conference on Intelligent Robots and Systems (IROS)* (pp. 6130-6135). IEEE.
11. Alfurati, I. S., & Rashid, A. T. (2020). Multi-Robot localization system using an array of LEDs and LDR sensors. *International Journal of Computer Applications, 975*, 8887.
12. Al-Furati, I. S., & Rashid, A. T. (2020, June). Shortest distance orientation algorithm for robot path planning using low-cost IR sensor system. In *2020 International Conference on Electrical, Communication, and Computer Engineering (ICECCE)* (pp. 1-6). IEEE.
13. Al-Forati, I. S. A., & Rashid, A. (2019, April). Design and implementation an indoor robot localization system using minimum bounded circle algorithm. In *2019 8th International Conference on Modeling Simulation and Applied Optimization (ICMSAO)* (pp. 1-6). IEEE.
14. Alfurati, I. S., & Rashid, A. T. (2020). Multi-Robot localization system using an array of LEDs and LDR sensors. *International Journal of Computer Applications, 975*, 8887.
15. Al-Furati, I., Rashid, A. T., & Al-Ibadi, A. (2019). IR sensors array for robots localization using K means clustering algorithm. *In-*

ternational Journal of Simulation Systems, Science & Technology, 12-1.

16. Wang, S., Kobayashi, Y., Ravankar, A. A., Ravankar, A., & Emaru, T. (2019). A novel approach for lidar-based robot localization in a scale-drifted map constructed using monocular slam. *Sensors, 19*(10), 2230.
17. Olcay, E., Schuhmann, F., & Lohmann, B. (2020). Collective navigation of a multi-robot system in an unknown environment. *Robotics and Autonomous Systems, 132*, 103604.
18. Dang, A. D., La, H. M., Nguyen, T., & Horn, J. (2017). Distributed formation control for autonomous robots in dynamic environments. *arXiv preprint arXiv:1705.02017.*
19. Olcay, E., Lohmann, B., & Akella, M. R. (2019, October). An information-driven algorithm in flocking systems for an improved obstacle avoidance. In *IECON 2019-45th Annual Conference of the IEEE Industrial Electronics Society* (Vol. 1, pp. 298-304). IEEE.
20. Kolling, A., Walker, P., Chakraborty, N., Sycara, K., & Lewis, M. (2015). Human interaction with robot swarms: A survey. *IEEE Transactions on Human-Machine Systems, 46*(1), 9-26.
21. Endsley, M. R. (2017). From here to autonomy: lessons learned from human–automation research. *Human factors, 59*(1), 5-27.
22. Setter, T., Fouraker, A., Egerstedt, M., & Kawashima, H. (2015). Haptic interactions with multi-robot swarms using manipulability. *Journal of Human-Robot Interaction, 4*(1), 60-74.
23. Kapellmann-Zafra, G., Salomons, N., Kolling, A., & Groß, R. (2016, September). Human-robot swarm interaction with limited situational awareness. In *International Conference on Swarm Intelligence* (pp. 125-136). Springer, Cham.
24. Nagi, J., Giusti, A., Gambardella, L. M., & Di Caro, G. A. (2014, September). Human-swarm interaction using spatial gestures. In *2014 IEEE/RSJ International Conference on Intelligent Robots and Systems* (pp. 3834-3841). IEEE.
25. Alonso-Mora, J., Lohaus, S. H., Leemann, P., Siegwart, R., & Beardsley, P. (2015, May). Gesture based human-multi-robot swarm interaction and its application to an interactive display. In *2015 IEEE International Conference on Robotics and Automation (ICRA)* (pp. 5948-5953). IEEE.
26. Gromov, B., Gambardella, L. M., & Di Caro, G. A. (2016, October). Wearable multi-modal interface for human multi-robot interaction. In *2016 IEEE International Symposium on Safety, Security, and Rescue Robotics (SSRR)* (pp. 240-245). IEEE.
27. Podevijn, G., O'Grady, R., Nashed, Y. S., & Dorigo, M. (2013, August). Gesturing at subswarms: Towards direct human control of robot swarms. In *Conference Towards Autonomous Robotic*

Systems (pp. 390-403). Springer, Berlin, Heidelberg.

28. Ayanian, N., Spielberg, A., Arbesfeld, M., Strauss, J., & Rus, D. (2014, May). Controlling a team of robots with a single input. In *2014 IEEE International Conference on Robotics and Automation (ICRA)* (pp. 1755-1762). IEEE.
29. Diaz-Mercado, Y., Lee, S. G., & Egerstedt, M. (2015, July). Distributed dynamic density coverage for human-swarm interactions. In *2015 American control conference (ACC)* (pp. 353-358). IEEE.
30. Bashyal, S., & Venayagamoorthy, G. K. (2008, September). Human swarm interaction for radiation source search and localization. In *2008 IEEE Swarm Intelligence Symposium* (pp. 1-8). IEEE.
31. Bottino, A., Garbo, A., Loiacono, C., & Quer, S. (2016). Street viewer: An autonomous vision based traffic tracking system. *Sensors, 16*(6), 813.
32. Roldán-Gómez, J. J., Garcia-Aunon, P., Mazariegos, P., & Barrientos, A. (2020). SwarmCity project: monitoring traffic, pedestrians, climate, and pollution with an aerial robotic swarm: Data collection and fusion in a smart city, and its representation using virtual reality.
33. Aparicio, F., Arenas, B., Gómez, A., Jiménez, F., López, J. M., Martínez, L., & Páez, F. J. (2008). Ingeniería del transporte. *CIE Dossat: Madrid, Spain*.
34. Draganjac, I., Petrović, T., Miklić, D., Kovačić, Z., & Oršulić, J. (2020). Highly-scalable traffic management of autonomous industrial transportation systems. *Robotics and Computer-Integrated Manufacturing, 63,* 101915.
35. Zhu, J., Li, X., Jin, P., Xu, Q., Sun, Z., & Song, X. (2021). Mme-yolo: Multi-sensor multi-level enhanced yolo for robust vehicle detection in traffic surveillance. *Sensors, 21*(1), 27.
36. Hou, Y. C., Mohamed Sahari, K. S., Weng, L. Y., Foo, H. K., Abd Rahman, N. A., Atikah, N. A., & Homod, R. Z. (2020). Development of collision avoidance system for multiple autonomous mobile robots. *International Journal of Advanced Robotic Systems, 17*(4), 1729881420923967.
37. Bozcan, I., & Kayacan, E. (2020, May). Au-air: A multi-modal unmanned aerial vehicle dataset for low altitude traffic surveillance. In *2020 IEEE International Conference on Robotics and Automation (ICRA)* (pp. 8504-8510). IEEE.
38. Rahwan, T., Ramchurn, S. D., Jennings, N. R., & Giovannucci, A. (2009). An anytime algorithm for optimal coalition structure generation. *Journal of artificial intelligence research, 34,* 521-567.
39. Rahwan, T., Ramchurn, S. D., Dang, V. D., Giovannucci, A., & Jen-

nings, N. R. (2007, February). Anytime optimal coalition structure generation. In *AAAI* (Vol. 7, pp. 1184-1190).

40. Rahwan, T., Ramchurn, S. D., Dang, V. D., & Jennings, N. R. (2007, January). Near-Optimal Anytime Coalition Structure Generation. In *IJCAI* (Vol. 7, pp. 2365-2371).
41. Adams, J. A. (2011). Coalition formation for task allocation: Theory and algorithms. *Autonomous Agents and Multi-Agent Systems, 22*(2), 225-248.
42. Dutta, A., & Asaithambi, A. (2019, May). One-to-many bipartite matching based coalition formation for multi-robot task allocation. In *2019 International Conference on Robotics and Automation (ICRA)* (pp. 2181-2187). IEEE.
43. Geihs, K. (2020). Engineering challenges ahead for robot teamwork in dynamic environments. *Applied Sciences, 10*(4), 1368.
44. Dutta, A., Ufimtsev, V., & Asaithambi, A. (2019, April). Correlation clustering based coalition formation for multi-robot task allocation. In *Proceedings of the 34th ACM/SIGAPP Symposium on Applied Computing* (pp. 906-913).
45. Smirnov, A. V., & Teslya, N. (2020). Modelling and Visualization of Robot Coalition Interaction through Smart Space and Blockchain. In *ICEIS (2)* (pp. 278-286).
46. Roldán, J. J., del Cerro, J., Garzón-Ramos, D., Garcia-Aunon, P., Garzón, M., de León, J., & Barrientos, A. (2018). Robots in agriculture: State of art and practical experiences. *Service robots,* 67-90.

10

Robotic Garbage Collection System for Smart City

Ajay Sudhir Bale[1], Naveen Ghorpade[1], Pravesh P[1], Kishore P[1], Vinayak N[1], Rajesh Kumar N[1], and Manjunath V[1]

[1]*School of Engineering and Technology, CMR University, Bengaluru, Karnataka, India*

10.1 Introduction

Infrastructural developments, job opportunities, improved facilities, comfortable lifestyle, education, business developments are the reason for migrated population and accumulation of population in developing and developed cities. This leads to proliferation of garbage. Various types of waste materials which need to be segregated and disposed in a proper way so as to maintain tidiness as well as to avoid the formation of microorganisms and viral which lead to infections in human and other living organisms [1]. If it can be segregated in a proper way then further recycling, disposal or else preparation of composed fertilizers are possible. If the disposal method is not proper, then recycling is not possible. If waste management is done in an effective way, then it will result in the production of electricity. The conveyance and assortment of waste can be reduced if the domestic waste is segregated properly by the Municipality. The amount spent on the collection of waste material itself covers 40 to 60% of the total cost spent on Solid Waste Management (SWM) [2] [3] [4]. Weekly two times or three times the garbage bins are cleared.

As per an investigation mentioned in [5], 10,000 tons are being collected by the Municipality in Tapebasi as an average of 800 tons of packed material. Sustainability, Cost management, innovativeness, Aesthetics are the key goals in most of the research works. When the bin is filled with the maximum unwanted materials, the information is generated and automatically induces the server through a 2nd stage server. Then a route optimization task takes place in the server in order to check its availability of garbage trucks. Li-Fi-based SWC and GSM based modules are helpful in the segregation and availability of trucks at different locations and sensitivity of the weight [5].

Garbage monitoring is also equally important as that of Garbage collection. A system using IoT is responsible in sending SMS to the concerned authority about the condition of the bin if it is filled and also a Google map link to direct the assigned person to reach the filled bin [6] [7]. An intelligent system using GPRS technology in which the bins are GPS positioned. The system also supportive in sorting of garbage [8].

A deep learning algorithm is used in the segregation of biodegradable and non-biodegradable wastes [9]. IoT based ultrasonic sensors whether the garbage bin is filled or not is monitored, and the location of the garbage bin is shared and the nearby vehicle is directed to collect the garbage from the bin [10].

10.2 Methodology

10.2.1 Block Diagram

In the proposed system there are two units. One is Garbage Collection Unit and other one is Garbage Monitoring Unit. In both the units, Arduino Uno is used to monitor as well as to control the system.

In monitoring unit Arduino microcontroller is interfaced with the RF transmitter, three IR sensors and relays to activate motor. One IR sensor is used to detect the level of dumpster. If IR detects the full Garbage in the dumpster, then controller sends RF signal to the vehicle and at the same time buzzer starts beeping to indicate vehicle is coming to collect the Garbage. IR sensors are kept near dumpster to detect the vehicle when vehicle arrives near dumpster only when dumpster is full.

Vehicle starts moving using line following method and stops automatically below the first dumpster, then the dumpster is emptied to the vehicle and the vehicle then moves to the next dumpster. The entire vehicle is controlled by monitoring system using RF technology. The vehicle can also be controlled by Bluetooth in manual mode. The vehicle setup and garbage monitoring system are shown in Figure 10.1 and Figure 10.2 respectively.

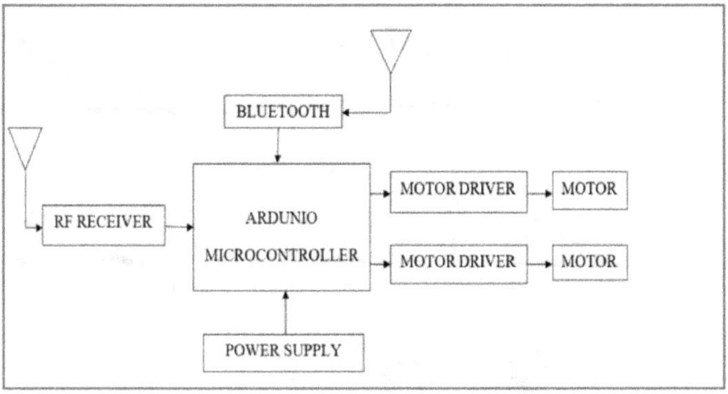

Figure 10.1 Block diagram showing the vehicle setup.

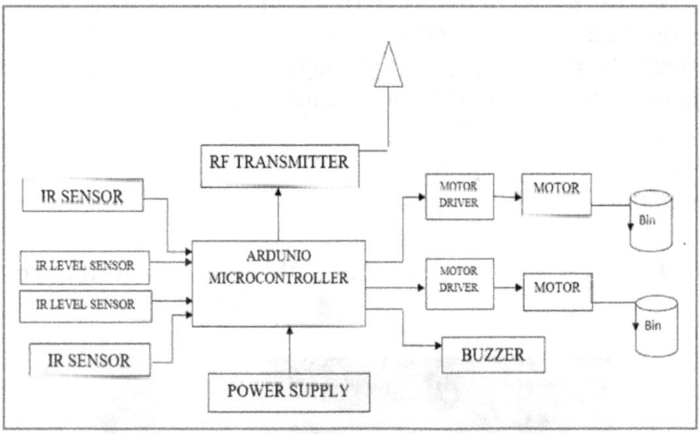

Figure 10.2 Block diagram showing the garbage monitoring system.

10.2.2 Software and Components Used

The software used in this work are Arduino v.6 and Embedded C. The hardware components used in this work are Arduino Uno, RF Transmitter and Receiver, Bluetooth, IR Sensor, Buzzer, Relays and DC motor.

10.2.3 Proposed Methodology Scenario

Figure 10.3 shows the scenario of automated garbage system in which the vehicle moves towards the dumpster when it is full.

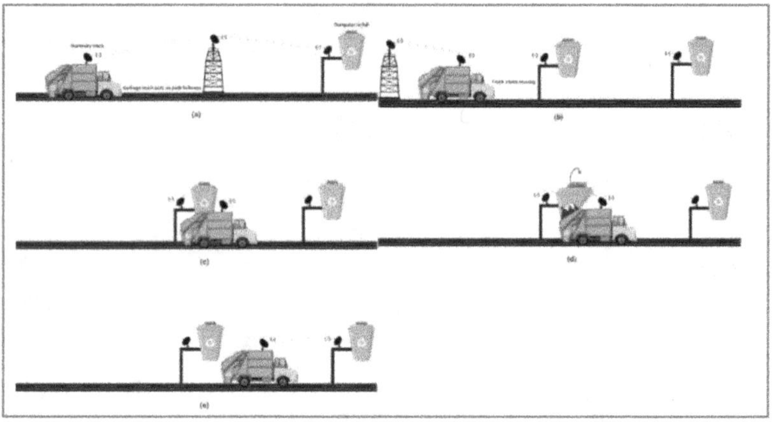

Figure 10.3 Garbage monitoring process. **(a):** Garbage dumpster getting filled and sending signal to the truck **(b):** Truck starts moving **(c):** Garbage truck has reached the dumpster and it halts for a few seconds and the sensors sense the garbage **(d):** The garbage pours the garbage into the truck **(e):** Truck moves on to the next dumpster.

10.3 Results

Figures 10.4 to 10.7 show the results of the automated garbage system.

Figure 10.4 Vehicle is at rest.

Here the vehicle is at rest and is waiting to receive the signal from the dumpster.

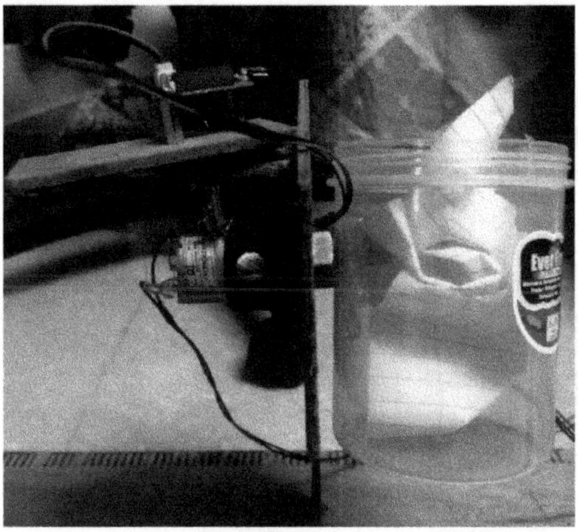

Figure 10.5 The dumpster has started filling.

The IR sensor will not detect the dumpster unless it is completely filled and hence does not send any signal to the vehicle, furthermore there is a 3 second delay to the IR sensor of in case there are unwanted disturbances cause to the sensor.

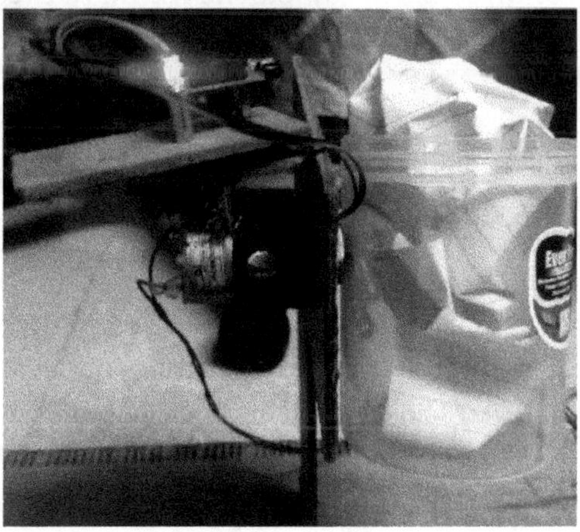

Figure 10.6 The dumpster is completely filled to its maximum level.

The IR sensor here detects the dumpster to be full as it is completely filled and is needed to be emptied and hence sends a signal through the transmitter to the vehicle and the vehicle upon receiving the signal, prepares itself to start moving and moves only when placed in the designed route.

Figure 10.7 The garbage is dumped onto the vehicle.

Once the signal is been received that the dumpster is full the vehicle starts moving towards the dumpster, using the path follower to stick to designed route and upon reaching the dumpster the second IR sensor detects the vehicle and the vehicle stops therefore leading to the emptying of the dumpster. Once the dumpster is emptied it goes back to its original position and the vehicle moves on to the next dumpster.

10.4 Conclusions

This paper involves the automation of garbage collection system. The results shows that a fully automatic system reduces man power as humans are not require to keep monitoring the dustbins. It always keeps the dustbin area clean as there is no overflow of waste from dustbin. This project can be used in all type of industries to keep industry environment clean. It can be used in apartments. This application can be used in university college campus for collection of garbage from dustbins. The system is designed only for automatic collection so segregation of waste must be done at the initial stage itself.

References

1. Jayson, M., Hiremath, S., & Lakshmi, H. R. (2018, February). Smart-Bin-Automatic waste segregation and collection. In *2018 Second international conference on advances in electronics, computers and communications (ICAECC)* (pp. 1-4). IEEE. (ICAECC). doi:10.1109/icaecc.2018.8479531
2. Jouhara, H., Czajczyńska, D., Ghazal, H., Krzyżyńska, R., Anguilano, L., Reynolds, A. J., & Spencer, N. (2017). Municipal waste management systems for domestic use. *Energy, 139*, 485-506.
3. Czajczyńska, D., Anguilano, L., Ghazal, H., Krzyżyńska, R., Reynolds, A. J., Spencer, N., & Jouhara, H. (2017). Potential of pyrolysis processes in the waste management sector. *Thermal Science and Engineering Progress, 3*, 171-197.
4. McCormick, K., Anderberg, S., Coenen, L., & Neij, L. (2013). Advancing sustainable urban transformation. *Journal of cleaner production, 50*, 1-11. https://doi.org/10.1016/j.jclepro.2013.01.003.
5. Topaloglu, M., Yarkin, F., & Kaya, T. (2018). Solid waste collection system selection for smart cities based on a type-2 fuzzy multi-criteria decision technique. *Soft Computing, 22*(15), 4879-4890.
6. Patel, A., & Patel, N. (2019). Garbage monitoring system using internet of things. In *Soft Computing and Signal Processing* (pp. 291-298). Springer, Singapore.
7. Soni, G., & Kandasamy, S. (2017, December). Smart garbage bin systems–a comprehensive survey. In *International Conference on Intelligent Information Technologies* (pp. 194-206). Springer, Singapore.
8. Zhao, Y., Wang, Q., Zang, Y., & Wan, J. (2019, June). Design of intelligent garbage collection system. In *International Conference on Applications and Techniques in Cyber Security and Intelligence* (pp. 542-547). Springer, Cham.
9. Sudha, S., Vidhyalakshmi, M., Pavithra, K., Sangeetha, K., & Swaathi, V. (2016, July). An automatic classification method for environment: Friendly waste segregation using deep learning. In *2016 IEEE Technological Innovations in ICT for Agriculture and Rural Development (TIAR)* (pp. 65-70). IEEE.
10. Kumar, S. V., Kumaran, T. S., Kumar, A. K., & Mathapati, M. (2017, August). Smart garbage monitoring and clearance system using internet of things. In *2017 IEEE International Conference on Smart Technologies and Management for Computing, Communication, Controls, Energy and Materials (ICSTM)* (pp. 184-189). IEEE.

11

Application of Flower Pollination Based Optimization Method for Parametric Appraisal of Bio-Diesel Powered Compression Ignition Engine

Ankita Barua[1], Abhishek Barua[2], Sangam Kumar Singh[3], Siddharth Jeet[2], Dilip Kumar Bagal[4], and Siba Sankar Mahapatra[5]

[1]*College of Agricultural Engineering and Technology, OUAT, Bhubaneswar, Odisha, India*
[2]*Centre for Advanced Post Graduate Studies, BPUT, Rourkela, Odisha, India*
[3]*Delhi Technical Campus, Bahadurgarh, Haryana, India*
[4]*Government College of Engineering, Kalahandi, Bhawanipatna, Odisha, India*
[5]*National institute of Technology, Rourkela, Odisha India*

11.1 Introduction

Practical design challenges in technology and economy are typically multi-objective or multi-criteria, and these numerous objectives are frequently at odds with one another, making it hard to adopt a single design choice without compromising [1]. Over the years, nature has solved difficult issues, and numerous biological arrangements have developed with remarkable and unexpected efficiency in optimizing evolutionary intents for instance reproduction [2].

Many algorithms inspired from the nature have been created all through the preceding few span of years decades were fully based on the effective abilities of biological classifications. All of these methods have been used in a variety of applications. Various project presentations in technology and industry need us to strive to identify the best explanation to a given issue while dealing with very complicated limitations. Because such limited optimization problems are sometimes very nonlinear, finding the best solutions is sometimes a difficult, if not impossible, effort. Most standard optimization methods do not function well for nonlinear and multimodal issues. The current tendency is to employ the nature-inspired metaheuristic procedures for solving such tough issues, and metaheuristics have been demonstrated to be surprisingly efficient for this purpose. As a result, the literature on metaheuristics has grown enormously in the recent two decades. Until date, researchers have only used a few natural-inspired qualities, and there is still opportunity for greater algorithm development [3].

11.2 Features of Flower Pollination Method

The primary function of a flower is eventually reproductive through pollination which is often related with pollen transfer, which is frequently related with pollinators such as birds, insects, bats, etc. [4]. Pollination may be classified into two types: biotic and abiotic. Approximately 90% of blooming plants are pollinated biotically where pollen is conveyed by a pollinator such as animals or insects. Whereas, Abiotic pollination does not require any pollinators and pollination occurs with help of wind, dispersion in water etc. Pollinators, also known as pollen carriers, come in a wide range of forms [5-8].

In actuality, one plant can have numerous flowers which can produce several no. of pollen gametes. We may simply expand future research to numerous pollen gametes for each blossom and numerous blooms for multi-objective optimization issues. All of the following flower pollination elements and processes cooperate to ensure that flowering plants reproduce optimally. As a result, this may stimulate the development of a novel optimization method.

11.3 Sunflower Optimization

A sunflower's life cycle is consistent: like the needles of a clock, they arise and accompany the sun every day. They go in the other way at night to await their disappearance the following morning [9-12].

The inverse square law radiation is another key nature-based optimization here. The quantity of heat Q received by each plant is thus given by:

$$Q_i = \frac{P}{4\pi r_i^2} \tag{1}$$

where P denotes the source's power and r_i denotes the distance amid the present paramount and the plant i.

The path of sunflowers towards the sun is:

$$\vec{S_i} = \frac{X^* - X_i}{\|X^* - X_i\|}, i = 1,2, \dots, n_p \tag{2}$$

The sunflowers' stride in the direction s is calculated by:

$$d_i = \lambda \times P_i(\|X_i - X_{i-1}\|) \times \|X_i - X_{i-1}\| \tag{3}$$

where λ is the perpetual value which defines a "inertial" dislocation of the plants, $P_i(\|X_i - X_{i-1}\|)$ is the probability of pollination.

The maximum step is defined as follows:

$$d_{max} = \frac{\|X_{max} - X_{min}\|}{2 \times N_{pop}} \tag{4}$$

where X_{max} and X_{min} are the upper and lower boundaries, and N_{pop} is the total number of plants.

The newest plantation would be:

$$\vec{X}_{i+1} = \vec{X}_i + d_i \times \vec{s}_i \tag{5}$$

The algorithm starts by creating a population of people which might be even or random. Respective individual's rating helps us to choose which one will be changed into the sun. Though it is intended to include the ability to function with several suns in a future edition, it is now limited to only one in this research. Then, like the sunflowers, all the additional entities will position themselves towards the sun and move in a random controlled manner. Paramount plants will pollinate around the sun.

11.4 Optimization of Compression Ignition Engine Parameters Using Sunflower Optimization

To study the effectiveness of Sunflower optimization, parametric appraisal of Karanja Bio-diesel fueled Compression Ignition Engine has been piloted. At first Karanja Bio-Diesel was prepared from the Karanja seeds using trans-esterification process. As per previous studies conducted by Bagal et al. [13], the operating parameters has been chosen as shown in Table 11.1. Bagal et al. [13] used Taguchi-based WASPAS method, Genetic Algorithm, Simulated Annealing Algorithm and Particle Swarm Optimization approach for solving multiple-criteria optimization problem for improvement in Brake Thermal efficiency (BTE), Exhaust Gas Temperature (EGT) and Brake Specific Fuel Consumption (BSFC) of Karanja Bio-Diesel powered Compression Ignition engine. Table 11.1 shows the operating parameters and Table 11.2 shows L_{16} orthogonal array with output responses.

Table 11.1 Parameters, codes, and level values used for orthogonal array

Factor	Unit	Symbol	L1	L2	L3
Compression Ratio		A	17.5	17.7	17.9
Blend	%	B	B10	B15	B20
Load	%	C	20	60	80

Table 11.2 Experimental runs and output responses

Run No.	A	B	C	BTE (%)	BSFC (kg/kWh)	EGT (⁰K)
1	17.5	10	40	32.34	0.25	457
2	17.5	15	60	30.19	0.26	485
3	17.5	20	80	30.51	0.29	496
4	17.5	25	100	29.10	0.29	534
5	17.7	10	60	32.34	0.26	473
6	17.7	15	40	31.13	0.26	470
7	17.7	20	100	30.77	0.30	524
8	17.7	25	80	29.77	0.28	507
9	17.9	10	80	30.03	0.30	475
10	17.9	15	100	28.14	0.31	515
11	17.9	20	40	30.06	0.31	473
12	17.9	25	60	28.39	0.31	499
13	18.1	10	100	32.93	0.33	510
14	18.1	15	80	31.46	0.32	496
15	18.1	20	60	32.71	0.34	497
16	18.1	25	40	31.97	0.32	491

11.5 Results and Discussion

As maximization of Brake Thermal efficiency (BTE) is the objective of the problem, the output responses were analyzed using larger the better formula of Taguchi technique. Similarly, minimization of Exhaust Gas Temperature (EGT) and Brake Specific Fuel Consumption (BSFC) is the objective of the problem, hence the output responses were analyzed using smaller the better formula of Taguchi technique. Figure 11.1-11.3 show the main effect plot for respective output responses.

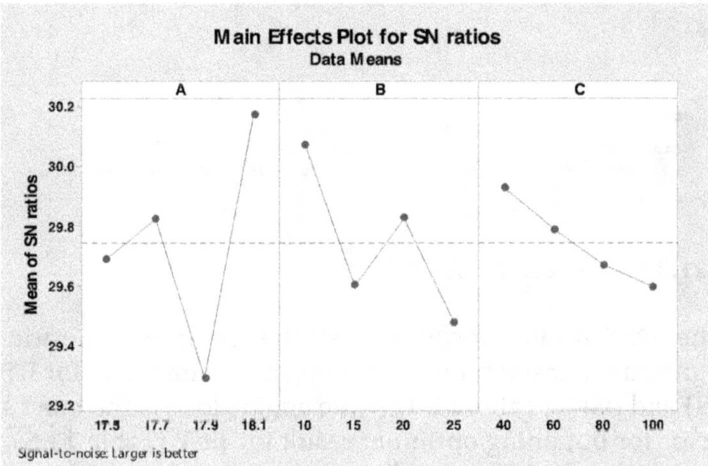

Figure 11.1 Main effect plot for BTE.

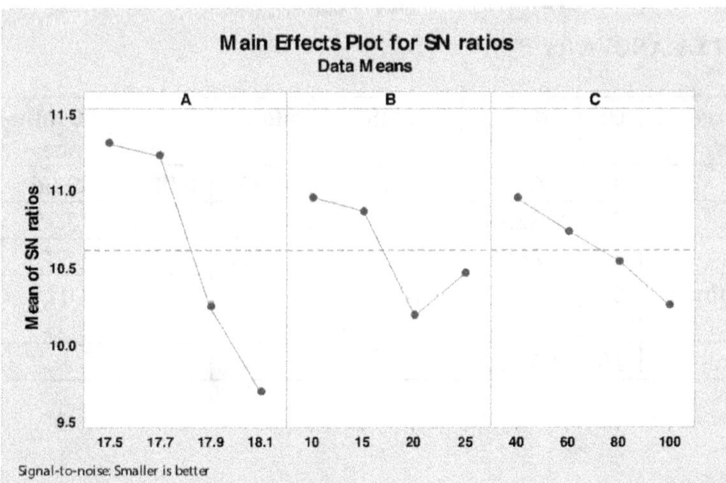

Figure 11.2 Main effect plot for BSFC.

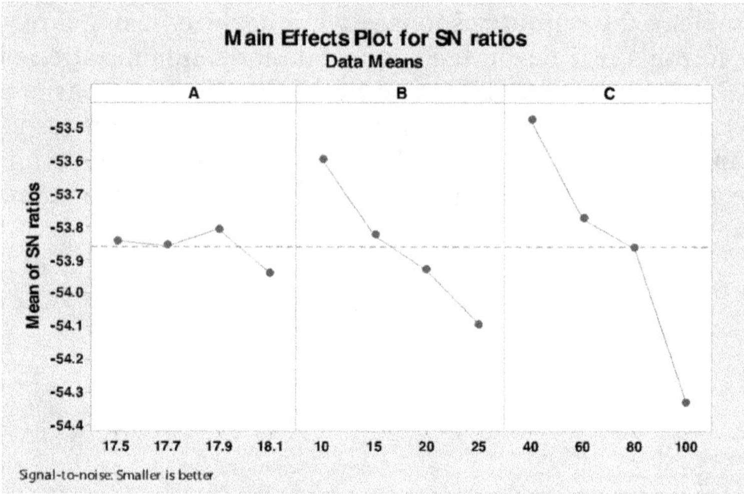

Figure 11.3 Main effect plot for EGT.

From the ANOVA table it can be seen that compression ratio is the most significant parameter for obtaining optimum result for BTE (Table 11.3) and BSFC (Table 11.4) while engine load is the most significant factor for obtaining optimum result for EGT (Table 11.5). It can be also seen that fuel blend is the 2nd most significant parameter for all three engine output responses. The R-squared values of all the responses are more than 99% which shows the consistency of the experiment results.

Table 11.3 ANOVA for BTE

Source	DF	Adj SS	Adj MS	F value	P value	% Influence
A	3	1.6005	0.5335	6361.47	0.000	59.66
B	3	0.8258	0.2753	3282.37	0.000	30.79
C	3	0.2558	0.0853	1016.58	0.000	9.53
Residual Error	6	0.0005	0.0001			0.02
Total	15	2.6825				

Table 11.4 ANOVA for BSFC

Source	DF	Adj SS	Adj MS	F value	P value	% Influence
A	3	7.357	2.452	435.44	0.000	73.66
B	3	1.536	0.512	90.94	0.000	15.38
C	3	1.060	0.353	62.78	0.000	10.62
Residual Error	6	0.033	0.005			0.34
Total	15	9.988				

Table 11.5 ANOVA for EGT

Source	DF	Adj SS	Adj MS	F value	P value	% Influence
A	3	0.0373	0.0124	71.75	0.000	1.80
B	3	0.5264	0.1755	1013.45	0.000	25.43
C	3	1.5054	0.5018	2898.30	0.000	72.72
Residual Error	6	0.0010	0.0002			0.05
Total	15	2.0701				

The regression equations were created using values of the alternatives, which will be used as a fitness function in Sunflower optimization algorithm:

Maximize:

$$BTE = 5162 - 582x_1 - 0.75x_2 + 0.81x_3 + 16.53x_1^2 + 0.0048x_2^2 + 0.000162x_3^2 + 0.012x_1x_2 - 0.051x_1x_3 + 0.00332x_2x_3 \quad (1)$$

Minimize:

$$BSFC = 32.3 - 3.75x_1 + 0.0208x_2 + 0.0111x_3 + 0.1094x_1^2 - 0.000125x_2^2 + 0.000002x_3^2 - 0.00080x_1x_2 - 0.00060x_1x_3 - 0.000019x_2x_3 \quad (2)$$

Minimize:

$$EGT = 14230 - 1526x_1 - 19.0x_2 - 3.93x_3 + 42.2x_1^2 - 0.0325x_2^2 + 0.00672x_3^2 + 1.19x_1x_2 + 0.202x_1x_3 + 0.0119x_2x_3 \quad (3)$$

According to previous studies, meta-heuristic optimization techniques provide more accurate outcomes than mathematical optimization techniques. For performing a concurrent optimization of output parameters, Sunflower optimization algorithms was employed for obtaining optimal engine running conditions The algorithm was coded in MATLAB R2018a software and ran on Lenovo Ideapad 330 Laptop with Intel Core i5- 8th Gen Processor with 1.60 GHz speed, RAM of 12GB and Windows 10 Pro Operating platform.

According to the problem statement BTE has to be maximized while BSFC and EGT has to be minimized. To achieve optimal running parameter, some fixed parameters have been defined in the Sunflower optimization algorithm where total no. of sunflower population has been taken as 100; pollination rate as 0.05; mortality rate as 0.1 and iteration limit was set to 100. Figures 11.4, 11.6 and 11.8 show the convergence of the population with generations for BTE, BSFC and EGT respectively. Figures 11.5, 11.7 and 11.9 show the convergence curve for BTE, BTE, BSFC and EGT respectively. Optimal factor setting along with predicted fitness values has been tabulated in Table 11.6.

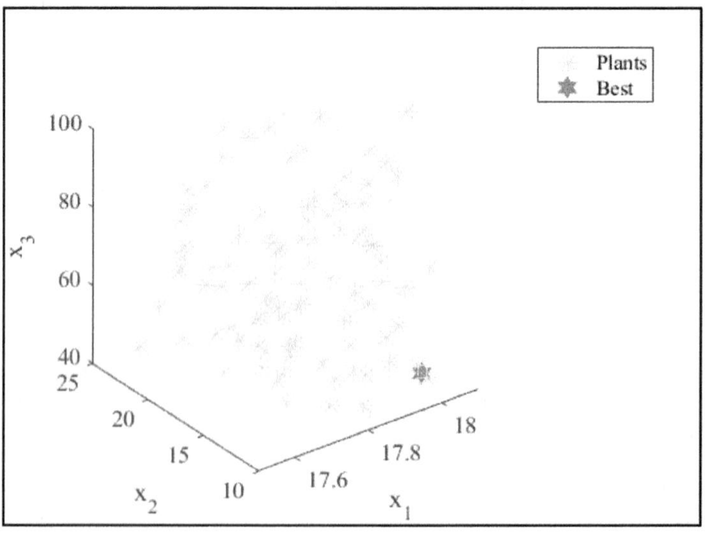

Figure 11.4 Convergence of the population with generations for BTE.

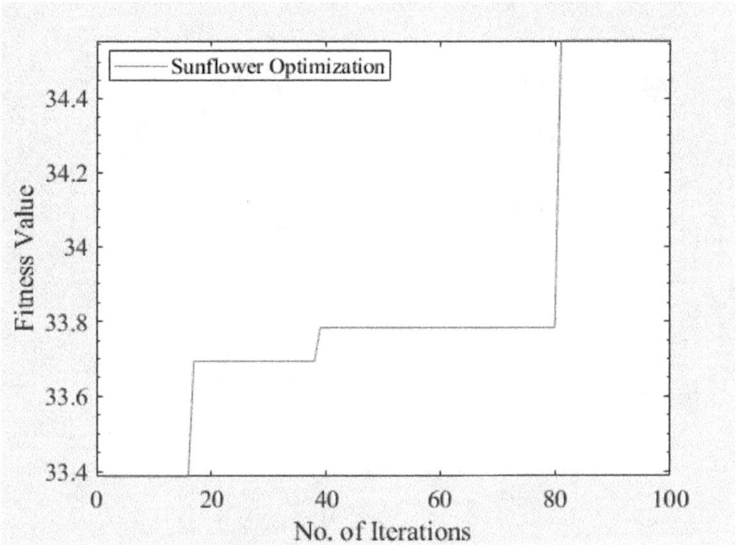

Figure 11.5 Convergence curve for BTE.

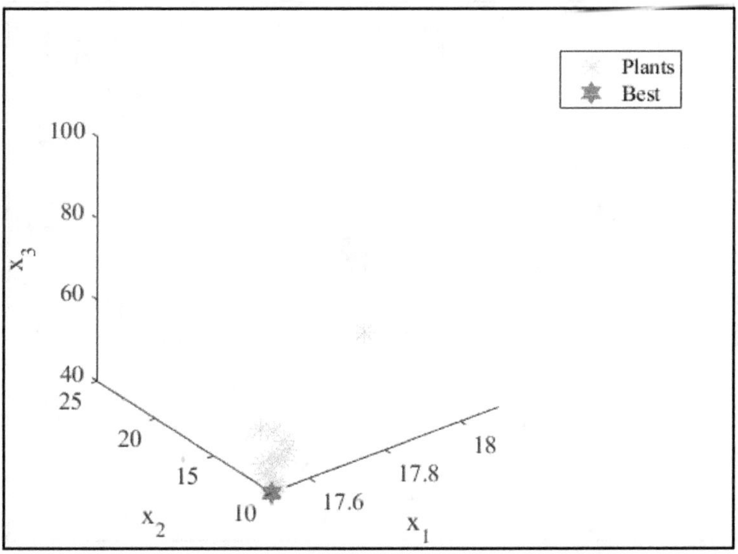

Figure 11.6 Convergence of the population with generations for BSFC.

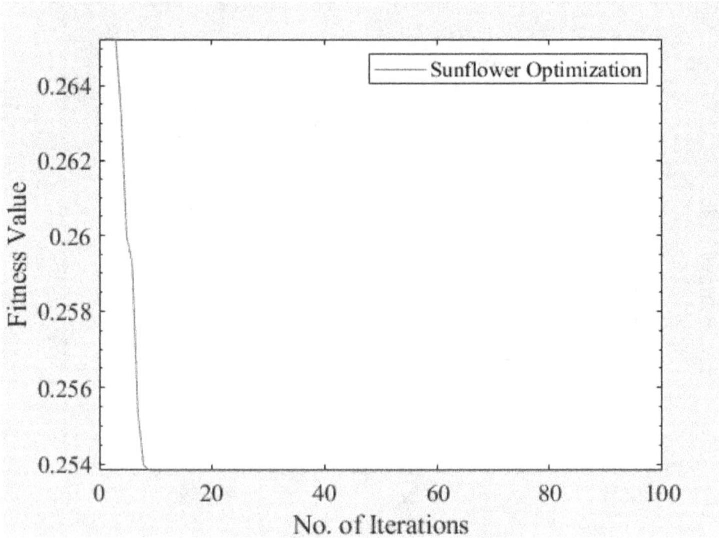

Figure 11.7 Convergence curve for BSFC.

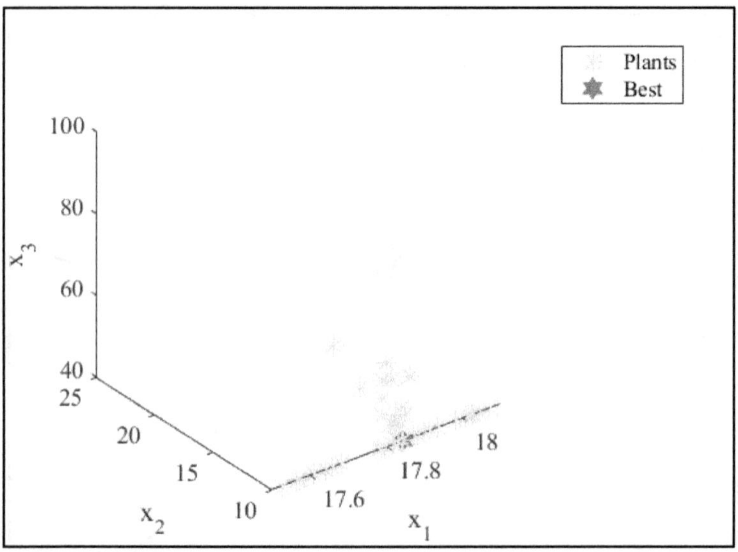

Figure 11.8 Convergence of the population with generations for EGT.

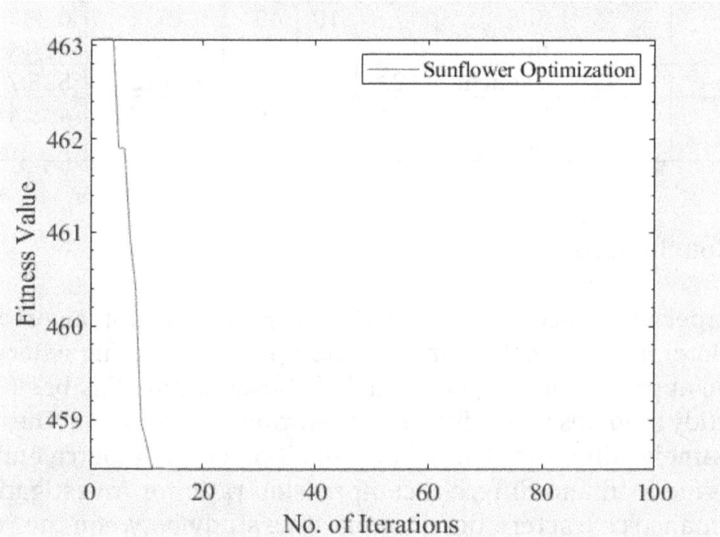

Figure 11.9 Convergence curve for EGT.

It is interesting to notice that results obtained from Sunflower optimization are precise than the values obtained by Taguchi method. It can be clearly seen in Table 11.6 that experimental results of BTE obtained using predicted factor setting by Sunflower optimization algorithm is more than that of predicted factor setting by Taguchi method as maximization of BTE is needed. Similarly experimental results of BSFC and EGT obtained using predicted factor setting by Sunflower optimization algorithm is less than that of predicted factor setting by Taguchi method as minimization of BSFC and EGT are needed for optimum running of the Diesel engine using Bio-Diesel. The results were in good agreement with the previous studies [13-15].

Table 11.6 R-squared value and optimal control parameters obtained from different optimization technique

Responses	R-sq value	Method	A	B	C	Predicted Value	Experimental Value
BTE	99.98	Taguchi	18.1	10	40	34.07 %	32.8 %
		Sunflower	18.1	15	40	33.72 %	33.7 %
BSFC	99.66	Taguchi	18.1	20	100	0.35 kg/kWh	0.31 kg/kWh

		Sun-flower	17.5	10	40	0.25 kg/kWh	0.25 kg/kWh
EGT	99.95	Taguchi	18.1	20	100	538.12 K	525.71 K
		Sun-flower	17.9	10	40	458.47 K	458.4 K

11.6 Conclusions

This paper introduced usage of Sunflower optimization algorithm in the field of multi-objective optimization problem solving where parametric appraisal of Bio-Diesel fueled Diesel Engine has been done. This study involves three different operating parameters of Diesel engine, namely, different blend of Karanja Biodiesel and petroleum diesel, engine load and different compression ratio for investigation of performance characteristics. Comparative study between the results obtained using Taguchi technique and Sunflower Algorithm has been carried to predict optimal operating parameter setting by using the regression analysis. According to ANOVA, compression ratio is the most significant parameter for obtaining optimum result for BTE and BSFC while engine load is the most significant factor for obtaining optimum result for EGT. It can be also seen that fuel blend is the 2nd most significant parameter for all three engine output responses. The results Confirmatory test showed that the Sunflower optimization algorithm method was able to find better prediction of the parameters, which proved its good performance. It can be concluded Sunflower Optimization algorithm which is a flower pollination-based algorithm can be employed for optimization of multi-parameter-based investigation in any field.

References

1. Yang X. S. (2012) Flower Pollination Algorithm for Global Optimization. In: Durand-Lose J., Jonoska N. (eds) Unconventional Computation and Natural Computation. UCNC 2012. Lecture Notes in Computer Science, vol 7445. Springer, Berlin, Heidelberg. https://doi.org/10.1007/978-3-642-32894-7_27.
2. Song, Y., Zhang, K., Hong, X., & Li, X. (2021). A novel multi-objective mutation flower pollination algorithm for the optimization of industrial enterprise R&D investment allocation. Applied Soft Computing, 107530.

3. Yang, X. S., Karamanoglu, M., & He, X. (2014). Flower pollination algorithm: a novel approach for multiobjective optimization. Engineering optimization, 46(9), 1222-1237.
4. Nabil, E. (2016). A modified flower pollination algorithm for global optimization. Expert Systems with Applications, 57, 192-203.
5. Gomes, G. F., da Cunha, S. S., & Ancelotti, A. C. (2019). A sunflower optimization (SFO) algorithm applied to damage identification on laminated composite plates. Engineering with Computers, 35(2), 619-626.
6. Yuan, Z., Wang, W., Wang, H., & Razmjooy, N. (2020). A new technique for optimal estimation of the circuit-based PEMFCs using developed sunflower optimization algorithm. Energy Reports, 6, 662-671.
7. Gomes, G. F., & de Almeida, F. A. (2020). Tuning metaheuristic algorithms using mixture design: Application of sunflower optimization for structural damage identification. Advances in Engineering Software, 149, 102877.
8. Gomes, G. F., de Almeida, F. A., Ancelotti, A. C., & da Cunha, S. S. (2020). Inverse structural damage identification problem in CFRP laminated plates using SFO algorithm based on strain fields. Engineering with Computers, 1-21.
9. Gomes, G. F., & Giovani, R. S. (2020). An efficient two-step damage identification method using sunflower optimization algorithm and mode shape curvature (MSDBI–SFO). Engineering with Computers, 1-20.
10. Francisco, M. B., Pereira, J. L. J., Oliver, G. A., da Silva, F. H. S., da Cunha Jr, S. S., & Gomes, G. F. (2021). Multiobjective design optimization of CFRP isogrid tubes using sunflower optimization based on metamodel. Computers & Structures, 249, 106508.
11. El-Sehiemy, R. A., Hamida, M. A., & Mesbahi, T. (2020). Parameter identification and state-of-charge estimation for lithium-polymer battery cells using enhanced sunflower optimization algorithm. International Journal of Hydrogen Energy, 45(15), 8833-8842.
12. Qais, M. H., Hasanien, H. M., & Alghuwainem, S. (2019). Identification of electrical parameters for three-diode photovoltaic model using analytical and sunflower optimization algorithm. Applied Energy, 250, 109-117.
13. Bagal, D. K., Patra, A. K., Jeet, S., Barua, A., Pattanaik, A. K., and Pattnaik, D. (2021) MCDM Optimization of Karanja Biodiesel Powered CI Engine to Improve Performance Characteristics Using Super Hybrid Taguchi-Coupled WASPAS-GA, SA, PSO Method. In Next Generation Materials and Processing Technologies: Select Proceedings of RDMPMC 2020. Springer Singapore. DOI: 10.1007/978-981-16-0182-8_37.
14. Ganapathy, T., Murugesan, K., and Gakkhar, R. P. (2009) Performan-

ce optimization of Jatropha Biodiesel engine model using Taguchi approach. Applied Energy, 86, pp 2476–2486.

15. Sharma, A., Singh, Y., Singh, N. K., and Singla, A. (2019) Sustainability of jojoba Biodiesel/diesel blends for DI diesel engine applications-taguchi and response surface methodology concept. Industrial Crops & Products, 139, pp 1-13.

12

Application of Taguchi and Whale Optimization Algorithm for Parametric Optimization of CNC Milling of Aluminum Alloy

Siddharth Jeet[1], Deepak Kumar Bagal[2], Kanchan Kumari[3], Sandeep Bhoi[3], Surya Narayan Panda[4], Ajit Kumar Pattanaik[5]

[1]*Centre for Advanced Post Graduate Studies, BPUT, Rourkela, Odisha, India*
[2]*Veer Surendra Sai University of Technology, Burla, Odisha, India*
[3]*Parala Maharaja Engineering College, Berhampur, Odisha*
[4]*Birsa Institute of Technology, Sindri, Dhanbad, Jharkhand, India*
[5]*Government College of Engineering, Kalahandi, Bhawanipatna, Odisha, India*

12.1 Introduction

Milling is the removal of material from workpiece surface by employing milling cutter which is a rotational cutting instruments with many cutting points. The tool's cutting blades machines the material by shaving off chips with each pass as the milling cutter penetrates the workpiece which is termed as Shear deformation. CNC (Computer Numerical Control) machine tools need fewer user inputs, enhance productivity, and improve machined item quality. Among the several industrial CNC machining techniques, milling is the most basic machining operation.

Milling is a widely used in industry as a fundamental machining operation. Surface roughness with material removal rate is a chief measure of a product's scientific quality as well as a component with a substantial influence on manufacturing costs. For attaining precise surface roughness is very vigorous, problematical, and process-driven; it is extremely demanding to quantify its value by theoretical study. For years, academics have been attempting to discover the optimal collection of input parameters as well as the influence of each on a certain output. Figure 12.1 shows the interaction of milling cutter over the metal surface which will be machined.

Figure 12.1 Interaction of milling cutter over the metal surface.

12.2 Literature Survey

Chandel and Giri reported depth of cut as most significant factor in CNC milling of Aluminium alloy 6061 by employing Taguchi Based Grey Relational analysis [1]. Similar kind of result using Aluminium alloy 6061 and Taguchi process was also reported by Rao [2]. Roy et. al. also studied the effect of CNC milling parameters on Aluminium alloy and reported depth of cut as the most significant factor for optimizing material removal rate along with surface roughness with the help of Taguchi coupled TOPSIS method [3]. Titu and Pop [4] studied the effect of CNC milling parameters on Aluminium alloy 7136 and reported cutting speed as the most significant factor for optimizing surface finish. Abdullah et. al. [5] optimized the surface finish of Aluminium alloy 7024 machined using CNC milling machine with the help of Taguchi process and reported tool path as the most influencing parameter. Sakthivelu et. al. optimized the surface finish and material removal rate of Aluminium alloy 7075 T6 machined using CNC milling machine with the help of Taguchi process. They reported Feed rate as the most influencial factor for controlling surface finish while depth of cut as the most influential parameter for controlling material removal rate [6]. Joshi and Kothiyal et. al. [7] reported depth of cut as the supreme factor which effects the surface finish of CNC milling of Aluminium alloy using Taguchi technique. Vardhan et. al. employed response surface methodology and Taguchi technique for optimizing the CNC milling parameters and reported depth of cut as the most significant factor for controlling surface finish and material removal rate of P20 steel [8]. Vishnu et. al. [9] reported coolant flow as the supreme factor which affected the surface finish of Aluminium alloy 6351 machined by CNC milling. Abdulridha [10] reported spindle speed as the most supreme parameter which affected surface finish of mild steel machined using CNC Milling and optimized using Genetic Algorithm and Artificial Neural network. Azim et. al. [11] studied the effect of CNC milling parameters on grade 60 mild steel and reported spindle speed as the most significant factor for optimizing surface finish [12-14].

It can be noted that feed rate is the most significant in milling process though less attention has been given in refining the results using meta-heuristic optimization method. This chapter comprises of end-ball milling operation of Aluminium alloy and optimization of machining parameters using Taguchi as well as Whale optimization method.

12.3 Background of Taguchi Design

Taguchi techniques were developed by Genichi Taguchi for the improvement of manufactured items quality. It's a technique that uses Orthogonal Array (OA) trials that greatly reduces experiment variance and attaining optimal selection of process control parameters. This approach is useful for data analysis and predicting optimal results.

12.4 Whale Optimization Algorithm

Whale Optimization Algorithm is a novel metaheuristic algorithm proposed by Mirjalili and Lewis that resembles humpback whale hunting [15, 16]. Humpback whales hunt schools of shrimp or tiny fish near the surface by swimming around them in a diminishing circle and producing unique bubbles along a circle or '9'-shaped route. The first phase of the algorithm represents encircling prey and spiral bubble-net assaulting approach; exploited phase, the later phases where seek arbitrarily for a predator (Figure 12.2). Figure 12.3 shows the algorithm of Whale optimization.

Figure 12.2 WOA Bubble net searching mechanism.

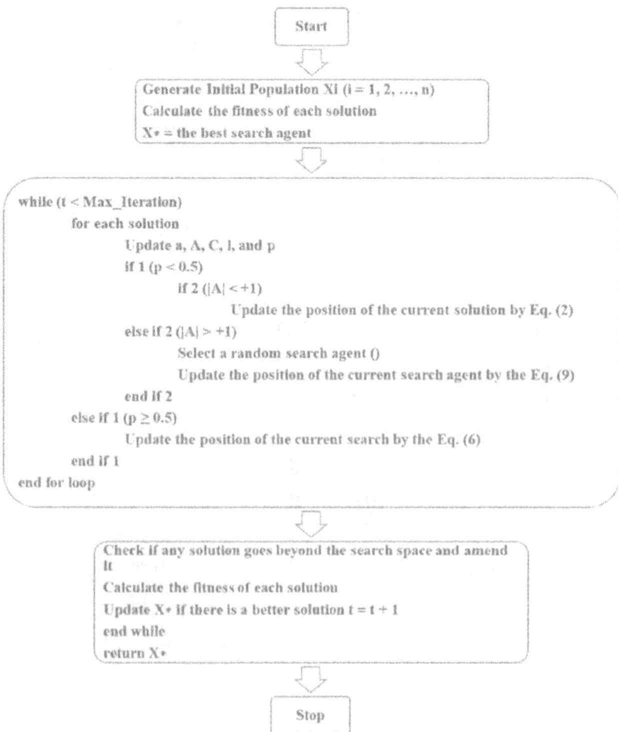

Figure 12.3 Algorithm of Whale optimization.

12.5 Experimental Design

Aluminum alloy-8011 (50 mm x 50 mm x 50 mm) was machined using Makino FDNC 86 mill under varying end-milling parameter conditions (Table 12.1). Experimentation will be carried out using Taguchi's L9 orthogonal array.

Table 12.1 Orthogonal array parameters, codes, and level values

Parameter	Code	Lvl 1	Lvl 2	Lvl 3
Spindle Speed	A	800 rpm	900 rpm	1000 rpm
Feed Rate	B	50 mm/min	100 mm/min	150 mm/min
Depth of cut	C	0.1 mm	0.2 mm	0.3 mm

12.6 Results and Discussion

Table 12.2 shows the Taguchi L_9 orthogonal array along with the material removal rate and surface roughness of the respective experimental run. Here, the MRR is needed to be maximized along with the minimization of roughness of the surface as per the objective of the problem. Hence the output responses were analyzed using larger the better formula for MRR and smaller the better formula for SR of Taguchi technique. Figures 12.4-12.5 show the main effect plot for respective output responses.

Table 12.2 MRR and SR results

Run No	A	B	C	MRR (mm^3/min)	SR (μm)
1	800	50	0.1	18.06	1.43
2	800	100	0.2	98.85	1.4
3	800	150	0.3	163.09	1.77
4	900	50	0.1	23.09	0.91
5	900	100	0.2	118.06	1.67
6	900	150	0.3	78.85	2.01
7	1000	50	0.1	60.90	1.15
8	1000	100	0.2	70.37	1.88
9	1000	150	0.3	120.43	1.46

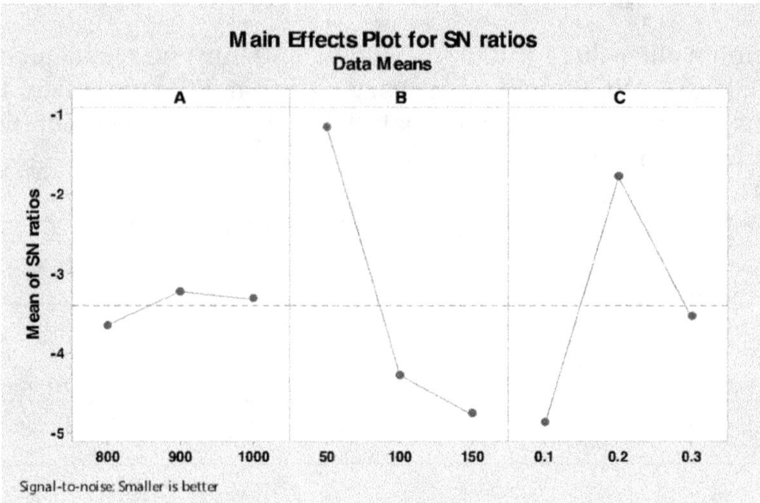

Figure 12.4 Main effect plot for surface roughness.

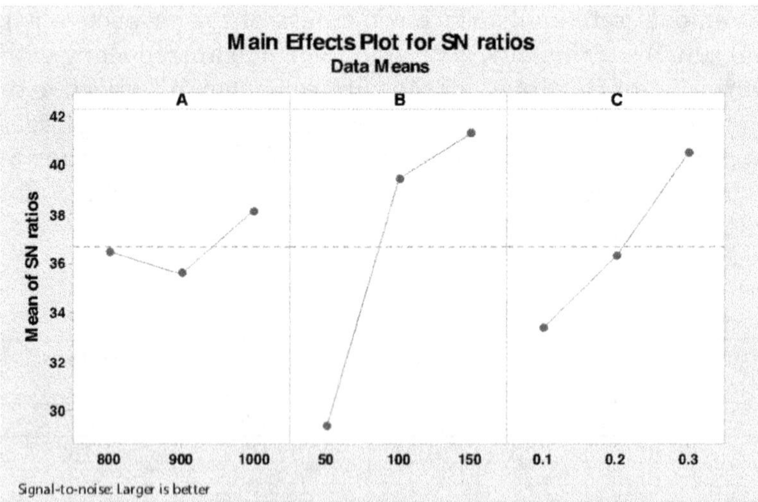

Figure 12.5 Main effect plot for material removal rate.

From the ANOVA table it can be seen that feed rate is the most significant parameter for obtaining optimum result for surface finish as well as MRR (Tables 12.3 and 12.4). It can be also seen that the depth of the cut is the second most significant parameter. The R-squared values of the responses are more than 98% which shows the consistency of the experiment results.

Table 12.3 Analysis of Variance for SN ratios of surface roughness

Source	DF	Adj SS	% Influence	Adj MS	F test	P test
A	2	0.306	0.81	0.153	1.22	0.45
B	2	22.946	60.57	11.473	91.70	0.011
C	2	14.383	37.96	7.191	57.48	0.017
Residual Error	2	0.250	0.66	0.125		
Total	8	37.885				

Table 12.4 Analysis of Variance for SN ratios material removal rate

Source	DF	Adj SS	% Influence	Adj MS	F test	P test
A	2	9.94	2.93	4.97	1.65	0.377
B	2	246.30	72.62	123.15	40.91	0.024
C	2	76.89	22.67	38.44	12.77	0.073
Residual Error	2	6.02	1.78	3.01		
Total	8	339.17				

The regression equations were created using values of the alternatives, which will be used as a fitness function in Whale optimization algorithm:

Maximize:

$$MRR = 1181 - 2.798x_1 + 0.9601x_2 + 576.1x_3 + 0.001528x_1^2 - 0.006146x_2^2 + 709.1x_3^2 + 0.001197x_1x_2 - 0.5983x_1x_3 \quad (1)$$

Minimize:

$$SR = 0.8067 + 0.002517x_1 + 0.02143x_2 - 17.02x_3 - 0.000001x_1^2 - 0.000078x_2^2 + 39.50x_3^2 + 0.000000x_1x_2 - 0.000000x_1x_3 \quad (2)$$

According to the problem statement MRR has to be maximized while surface roughness has to be minimized. To achieve optimal parameter, some fixed parameters have been defined in the Whale optimization algorithm where total no. of whale population has been taken as 30 and iteration limit was set to 50. Figures 12.6 and 12.7 show the convergence of MRR and surface roughness respectively. Optimal factor setting along with predicted fitness values has been tabulated in Table 12.5.

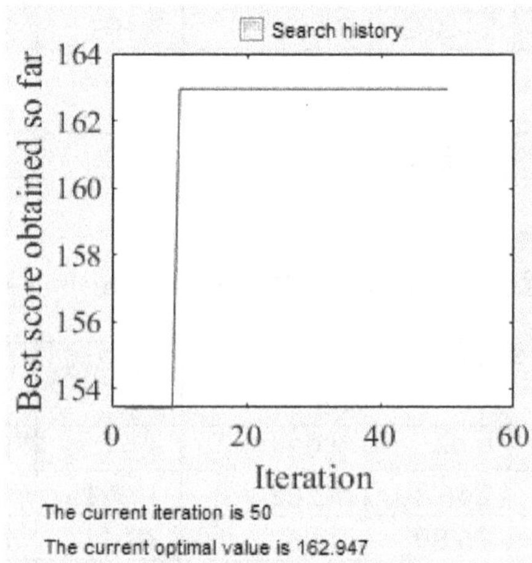

Figure 12.6 Convergence plot for MRR.

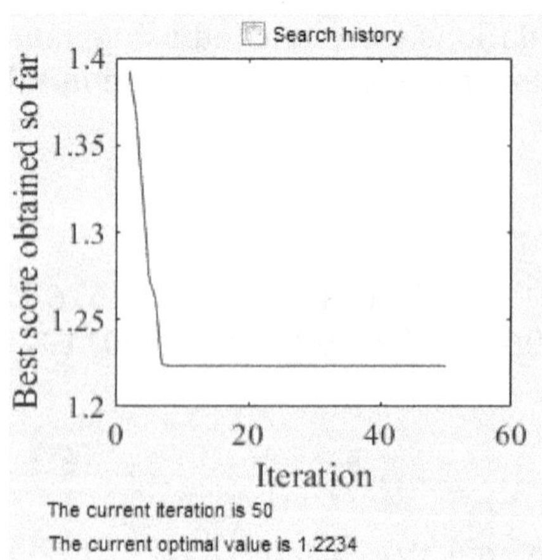

Figure 12.7 Convergence plot for surface roughness.

It is interesting to notice that results obtained from Whale optimization are more precise than the values obtained by Taguchi method. It can be clearly seen in Table 5 that experimental results of MRR obtained using predicted factor setting by Whale optimization algorithm is more than that of predicted factor setting by Taguchi method as maximization of MRR is needed. Similarly experimental results of surface roughness obtained using predicted factor setting by Whale optimization algorithm is less than that of predicted factor setting by Taguchi method as minimization of surface roughness is needed. The results were in good agreement with the previous studies [13-15].

Table. 12.5 R-squared value and optimal control parameters obtained from different optimization technique

Responses	R-sq value	Method	A	B	C	Predicted	Experimental
MRR	98.22	Taguchi	1000	150	0.3	151.66 mm^3/min	150.7 mm^3/min
		Whale	800	150	0.3	162.94 mm^3/min	162.01 mm^3/min
SR	99.34	Taguchi	800	150	0.1	2.01 μm	1.94 μm
		Whale	800	50	0.2	1.22 μm	1.21 μm

12.7 Conclusions

The Taguchi techniques offer the benefit of examining the impact of several variables and their interactions in a single experiment with only a few responses. Given the number of control and noise factors, the most significant factor was feed rate, which was achieved with a relatively small number of experimental runs, implying that Taguchi parameter design is an efficient and effective method for optimizing surface roughness and MRR in a milling operation. The following findings have been obtained from studies done on a CNC end milling machine:

It is interesting to notice that results obtained from Whale optimization are more precise than the values obtained by Taguchi method. It can be clearly seen in Table 6 that experimental results of MRR obtained using predicted factor setting by Whale optimization algorithm is more than that of predicted factor setting by Taguchi method as maximization of MRR is needed. Similarly experimental results of surface roughness obtained using predicted factor setting by Whale optimization algorithm is less than that of predicted factor setting by Taguchi method as minimization of surface roughness is needed. The percentage contributions of feed rate are more followed by depth of cut and spindle speed for milling process using Taguchi technique.

In future view, this approach of Hybrid Taguchi Methods can be applied in all engineering applications to achieve the best optimal values of process parameters. Again, the materials of workpiece can be extended to other types as well in which the CNC End-Milling process happens. The emerging optimization techniques can be applied to this experimental data set for better accuracy of optimal setting. In addition, this type of approach can enhance the better productivity in an industry where machining processes are carried out. This will also help to Research scholars as well as CNC operators to obtain the best optimal parametric setting with a smaller number of experimentations.

References

1. Chandel, D. K., & Giri, P. K. (2020) Optimization of Cutting Parameters in End Milling Process using Grey Taguchi Method. International Journal for Research in Applied Science & Engineering Technology 8(XII), 523-528.
2. Rao, A. D. (2020) Experimental Analysis on Multi Axis Spindle for Milling Operation by Taguchi Method. JAC: A Journal of Composition Theory XIII(XII), 173-180.
3. Roy, M. R. (2020) Optimization of Cutting Parameters to obtain the Optimum Surface Roughness and Metal Removal Rate in CNC End Milling using TOPSIS based Taguchi Method. Mukt Shabd Journal IX(IV), 954-960.
4. Țîțu, M. A., & Pop, A. B. (2020) CNC End Milling Process Parameters Optimization on Aluminium Alloy by Using Taguchi and ANOVA Methods. In IOP Conference Series: Materials Science and Engineering 877(1), 012021.
5. Abdullah, Z. K., Jaleel, A. K., & Abdulrazaq, M. M. (2020). Using of Taguchi Method to Improve the Surface Roughness in CNC Milling Machine. Solid State Technology, 63(2).
6. Sakthivelu S., Anandaraj T., Selwin M. (2017) Multi–Objective Optimization of Machining Conditions on Surface Roughness and MRR during CNC End Milling of Aluminium Alloy 7075 Using Taguchi Design of Experiments. Mechanics and Mechanical Engineering 21(1),

95–103.

7. Joshi A., Kothiyal P. (2021) Investigating Effect of Machining Parameters of CNC Milling on Surface Finish by Taguchi Method. International Journal on Theoretical and Applied Research in Mechanical Engineering 1(2), 60-65.
8. Vardhan, M. V., Sankaraiah, G., Yohan, M., & Rao, H. J. (2017). Optimization of Parameters in CNC milling of P20 steel using Response Surface methodology and Taguchi Method. Materials Today: Proceedings, 4(8), 9163-9169.
9. Vishnu, A. V., Tilak, K. B. G., Naidu, G. G., & Raju, D. G. J. (2015). Optimization of different process parameters of aluminium alloy 6351 in CNC milling using Taguchi method. International Journal of Engineering Research and General Science, 3(2), 407-413.
10. Abdulridha, H. H. (2020). Prediction of Surface Roughness of Mild Steel Alloy in CNC Milling Process Using ANN and GA Technique. Engineering and Technology Journal, 38(12A), 1842-1851.
11. Azim, S., Noor, S., Khalid, Q. S., Khan, A. M., Pimenov, D. Y., Ahmad, I. Babar, A.R., Pruncu, C. I. (2020). Sustainable manufacturing and parametric analysis of mild steel grade 60 by deploying CNC milling machine and Taguchi method. Metals, 10(10), 1303.
12. Mirjalili, S., & Lewis, A. (2016). The whale optimization algorithm. Advances in engineering software, 95, 51-67.
13. Aljarah, I., Faris, H., & Mirjalili, S. (2018). Optimizing connection weights in neural networks using the whale optimization algorithm. Soft Computing, 22(1), 1-15.
14. Mafarja, M. M., & Mirjalili, S. (2017). Hybrid whale optimization algorithm with simulated annealing for feature selection. Neurocomputing, 260, 302-312.
15. Bagal, D. K., Parida, B., Barua, A., Naik, B., Jeet, S., Singh, S. K., & Pattanaik, A. K. (2020). Mechanical Characterization of Hybrid Polymer SiC Nano Composite Using Hybrid RSM-MOORA-Whale Optimization Algorithm. In IOP Conference Series: Materials Science and Engineering, 970(1).
16. Mohanty, A., Nag, K.S., Bagal, D.K., Barua, A., Jeet, S., Mahapatra, S.S., Cherkia, H. (2021). Parametric optimization of parameters affecting dimension precision of FDM printed part using hybrid Taguchi-MARCOS-nature inspired heuristic optimization technique. Materials Today: Proceedings.

13

Comparative Analysis of Some Popular Multi-Objective Statistical Optimization Methods for solving MCDM Problems: A Case Study on FDM Parameters

Siddharth Jeet[1], Swarnima Subhadarsini[2], Abhishek Samantaray[3], Biswajit Parida[1], Bibekananda Naik[4], and Sriparna Paul[5]

[1]*Centre for Advanced Post Graduate Studies, BPUT, Rourkela, Odisha, India*
[2]*Veer Surendra Sai University of Technology, Burla, Odisha, India*
[3]*Scan Steels Limited, Rajgangpur, Odisha, India, India*
[4]*CSIR - Institute of Minerals & Materials Technology, Bhubaneswar, Odisha, India*
[5]*Council of Scientific and Industrial Research–Advanced Materials and Process Research, Bhopal, Madhya Pradesh, India*

13.1 Introduction

Multiple decision-making groups become active in the substantive problem-solving process, each having different parameters and points of view that must be resolved through a system of understanding and constructive agreement (concessions). The challenges are always distinguished by many incommensurable and contradictory conditions, and there can be no solution that meets these specifications at the same time. Robust choices are those that reflect a method outcome or appear after some algorithmic application. Both the method and the algorithm must guide and assist the decision-maker in the daunting challenge of selecting the right compromise solution to the decision problem at hand. However, the decision-maker must ultimately consider only one solution out of the multiple possible solutions, taking into account the problem's goals and objectives [1,2].

Continuous methods, also known as multi-objective decision-making methods, seek to find an ideal quantity in a decision problem that can differ indefinitely. Discrete MCDM methods, also known as multi-attribute decision-making (MADM) methods, are decision-support approaches that have a finite range of alternatives, a set of priorities, parameters for judging the options, and a system for rating choices based on how well they meet the aims and steps. Weighting methods and ranking methods are two types of discrete methods. Empirical MCDM methods are still being fine-tuned, and their application to new problems is expanding. If the number of applications increases, new ideas into how to develop MADM methods emerge [3,4].

(MCDM) approaches are gaining popularity as potential techniques for evaluating and solving complex real-time problems due to their intrinsic capacity to compare multiple options against varying parameters in order to choose the best alternative. The existence of several non-commensurable and overlapping parameters, separate units of measurement among the criteria, and the presence of very different alternatives distinguishes MCDM issues. Various MCDM approaches are used to solve these decision-making challenges describing multidimensional scenarios. The primary goal of the MCDM approaches is to evaluate and rate the possible alternatives. There are many instances where various MCDM approaches produce disparate effects (i.e., ranks of the same alternatives differ depending on the methods adopted). However, the issue of selecting a suitable MCDM approach in a given scenario persists [5,6].

The method of evaluating the best feasible solution based on the defined parameters is known as multi-criteria optimization (representing different effects). In MCDM, algorithmic reasoning and model construction offer a modern way to understanding some types of human behaviour and decision-making. MCDM is one of the most appealing interdisciplinary study areas in management science and operations research.

In most management decision-making problems, four assumptions are made:

(1) Prior to assessing solutions in a given problem domain, one must first acknowledge that a problem exists and explore innovative methods for evaluation.

(2) There is a need to choose the right choice from a limited range of possible alternatives. A decision-maker is aware of or has the ability to decide the other options; he or she studies the dilemma, where each alternative is represented by a set of parameters that are interdependent in different ways.

(3) In those cases, most decision-makers would choose a tool that would assist them in processing the criterion significance information for each option. In certain ways, the chosen solution is ideal or perfect.

(4) Even though the constraints do not actually simplify the analysis, a clear description of constraints for a problem can be useful.

Problem solving can be viewed in a variety of ways:

(1) Choosing the right or more preferable decision-maker alternative from a list of possible choices

(2) Selecting a limited set of appropriate options, or sorting alternatives into separate choice sets; and

(3) Identifying all effective or non-dominated options.

In certain situations, a system analyst can help with decision-making by doing a thorough review and listing the basic properties of non-inferior or compromise alternatives. The following are the key stages of MCDM:

a) Designing system assessment parameters that link system capabilities to objectives
b) Creating alternative systems for achieving the goals (generating alternatives)
c) Assessing choices concerning rules (the values of the criterion functions)
d) Using a normative multi-criteria design method
e) Approving one choice as "optimal" (preferred); and
f) If the final solution is rejected, collect new data and proceed to the next iteration of multi-criteria optimization.

For answering problems with different and non-commensurable parameters, dozens of MCDM models are created. There are times where the assessment of viable alternatives must deal with the compromising of existing standard principles, necessitating the implementation of a compromise MCDM model.

The development of emerging technology and tools in each field of research provides a viewpoint that helps consumers and experts to compare and select depending on relevant criteria. As a major cluster in process analysis, multiple decision-making theories are effectively addressed in order to reach more optimal solutions. As a significant part of decision science, the MCDM family has been broadly improved and implemented in related decision-making problems ranging from human capital, electricity, development, and manufacturing to air travel management and education. It is an excellent subject if one focuses on the central concept of MCDM and develops an enhanced algorithm that handles differently from others. An effort is already being made to restructure and remodel a prior paradigm using different techniques, even though the outcomes are unsatisfactory. In this area of knowledge, a study strategy includes the reduction, division, distinction, or aggregation of rational formulas. Academicians presented many exciting projects related to the growth, expansion, and incorporation of MCDM methods [4-7].

Various decision-making techniques have been developed and expanded through various implementations and case studies. When a particular MCDM method is finally recommended for a specific application, it is observed that its solution accuracy and ranking performance are seriously influenced by the value of its control parameter. In this paper, the applicability and usefulness of some popular methods will be explored and will be compared with some recently formulated methods while solving problem from real time manufacturing environment. Printing parameter of Fused Deposition Modelling (FDM) has be considered and solved using Desirability Function Approach, TOPSIS Method (Technique for Order of Preference by Similarity to Ideal Solution), Utility Concept, Grey Relational Analysis, WASPAS Method (Weighted aggregated sum product assessment), EDAS Method (Evaluation Based on Distance from Average Solution), COCOSO Method (Combined Compromise Solution), MV-COCOSO Method (Multi-Variance Combined Compromise Solution) and MARCOS Method (Measurement Alternatives and Ranking according to Compromise Solution) which are coupled with Taguchi Method. For each problem, the optimal control parameters are determined and their effects on alternatives are found using Analysis of Variance (ANOVA).

13.2 Optimization Methods

13.2.1 Desirability Function Approach

Derringer and Suich proposed formulae in 1980 to measure the desirability of each answer based on the target value criterion. To compute the person desirability index (di) for the respective responses, two types of desirability functions are used, depending on the response characteristics. [8-11].

Smaller-the better

Equation (i) can be used to describe the desirability property of the-smaller-the-better):

$$d_i = \begin{cases} 1 & \hat{y} \leq y_{min} \\ \left(\frac{\hat{y}-y_{max}}{y_{min}-y_{max}}\right)^r, y_{min} \leq y \leq y_{max}, r \geq 0 & r \geq 0 \\ 0 & \hat{y} \geq y_{max} \end{cases} \tag{i}$$

where the y_{min} signifies the lower tolerance limit of $\hat{y}$, the y_{max} signifies the upper tolerance limit of $\hat{y}$, and r denotes the weight.

Larger-the better

The larger-the-better desirability function can be written as Equation (ii):

$$d_i = \begin{cases} 0 & \hat{y} \leq y_{min} \\ \left(\frac{\hat{y}-y_{min}}{y_{max}-y_{min}}\right)^r, & y_{min} \leq \hat{y} \leq y_{max}, r \geq 0 \\ 1 & \hat{y} \geq y_{min} \end{cases} \tag{ii}$$

where the y_{min} represents the lower tolerance limit of $\hat{y}$, the y_{max} represents the upper tolerance limit of $\hat{y}$ and r represents the weight. Calculate the overall desirability value D_0 in the following stage. By using the following formula, the individual desirability indexes of all answers can be combined to create a single value known as composite desirability D_0. Formula (iii):

$$D_0 = (d_1^{w1} d_2^{w2} \dots d_n^{wn})^{\frac{1}{W}} \tag{iii}$$

where d_i is the individual desirability of the property y_i, w_i is the weight of the property y_i in the composite desirability and W is the sum of the individual weights.

13.2.2 TOPSIS Method (Technique for Order of Preference by Similarity to Ideal Solution)

TOPSIS Method (Technique for order of preference by similarity to ideal solution) is used for estimating substitutions prior to multiple attribute decision making; it is founded on the fact that the preferred substitute should be the closest to the positive ideal solution and the furthest away from the negative ideal solution. Positive ideal solution denotes the best performance values achieved by any substitution for each attribute, while negative ideal solution denotes the worst performance values. TOPSIS is comprised of the measures mentioned below [12-16]:

Step 1: Formation of decision matrix:

$$D = \begin{matrix} A_1 \\ A_2 \\ \cdot \\ A_i \\ \cdot \\ A_m \end{matrix} \begin{bmatrix} x_{11} & x_{12} & \cdot & x_{1j} & x_{1n} \\ x_{21} & x_{22} & \cdot & x_{2j} & x_{2n} \\ \cdot & \cdot & \cdot & \cdot & \cdot \\ x_{i1} & x_{i2} & \cdot & x_{1j} & \cdot \\ \cdot & \cdot & \cdot & \cdot & \cdot \\ x_{m1} & x_{m2} & \cdot & x_{mj} & x_{mn} \end{bmatrix} \quad \text{(iv)}$$

Here, A_i (i=1,2, ..., m) represents the possible alternatives; x_j (j=1,2, ..., n) represents the attributes relating to alternative performance, j=1,2, ..., n and x_{ij} is the performance of A_i with respect to attribute x_j

Step 2: Normalization of matrix:

$$r_{ij} = \frac{x_{ij}}{\sqrt{\sum_{i=1}^{m} x_{ij}^2}} \quad \text{(v)}$$

Here, r_{ij} represents the normalized performance of A_i with respect to attribute x_j.

Step 3: Weighted Decision matrix:

$$V = \lfloor v_{ij} \rfloor V = w_j r_{ij} \quad \text{(vi)}$$

$$D = \begin{bmatrix} y_{11} & y_{12} & \cdot & y_{1j} & y_{1n} \\ y_{21} & y_{22} & \cdot & y_{2j} & y_{2n} \\ \cdot & \cdot & \cdot & \cdot & \cdot \\ y_{i1} & y_{i2} & \cdot & y_{1j} & \cdot \\ \cdot & \cdot & \cdot & \cdot & \cdot \\ y_{m1} & y_{m2} & \cdot & y_{mj} & y_{mn} \end{bmatrix} \quad \text{(vii)}$$

Here, $\sum_{j=1}^{n} w_j = 1$

Step 4: Determine the positive ideal and negative ideal solutions:

a) The positive ideal solution:

$$A^+ = \left\{ \left(\max_i y_{ij} \,\middle|\, j \in J \right), \left(\min_i y_{ij} \,|\, j \in J | i = 1,2, \ldots, m \right) \right\} = \{y_1^+, y_2^+, \ldots \ldots .. y_j^+, \ldots \ldots \ldots y_n^+\} \quad \text{(viii)}$$

b) The negative ideal solution:

$$A^+ = \left\{ \left(\min_i y_{ij} \,\middle|\, j \in J \right), \left(\max_i y_{ij} \,|\, j \in J | i = 1,2, \ldots \ldots \ldots, m \right) \right\} = \{y_1^-, y_2^-, \ldots \ldots .. y_j^-, \ldots \ldots \ldots y_n^-\} \quad \text{(ix)}$$

Here,
$J = \{j = 1,2, \ldots \ldots .., n|j\}$: Associated with the beneficial attributes
$J' = \{j = 1,2, \ldots \ldots .., n|j\}$: Associated with non-beneficial attributes

Step 5: The n-dimensional Euclidean distance between each alternative and the ideal solution is given by the following equations:

$$S_i^+ = \sqrt{\sum_{j=1}^{n} (y_{ij} - y_j^+)^2 \; i = 1,2, \ldots \ldots \ldots, m} \quad \text{(x)}$$

$$S_i^- = \sqrt{\sum_{j=1}^{n} (y_{ij} - y_j^-)^2 \; i = 1,2, \ldots \ldots \ldots, m} \quad \text{(xi)}$$

Step 6: Determine the Overall efficiency coefficient that is nearest to the optimal solution:

$$C_i^+ = \frac{S_i^-}{S_i^+ + S_i^-}, i = 1,2, \ldots \ldots \ldots, m; 0 \le C_i^+ \le 1 \quad \text{(xii)}$$

13.2.3 Utility Concept

Utility can be described as the ease with which an object responds to the needs of its users. It is a test of an artifact's ability to satisfy the needs of its consumers. The cumulative usefulness of an item can be expressed by a single index called utility, which is the sum of a product's individual output characteristics.

If y_j is the measure of performance of jth attribute (criterion) and there are n alternatives in the entire selection space, then the joint utility function can be expressed as [17-19]:

$$U(y_1, y_2, \ldots, y_n) = f(U_1(y_1), U_2(y_2), \ldots, U_n(y_n)) \quad \text{(xiii)}$$

where $U_j(y_j)$ is the utility of jth attribute and n is the number of evaluation criteria. The overall utility function is the sum of individual utilities if the attributes are autonomous, and is given as follows:

$$U(y_1, y_2, \ldots, y_n) = \sum_{j=1}^{n} U_j(y_j) \quad \text{(xiv)}$$

The overall utility function after assigning weights to the attributes can be expressed as:

$$U(y_1, y_2, \ldots, y_n) = \sum_{j=1}^{n} w_j U_j(y_j) \quad \text{(xv)}$$

For each attribute, a preference scale is created to define its utility value. The preference numbers range from 0 to 9. (Denoting the lowest and highest performance value). Thus, preference number 0 denotes the most appropriate attribute, while preference number 9 denotes the best value of the attribute. The preference number (P_j) for j[th] attribute can be stated on a logarithmic scale as follows:

$$P_j = A_j \times \log\left(\frac{y_j}{y_j'}\right) \quad \text{(xvi)}$$

where y_j and y_j' are the observed and the just acceptable values, and A_j is a constant for j[th] attribute. For y_j', the minimum value is taken for beneficial criteria, while the maximum value is considered for non-beneficial criteria. The value of A_j can be found by the condition that if $y_j = y_j^*$ (where y_j^* is the best value for j[th] attribute), then $P_j = 9$.

Therefore,

$$A_j = \frac{9}{\log\left(\frac{y_j^*}{y_j'}\right)} \quad \text{(xvii)}$$

The overall utility value (U) can now be computed as follows:

$$U = \sum_{j=1}^{n} w_j P_j \quad \text{(xviii)}$$

subject to the constraint that $\sum_{j=1}^{n} w_j = 1$

13.2.4 Grey Relational Analysis

The grey relational analysis is a calculation of the absolute value of data discrepancy between the sequences, and it can be used to approximate the similarity between them. The parts that follow describe the grey relational analysis procedure that was used [20-22].

Data pre-processing

For "Larger the better", if the target value of original sequence is infinitely large then the normalized experimental results can be expressed as

$$x_i^*(t) = \frac{x_i^{(o)}(t) - \min(x_i^{(o)}(t))}{\max(x_i^{(o)}(t)) - \min(x_i^{(o)}(t))} \quad \text{(xix)}$$

For "Smaller the better", when the target value of original sequence is infinitely small then the normalized results is expressed as

$$x_i^*(t) = \frac{\max(x_i^{(o)}(t)) - x_i^{(o)}(t)}{\max(x_i^{(o)}(t)) - \min(x_i^{(o)}(t))} \quad \text{(xx)}$$

For "Nominal the best", if a defined target value j exists, when the target value is closer to desired value the normalization is done as

$$x_i^*(t) = 1 - \frac{|x_i^{(o)}(t) - j|}{\max\{\max(x_i^{(o)}(t)) - j, j - \min(x_i^{(o)}(t))\}} \quad \text{(xxi)}$$

Grey relational coefficients and Grey relational grades

After pre-processing the results, a grey relational coefficient is computed using the pre-processed sequences. The grey relational coefficient can be computed as follows:

$$\gamma(x_o^*(t),(x_i^*(t)) = \frac{\Delta_{min} + \zeta\Delta_{max}}{\Delta_{oi}(k) + \zeta\Delta_{max}}$$
and $0 < \gamma(x_o^*(t),(x_o^*(t)) \leq 1$ (xxii)

The grey relational grade is determined using the following equation after the grey relational coefficients have been calculated

$$\gamma(x_o^*,x_i^*) = \sum_{t=1}^{n} \beta_t \, \gamma(x_o^*(t),(x_i^*(t)) \quad \text{(xxiii)}$$

where $\sum_{t=1}^{n} \beta_t = 1$

The degree of similarity between the comparison and comparability sequences is represented by the grey relational grade $\gamma(x_o^*,x_i^*)$. [23-28].

13.2.5 WASPAS Method (Weighted Aggregated Sum Product Assessment)

Zavadskas et al. introduced the weighted aggregated sum product assessment (WASPAS) method for MCDM problems in 2012. The weighted sum method (WSM) and the weighted product method (WPM) are two well-known MCDM approaches that have been combined in this technique (WPM). The main method for solving MCDM problems using the WASPAS methodology is divided into several stages [24-27].

Step 1. The initial decision matrix has been developed.

Step 2. Normalization of the decision matrix using equations (xxiv) and (xxv), respectively, for maximization and minimization parameters:

$$\bar{x}_{ij} = x_{ij}/max_i x_{ij} \quad \text{(xxiv)}$$
$$\bar{x}_{ij} = min_i x_{ij}/x_{ij} \quad \text{(xxv)}$$

where x_{ij} is the assessment value of the ith alternative with respect to the jth criterion.

Step 3. Using equation (xxvi), calculate the overall relative value of the ith alternative using the weighted sum process (WSM):

$$Q_i^{(1)} = \sum_{j=1}^{n} \bar{x}_{ij} \cdot w_j \tag{xxvi}$$

Step 4. Using equation (xxvii), calculate the absolute relative value of the ith alternative using the weighted product process (WPM):

$$Q_i^{(2)} = \prod_{j=1}^{n} \bar{x}_{ij}^{w_j} \tag{xxvii}$$

Step 5: The WASPAS approach uses equation (xxviii) to determine the overall relative value of alternatives in order to improve rating accuracy and decision-making effectiveness:

$$Q_i = \lambda \, . \, Q_i^{(1)} + (1 - \lambda) \, . \, Q_i^{(2)} \tag{xxviii}$$

where λ=0,0.1,...,1 is coefficient of linear combination and usually takes value of 0.5.

13.2.6 EDAS Method (Evaluation Based on Distance from Average Solution)

Developed by Keshavarz Ghorabaee et al. [3,28,29], EDAS method was employed for multi-criteria inventory classification. The stages of EDAS method are presented as follows:

Step 1: Selection of important criteria for describing altenatives.

Step 2: Construct the decision-making matrix (X), shown as follows:

$$X = [x_{ij}]_{nXm} = \begin{bmatrix} x_{11} & x_{12} & \cdots & x_{1m} \\ x_{21} & x_{22} & \cdots & x_{2m} \\ \cdots & \cdots & \cdots & \cdots \\ x_{n1} & x_{n2} & \cdots & x_{nm} \end{bmatrix} \tag{xxix}$$

where X_{ij} denotes the performance value of ith alterative on j[th] criterion.

Step 3: Computation of average solution using equation (xxx):

$$AV = \left[AV_j\right]_{1Xm}$$

where,

$$AV_j = \frac{\sum_{i=1}^{n} X_{ij}}{n} \tag{xxx}$$

Step 4: Computation of positive distance from average (PDA) and the negative distance from average (NDA) matrixes in accordance to the type of measures (benefit and cost):

$$PDA = \left[PDA_{ij}\right]_{nXm} \tag{xxxi}$$

$$NDA = \left[NDA_{ij}\right]_{nXm} \tag{xxxii}$$

if j[th] criterion is beneficial,

$$PDA_{ij} = \frac{\max\left(0,\left(X_{ij} - AV_j\right)\right)}{AV_j} \tag{xxxiii}$$

$$NDA_{ij} = \frac{\max\left(0,\left(AV_j - X_{ij}\right)\right)}{AV_j} \tag{xxxiv}$$

and if j[th] criterion is non-beneficial,

$$PDA_{ij} = \frac{\max\left(0,\left(AV_j - X_{ij}\right)\right)}{AV_j} \tag{xiii}$$

$$NDA_{ij} = \frac{\max\left(0,\left(X_{ij} - AV_j\right)\right)}{AV_j} \tag{xxxv}$$

where PDA_{ij} and NDA_{ij} denote the positive and negative distance of ith altemative from average solution in terms of jth criterion, respectively.

Step 5: Computation of weighted sum of PDA and NDA for all alternatives using equation (xxxvi) and (xxxvii):

$$SP_i = \sum_{j=1}^{m} w_j PDA_{ij} \quad \text{(xxxvi)}$$

$$SN_i = \sum_{j=1}^{m} w_j NDA_{ij} \quad \text{(xxxvii)}$$

where w_j is the weight of jth criterion.

Step 6: Normalization of SP and SN values for all alterative using equation (xxxvii):

$$NSP_i = \frac{SP_i}{max_i(SP_i)} \quad \text{(xvii)}$$

$$NSN_i = 1 - \frac{SN_i}{max_i(SN_i)} \quad \text{(xviii)}$$

Step 7: Calculation of appraisal score (AS) for all alterative using equation (xxxviii):

$$AS_i = \frac{1}{2}(NSP_i + NSN_i) \quad \text{(xxxviii)}$$

where $0 \leq AS_i \leq 1$.

13.2.7 COCOSO Method (Combined Compromise Solution)

Formulated by Yazdani et. al., Combined Compromise Solution (COCOSO) Method is one of the recently formulated method used for solving decision making problems in a simplified manner.
The following steps are used to solve COCOSO decision problem [5, 30,31]:

1) Computation of initial decision-making matrix using equation (xxxix):

$$X_{ij} = \begin{bmatrix} x_{11} & x_{12} & \cdots & x_{1n} \\ x_{21} & x_{22} & \cdots & x_{2n} \\ \cdots & \cdots & \cdots & \cdots \\ x_{m1} & x_{m2} & \cdots & x_{mn} \end{bmatrix}; \quad \text{(xxxix)}$$

2) The criteria values are normalised using a consensus normalisation equation:

$$r_{ij} = \frac{x_{ij}-minx_{ij}}{maxx_{ij}-minx_{ij}} \quad \text{; for benefit criterion;} \tag{xl}$$

$$r_{ij} = \frac{maxx_{ij}-x_{ij}}{maxx_{ij}-minx_{ij}} \quad \text{; for cost criterion:} \tag{xli}$$

3) Calculation of overall weighted comparability series and total power of weight of comparability sequences for respective alternates S_i and P_i:

$$S_i = \sum_{j=1}^{n}(w_j r_{ij}) \tag{xlii}$$

$$P_i = \sum_{j=1}^{n}(r_{ij})^{w_j} \tag{xliii}$$

4) Three appraisal score are used for generation of comparative weights of other options derived using equation (xliv), (xlv), (xlvi):

$$k_{ia} = \frac{P_i+S_i}{\sum_{i=1}^{m}(P_i+S_i)} \tag{xliv}$$

$$k_{ib} = \frac{S_i}{minS_i} + \frac{P_i}{minP_i} \tag{xlv}$$

$$k_{ic} = \frac{\lambda(S_i)+(1-\lambda)(P_i)}{(\lambda maxS_i+(1-\lambda)maxP_i)} \tag{xlvi}$$

5) Ranking of all alternatives is determined from higher to lower based on k_i values:

$$k_i = (k_{ia}k_{ib}k_{ic})^{\frac{1}{3}} + (k_{ia} + k_{ib} + k_{ic}) \tag{xlvii}$$

13.2.8 MV-COCOSO Method (Multi-Variance Combined Compromise Solution)

An improved CoCoSo method with a new integration function based on the nonlinear programming model with maximum variance is proposed as follows [6]:

Steps 1-4. The same as Steps 1-4 in the original CoCoSo method.

Step 5. Apply the formula of "Linear Sum Normalization", i.e., Eq. (10), to normalize the three subordinate compromise scores:

$$k'_{ih} = \frac{k_{ih}}{\sum_{i=1}^{m} k_{ih}}, for\ i = 1, 2.., m; h = 1,2,3 \quad \text{(xlviii)}$$

where k_{ih} is the evaluation score of the ith alternative under the hth aggregation strategy, k'_{ih} is the normalized value of k_{ih}, such that $k'_{ih} \in [0,1] and\ \sum_{i=1}^{m} k'_{ih} = 1$.

Step 6. Construct the objective function and its constraints. To make the differences among the evaluated objects as large as possible, the overall difference of the evaluated objects can be measured by the variance of the comprehensive evaluation value of the evaluated objects:

$$S^2 = \frac{1}{m}\sum_{i=1}^{m}\left(k''_i - \overline{k''}\right)^2 \quad \text{(xlix)}$$

where k''_i denotes the comprehensive score of each evaluated object after synthesizing the three normalized subordinate compromise scores.

Furthermore, using Eqs. (l) and (li), we can determine the reasonable range of the comprehensive score of each evaluated object:

$$k''_i \in [k'^{-}_i, k'^{+}_i] \quad \text{(l)}$$

$$\sum_{i=1}^{m} k''_i = 1 \quad \text{(li)}$$

where $k'^{-}_i = \min\{k'_{i1}, k'_{i2}, k'_{i3}\}\ and\ k'^{+}_i = \max\{k'_{i1}, k'_{i2}, k'_{i3}\}$.

According to Eq. (lii), the mean of the comprehensive scores of all alternatives can be obtained by:

$$\overline{k''} = (k''_1 + k''_2 +..+ k''_i)/m = {}^{1}\!/_{m} \quad \text{(lii)}$$

According to Eqs. (l) and (li), we can furtherly calculate the variance of comprehensive scores:

$$S^2 = \frac{1}{m}\sum_{i=1}^{m}\left(k''_i - \overline{k''}\right)^2 = \frac{1}{m}\sum_{i=1}^{m}\left(k''_i - \frac{1}{m}\right)^2 \quad \text{(liii)}$$

Finally, we can construct an objective function and its constraint by synthesizing Eqs. (li)-(liii) as following:

$$max \frac{1}{m}\sum_{i=1}^{m}\left(k_i'' - \frac{1}{m}\right)^2 \ subjected\ to \begin{cases} \sum_{i=1}^{m} k_i'' = 1 \\ k_i'^{-} \le k_i'' \le k_i'^{+} \end{cases} \quad \text{(liv)}$$

Step 7. Rank all the alternatives. The alternative with the maximum k_i'' is chosen as the optimal alternative.

13.2.9 Measurement Alternatives and Ranking according to Compromise Solution (MARCOS)

Measurement Alternatives and Ranking according to Compromise Solution (MARCOS) was proposed by Željko Stević et. al in 2020 [7]. Initially, MARCOS method was developed for a sustainable supplier selection in the healthcare industry present in Bosnia and Herzegovina.

This method is based on the basis of defining the relationship between alternatives and reference values and determination of utility functions of alternatives along with compromise ranking with respect to ideal solution as well as anti-ideal solutions.

Utility functions are used to describe decision priorities. Utility functions reflect an alternative's location in relation to an optimal and anti-ideal solution. The optimal solution is the one that is nearest to the ideal while still being the furthest away from the anti-ideal reference point. The MARCOS process is carried out in the following steps: Having an initial decision-making matrix, which includes defining a set of n parameters and m alternatives. The formation of a longer initial matrix. The initial matrix is extended in this step by specifying the ideal solution (IS) and anti-ideal solution (AIS), where 'IS' is the best alternative and 'AIS' is the worst.

$$X = \begin{bmatrix} x_{IS1} & x_{IS2} & .. & x_{ISj} \\ x_{11} & x_{12} & .. & x_{1j} \\ x_{21} & x_{22} & .. & x_{2j} \\ .. & .. & .. & .. \\ x_{i1} & x_{i2} & .. & x_{ij} \\ x_{AIS1} & x_{AIS1} & .. & x_{AISj} \end{bmatrix} \quad \text{(lv)}$$

Depending on the nature of the criteria, IS and AIS are defined by applying equations (lvi) and (lvii):

$$IS = maxx_{ij} \text{ for benefit and } minx_{ij} \text{ for cost} \quad \text{(lvi)}$$

$$AIS = minx_{ij} \text{ for benefit and } maxx_{ij} \text{ for cost} \quad \text{(lvii)}$$

Normalization of the extended initial matrix (X) by applying equations (lviii) and (lix):

$$n_{ij} = \frac{x_{ij}}{x_{is}} \quad \text{for benefit} \quad \text{(lviii)}$$

$$n_{ij} = \frac{x_{is}}{x_{ij}} \quad \text{for cost} \quad \text{(lix)}$$

Determination of the weighted matrix $V = [v]_{m \times n}$

$$v_{ij} = n_{ij} \times w_j \quad \text{(lx)}$$

Calculation of the utility degree of alternatives k_i by applying equations (lxi) and (lxii),

$$k_i^+ = \frac{S_i}{S_{ISi}} \quad \text{(lxi)}$$

$$k_i^- = \frac{S_i}{S_{AISi}} \quad \text{(lxii)}$$

where Si (i=1,2,..,m) represents the sum of the elements of the weighted matrix V, equation (lxiii).

$$S_i = \sum_{i=0}^{n} v_{ij} \quad \text{(lxiii)}$$

Determination of the utility function of alternatives f(Ki) by equation (lxiv).

$$f(k_i) = \frac{k_i^+ + k_i^-}{1 + \frac{1-f(k_i^+)}{f(k_i^+)} + \frac{1-f(k_i^-)}{f(k_i^-)}} \quad \text{(lxiv)}$$

where $f(k_i^+)$ represents the utility function in relation to the ideal solution, while $f(k_i^-)$ represents the utility function in relation to the anti-ideal solution which are determined by equations (lxv) and (lxvi).

$$f(k_i^+) = \frac{k_i^+}{k_i^+ + k_i^-} \tag{lxv}$$

$$f(k_i^-) = \frac{k_i^-}{k_i^+ + k_i^-} \tag{lxvi}$$

13.2.10 Calculation of Weights between Criterions by Pair-Wise Comparison

In this analysis, the geometric mean methodology of the AHP system was used to assess the relative value of performance responses dependent on practise. The pair-wise contrast matrix is constructed using the Saaty nine-point choice scale. The following calculations is used to calculate the criteria weights of fabricated composite sample output characteristics [23]

$$GM_i = \left(\prod_{j=1}^{n} b_{ij}\right)^{1/n} \tag{lxvii}$$

$$w_j = GM_i / \sum_{j=1}^{n} GM_i \tag{lxviii}$$

Using the above calculations, the criterion weights were calculated as w = [0.60, 0.13, 0.27]. Despite the fact that this is a subjective approach to determining the relative value of parameters, a quality review of the calculated criteria weights was conducted. For the three metrics considered, random index of 0.58, consistency index, and consistency ratio values of 0.037 and 0.064, respectively, were obtained. These values indicate that determining criterion weights is rational, as CR = 0.10 for accuracy. The weights assigned to each consistency characteristic would be used to optimize process parameters.

13.3 Comparative Study of the effectiveness of MCDM optimization

For studying the effectiveness of the MCDM optimization, Example of Fused Deposition Modelling printed part has been taken where parametric optimization will be done for minimization of the surface roughness of the part. Barua et. al. [19] suggested MOORA and

Taguchi-based utility approach for solving multiple-criteria optimization problem in Fused Deposition Modelling. Improvement in surface roughness was taken as output measures in this study, as shown in Table 13.1 with L_{27} orthogonal array using control parameters (Table 13.2).

Table 13.1 Control parameters with their levels [1-7]

Control Parameters	Symbol	Levels		
		1	2	3
Thickness of layer	A	0.127 mm	0.178 mm	0.254 mm
Orientation of part	B	0^0	15^0	30^0
Raster angle	C	0^0	30^0	60^0
Raster width	D	0.4064 mm	0.4654 mm	0.5064 mm
Air Gap	E	0.000 mm	0.004 mm	0.008 mm

Table 13.2 Orthogonal array of the experiment with responses

Run No.	A	B	C	D	E	Top (µm)	Bottom (µm)	Side (µm)
1.	1	1	1	1	1	4.49	5.35	1.10
2.	1	1	1	1	2	4.22	5.43	1.14
3.	1	1	1	1	3	4.74	6.99	0.94
4.	1	2	2	2	1	5.62	4.30	0.65
5.	1	2	2	2	2	5.34	4.38	0.69
6.	1	2	2	2	3	5.86	5.94	0.50
7.	1	3	3	3	1	3.40	4.45	0.75
8.	1	3	3	3	2	3.13	4.54	0.79
9.	1	3	3	3	3	3.65	6.09	0.59
10.	2	1	2	3	1	5.83	6.34	0.81
11.	2	1	2	3	2	5.55	6.42	0.85
12.	2	1	2	3	3	6.08	7.98	0.65
13.	2	2	3	1	1	5.53	5.70	0.66
14.	2	2	3	1	2	5.25	5.78	0.70
15.	2	2	3	1	3	5.78	7.34	0.50
16.	2	3	1	2	1	4.54	5.14	0.59
17.	2	3	1	2	2	4.26	5.22	0.63
18.	2	3	1	2	3	4.79	6.78	0.44

19.	3	1	3	2	1	4.69	5.51	0.40
20.	3	1	3	2	2	4.42	5.59	0.44
21.	3	1	3	2	3	4.94	7.15	0.25
22.	3	2	1	3	1	4.36	4.78	0.67
23.	3	2	1	3	2	4.09	4.87	0.71
24.	3	2	1	3	3	4.61	6.42	0.51
25.	3	3	2	1	1	3.21	4.14	0.86
26.	3	3	2	1	2	2.94	4.23	0.90
27.	3	3	2	1	3	3.46	5.78	0.70

Table 13.3 shows the computational value obtained using Desirability Function Approach where at first desirability index was calculated based on smaller the better criteria. Finally, a weighted desirability value was calculated which was used in significant factor prediction using Taguchi method and ANOVA.

Table 13.3 Computational values of desirability function approach

Expt. No.	Top D_i	Bottom D_i	Side D_i	D_0
1.	0.5042	0.6861	0.0442	0.2720
2.	0.5924	0.6638	0.0000	0.0000
3.	0.4248	0.2587	0.2207	0.3338
4.	0.1474	0.9594	0.5438	0.2675
5.	0.2355	0.9370	0.4996	0.3453
6.	0.0679	0.5320	0.7204	0.1679
7.	0.8514	0.9187	0.4338	0.7167
8.	0.9395	0.8964	0.3896	0.7363
9.	0.7720	0.4913	0.6103	0.6832
10.	0.0794	0.4274	0.3698	0.1497
11.	0.1676	0.4051	0.3256	0.2248
12.	0.0000	0.0000	0.5463	0.0000
13.	0.1746	0.5941	0.5375	0.2774
14.	0.2627	0.5718	0.4933	0.3446
15.	0.0952	0.1667	0.7140	0.1764
16.	0.4904	0.7401	0.6110	0.5490
17.	0.5785	0.7178	0.5668	0.5917
18.	0.4109	0.3128	0.7875	0.4728
19.	0.4405	0.6444	0.8235	0.5480
20.	0.5286	0.6221	0.7793	0.5996
21.	0.3611	0.2170	1.0000	0.4449
22.	0.5460	0.8332	0.5279	0.5716
23.	0.6341	0.8108	0.4836	0.6085

24.	0.4666	0.4058	0.7044	0.5121
25.	0.9119	1.0000	0.3179	0.6943
26.	1.0000	0.9777	0.2737	0.7027
27.	0.8324	0.5726	0.4944	0.6889

13.4 Results and Discussions

Table 13.4 shows the computational value obtained using TOPSIS method where a weight normalization of response value was determined along with computation of positive and negative ideal solution. These values were taken for evaluation of distance measure and overall performance coefficient.

Table 13.4 Computational values of TOPSIS method

Expt. No.	S^-	S^+	C^+
1.	0.1956	0.0413	0.8256
2.	0.1940	0.0476	0.8028
3.	0.1883	0.0333	0.8499
4.	0.2079	0.0233	0.8993
5.	0.2054	0.0268	0.8844
6.	0.2017	0.0140	0.9350
7.	0.1968	0.0684	0.7421
8.	0.1960	0.0748	0.7239
9.	0.1885	0.0606	0.7566
10.	0.1995	0.0113	0.9465
11.	0.1967	0.0157	0.9261
12.	0.1937	0.0054	0.9728
13.	0.2002	0.0189	0.9137
14.	0.1977	0.0239	0.8921
15.	0.1940	0.0108	0.9475
16.	0.1967	0.0412	0.8270
17.	0.1950	0.0473	0.8049
18.	0.1895	0.0333	0.8507
19.	0.1956	0.0373	0.8397
20.	0.1937	0.0434	0.8171
21.	0.1886	0.0299	0.8632
22.	0.1978	0.0457	0.8123
23.	0.1962	0.0518	0.7911
24.	0.1904	0.0376	0.8352
25.	0.1982	0.0732	0.7302
26.	0.1977	0.0797	0.7128

27.	0.1898	0.0654	0.7437

Table 13.5 shows the computational value obtained using Utility method where overall utility function was computed based of preference scale of output response.

Table 13.5 Computational values of utility method

Expt. No.	Top P	Bottom P	Side P	U_i
1.	4.6770	3.3211	9.0675	5.6862
2.	3.9779	3.5278	9.2810	5.3512
3.	5.2709	6.8015	8.1311	6.2421
4.	7.1254	0.4811	5.9171	5.9354
5.	6.5695	0.7376	6.2723	5.7311
6.	7.6033	4.6861	4.2455	6.3175
7.	1.6216	0.9452	6.7652	2.9224
8.	0.6888	1.1928	7.0751	2.4786
9.	2.3991	5.0236	5.3404	3.5345
10.	7.5355	5.5371	7.2090	7.1876
11.	7.0005	5.7116	7.4974	6.9671
12.	7.9963	8.5300	5.8968	7.4988
13.	6.9565	4.1530	5.9693	6.3255
14.	6.3918	4.3469	6.3215	6.1070
15.	7.4416	7.4427	4.3142	6.5974
16.	4.7832	2.8067	5.3343	4.6750
17.	4.0910	3.0216	5.7243	4.3930
18.	5.3715	6.4096	3.4646	4.9916
19.	5.1562	3.7052	3.0007	4.3856
20.	4.4878	3.9058	3.5660	4.1633
21.	5.7254	7.0964	0.0000	4.3578
22.	4.3514	1.8701	6.0479	4.4869
23.	3.6306	2.1009	6.3957	4.1783
24.	4.9627	5.7060	4.4174	4.9121
25.	0.9903	0.0001	7.5464	2.6317
26.	-0.0002	0.2662	7.8195	2.1458
27.	1.8121	4.3398	6.3132	3.3560

Table 13.6 shows the computational value obtained using Grey Relational Analysis where responses were normalized as per lower the better criteria followed by computation of grey relational coefficient and grades.

Table 13.6 Computational values of Grey relational analysis

Run No.	Grey Relational Coefficient			Grey Relational Grade
	Top GRC	Bottom GRC	Side GRC	
1.	0.5021	0.6143	0.3434	0.4866
2.	0.5509	0.5979	0.3333	0.4941
3.	0.4650	0.4028	0.3908	0.4196
4.	0.3696	0.9248	0.5229	0.6058
5.	0.3954	0.8882	0.4998	0.5945
6.	0.3491	0.5165	0.6413	0.5023
7.	0.7709	0.8601	0.4690	0.7000
8.	0.8921	0.8283	0.4503	0.7236
9.	0.6868	0.4957	0.5620	0.5815
10.	0.3520	0.4661	0.4424	0.4202
11.	0.3753	0.4566	0.4257	0.4192
12.	0.3333	0.3333	0.5243	0.3970
13.	0.3772	0.5519	0.5195	0.4829
14.	0.4041	0.5387	0.4967	0.4798
15.	0.3559	0.3750	0.6361	0.4557
16.	0.4952	0.6580	0.5624	0.5719
17.	0.5426	0.6392	0.5358	0.5725
18.	0.4591	0.4211	0.7018	0.5273
19.	0.4719	0.5844	0.7391	0.5985
20.	0.5148	0.5695	0.6937	0.5927
21.	0.4390	0.3897	1.0000	0.6096
22.	0.5241	0.7498	0.5143	0.5961
23.	0.5775	0.7255	0.4919	0.5983
24.	0.4838	0.4569	0.6284	0.5231
25.	0.8502	1.0000	0.4230	0.7577
26.	1.0000	0.9573	0.4077	0.7883
27.	0.7490	0.5392	0.4972	0.5951

Table 13.7 shows the computational value obtained using WASPAS method where normalization of the responses was conducted based on cost criteria followed by evaluation of total relative importance using weighted sum and weighted product method.

Table 13.7 Computational values of WASPAS method

Run No.	Top $\bar{x}_{ij}$	Bottom $\bar{x}_{ij}$	Side $\bar{x}_{ij}$	Q_1	Q_2	Q_i
1.	0.6535	0.7746	0.2233	0.5531	0.5000	0.5265
2.	0.6964	0.7624	0.2155	0.5752	0.5134	0.5443

3.	0.6192	0.5927	0.2607	0.5189	0.4874	0.5032
4.	0.5231	0.9637	0.3759	0.5406	0.5180	0.5293
5.	0.5502	0.9449	0.3545	0.5486	0.5242	0.5364
6.	0.5008	0.6974	0.4956	0.5250	0.5214	0.5232
7.	0.8629	0.9299	0.3267	0.7268	0.6703	0.6986
8.	0.9393	0.9124	0.3104	0.7660	0.6939	0.7300
9.	0.8040	0.6796	0.4135	0.6824	0.6573	0.6699
10.	0.5039	0.6533	0.3036	0.4692	0.4546	0.4619
11.	0.5290	0.6446	0.2895	0.4794	0.4612	0.4703
12.	0.4832	0.5190	0.3772	0.4592	0.4562	0.4577
13.	0.5311	0.7266	0.3727	0.5138	0.5028	0.5083
14.	0.5591	0.7159	0.3516	0.5235	0.5094	0.5165
15.	0.5082	0.5642	0.4900	0.5106	0.5101	0.5104
16.	0.6472	0.8059	0.4139	0.6049	0.5902	0.5976
17.	0.6893	0.7927	0.3881	0.6214	0.6011	0.6113
18.	0.6135	0.6109	0.5639	0.5998	0.5994	0.5996
19.	0.6256	0.7521	0.6089	0.6376	0.6361	0.6368
20.	0.6649	0.7406	0.5545	0.6449	0.6420	0.6435
21.	0.5941	0.5795	1.0000	0.7018	0.6816	0.6917
22.	0.6732	0.8661	0.3679	0.6158	0.5909	0.6033
23.	0.7188	0.8508	0.3473	0.6356	0.6037	0.6197
24.	0.6368	0.6448	0.4817	0.5959	0.5915	0.5937
25.	0.9139	1.0000	0.2871	0.7558	0.6764	0.7161
26.	1.0000	0.9797	0.2745	0.8015	0.7035	0.7525
27.	0.8480	0.7163	0.3521	0.6970	0.6543	0.6757

Table 13.8 shows the computational value obtained using EDAS method. At first average solution of the output are computed followed by evaluation of positive and negative distance from average in accordance to cost criteria. Finally, an appraisal score was computed using normalized weighted sum of positive and negative distance from average.

Table 13.8 Computational values of EDAS method

Run No.	SP	SN	NSP	NSN	AS
1.	0.0234	0.1673	0.0929	0.3114	0.2022
2.	0.0573	0.1830	0.2282	0.2467	0.2374
3.	0.0000	0.1513	0.0000	0.3770	0.1885
4.	0.0414	0.1292	0.1648	0.4679	0.3164
5.	0.0292	0.0987	0.1161	0.5936	0.3548
6.	0.0730	0.1682	0.2905	0.3074	0.2989

7.	0.1855	0.0288	0.7382	0.8813	0.8097
8.	0.2195	0.0445	0.8734	0.8166	0.8450
9.	0.1595	0.0102	0.6345	0.9580	0.7963
10.	0.0000	0.2244	0.0000	0.0763	0.0381
11.	0.0000	0.2061	0.0000	0.1515	0.0758
12.	0.0111	0.2429	0.0443	0.0001	0.0221
13.	0.0080	0.1192	0.0319	0.5091	0.2705
14.	0.0000	0.0929	0.0000	0.6174	0.3087
15.	0.0707	0.1894	0.2815	0.2204	0.2509
16.	0.0566	0.0000	0.2252	1.0000	0.6126
17.	0.0749	0.0000	0.2979	1.0000	0.6490
18.	0.0969	0.0477	0.3854	0.8037	0.5946
19.	0.1130	0.0097	0.4495	0.9603	0.7049
20.	0.1216	0.0000	0.4838	1.0000	0.7419
21.	0.1724	0.0765	0.6859	0.6852	0.6855
22.	0.0580	0.0000	0.2306	1.0000	0.6153
23.	0.0874	0.0111	0.3476	0.9542	0.6509
24.	0.0683	0.0177	0.2718	0.9269	0.5994
25.	0.2173	0.0700	0.8649	0.7117	0.7883
26.	0.2513	0.0857	1.0001	0.6470	0.8236
27.	0.1502	0.0103	0.5978	0.9575	0.7777

Table 13.9 shows the computational value obtained using COCOSO method where normalization of the outputs was performed based on cost criterion followed by computation of total weighted comparability sequence. Finally total appraisal score was calculated.

Table 13.9 Computational values of COCOSO method

Run No.	S_i	P_i	k_{ia}	k_{ib}	k_{ic}	k_i
1.	0.4037	2.0460	0.0326	4.7026	0.6949	2.2839
2.	0.4417	1.6785	0.0282	4.6075	0.6014	2.1731
3.	0.3481	2.1022	0.0326	4.3799	0.6950	2.1653
4.	0.3600	2.1599	0.0335	4.5157	0.7148	2.2310
5.	0.3980	2.2407	0.0351	4.8512	0.7485	2.3814
6.	0.3044	2.0357	0.0311	4.0197	0.6638	2.0077
7.	0.7474	2.6951	0.0458	7.6566	0.9765	3.5923
8.	0.7854	2.7244	0.0467	7.9427	0.9956	3.7122
9.	0.6918	2.6431	0.0443	7.2300	0.9460	3.4119
10.	0.2030	1.8786	0.0277	3.1815	0.5905	1.6397
11.	0.2411	1.9701	0.0294	3.5274	0.6272	1.7968
12.	0.1475	1.0408	0.0158	2.0000	0.3371	1.0043

13.	0.3271	2.1312	0.0327	4.2654	0.6973	2.1249
14.	0.3652	2.2047	0.0342	4.5940	0.7290	2.2711
15.	0.2716	1.9492	0.0295	3.7140	0.6299	1.8681
16.	0.5554	2.4892	0.0405	6.1571	0.8636	2.9529
17.	0.5935	2.5358	0.0416	6.4598	0.8876	3.0831
18.	0.4999	2.3838	0.0383	5.6792	0.8180	2.7411
19.	0.5704	2.5049	0.0409	6.2739	0.8723	3.0027
20.	0.6085	2.5572	0.0421	6.5822	0.8980	3.1363
21.	0.5149	2.3626	0.0382	5.7606	0.8162	2.7695
22.	0.5784	2.5136	0.0411	6.3366	0.8771	3.0295
23.	0.6165	2.5559	0.0422	6.6352	0.8999	3.1572
24.	0.5229	2.4320	0.0393	5.8816	0.8382	2.8315
25.	0.7629	2.6800	0.0458	7.7474	0.9766	3.6254
26.	0.8010	2.7018	0.0466	8.0263	0.9936	3.7409
27.	0.7074	2.6527	0.0447	7.3446	0.9531	3.4595

Table 13.10 shows the computational value obtained using MV-COCOSO method where normalization of the outputs was performed based on cost criterion followed by computation of total weighted comparability sequence. Further normalization was done using linear sum normalization. Finally comprehensive scores were calculated.

Table 13.10 Computational values of MV-COCOSO method

Run No.	S_i	P_i	k'_{ia}	k'_{ib}	k'_{ic}	k''
1.	0.4037	2.0460	0.0326	0.0313	0.0326	0.0313
2.	0.4417	1.6785	0.0282	0.0307	0.0282	0.0282
3.	0.3481	2.1022	0.0326	0.0292	0.0326	0.0292
4.	0.3600	2.1599	0.0335	0.0301	0.0335	0.0301
5.	0.3980	2.2407	0.0351	0.0323	0.0351	0.0323
6.	0.3044	2.0357	0.0311	0.0268	0.0311	0.0268
7.	0.7474	2.6951	0.0458	0.0510	0.0458	0.0458
8.	0.7854	2.7244	0.0467	0.0529	0.0467	0.0467
9.	0.6918	2.6431	0.0443	0.0482	0.0443	0.0443
10.	0.2030	1.8786	0.0277	0.0212	0.0277	0.0212
11.	0.2411	1.9701	0.0294	0.0235	0.0294	0.0235
12.	0.1475	1.0408	0.0158	0.0133	0.0158	0.0133
13.	0.3271	2.1312	0.0327	0.0284	0.0327	0.0284
14.	0.3652	2.2047	0.0342	0.0306	0.0342	0.0306
15.	0.2716	1.9492	0.0295	0.0247	0.0295	0.0247
16.	0.5554	2.4892	0.0405	0.0410	0.0405	0.0405

17.	0.5935	2.5358	0.0416	0.0430	0.0416	0.0416
18.	0.4999	2.3838	0.0383	0.0378	0.0383	0.0378
19.	0.5704	2.5049	0.0409	0.0418	0.0409	0.0409
20.	0.6085	2.5572	0.0421	0.0439	0.0421	0.0421
21.	0.5149	2.3626	0.0382	0.0384	0.0382	0.0382
22.	0.5784	2.5136	0.0411	0.0422	0.0411	0.0411
23.	0.6165	2.5559	0.0422	0.0442	0.0422	0.0422
24.	0.5229	2.4320	0.0393	0.0392	0.0393	0.0392
25.	0.7629	2.6800	0.0458	0.0516	0.0458	0.0458
26.	0.8010	2.7018	0.0466	0.0535	0.0466	0.0466
27.	0.7074	2.6527	0.0447	0.0489	0.0447	0.0447

Table 13.11 shows the computational value obtained using MARCOS where a weighted extended normalization matrix was created using cost criteria followed by utility degree and utility function.

Table 13.11 Computational values of MARCOS method

Run No.	S_i	K_i^-	K_i^+	$f(K^-)$	$f(K^+)$	$f(K_i)$
1.	0.55	1.32	0.55	0.30	0.70	0.4932
2.	0.58	1.37	0.58	0.30	0.70	0.5128
3.	0.52	1.24	0.52	0.30	0.70	0.4630
4.	0.54	1.29	0.54	0.30	0.70	0.4826
5.	0.55	1.31	0.55	0.30	0.70	0.4897
6.	0.53	1.26	0.53	0.30	0.70	0.4692
7.	0.73	1.74	0.73	0.30	0.70	0.6483
8.	0.77	1.83	0.77	0.30	0.70	0.6831
9.	0.68	1.63	0.68	0.30	0.70	0.6091
10.	0.47	1.12	0.47	0.30	0.70	0.4189
11.	0.48	1.15	0.48	0.30	0.70	0.4278
12.	0.46	1.10	0.46	0.30	0.70	0.4103
13.	0.52	1.23	0.52	0.30	0.70	0.4588
14.	0.53	1.25	0.53	0.30	0.70	0.4673
15.	0.51	1.22	0.51	0.30	0.70	0.4564
16.	0.61	1.45	0.61	0.30	0.70	0.5400
17.	0.62	1.48	0.62	0.30	0.70	0.5547
18.	0.60	1.43	0.60	0.30	0.70	0.5361
19.	0.64	1.53	0.64	0.30	0.70	0.5699
20.	0.65	1.54	0.65	0.30	0.70	0.5762
21.	0.71	1.68	0.71	0.30	0.70	0.6285
22.	0.62	1.47	0.62	0.30	0.70	0.5496
23.	0.64	1.52	0.64	0.30	0.70	0.5672
24.	0.60	1.42	0.60	0.30	0.70	0.5324

25.	0.76	1.80	0.76	0.30	0.70	0.6740
26.	0.80	1.91	0.80	0.30	0.70	0.7145
27.	0.70	1.66	0.70	0.30	0.70	0.6218

Figures 13.1-13.9 show the SN ratio plot computed using respective optimization. Tables 13.12-13.20 show the ANOVA results of the input parameters obtained using respective optimization process. It is interesting to note that Factor B i.e part orientation is the most significant input parameters among the other input parameters for minimization of surface roughness of the FDM printed part. All the optimal factor settings obtained from the SN ratio plot for different optimization methods are stated in Table 13.21 along with the R-squared values.

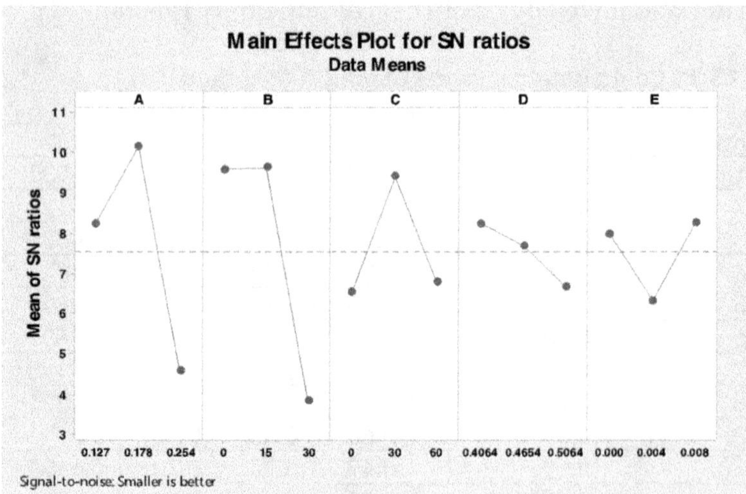

Figure 13.1 Main effect plot for overall desirability value.

Table 13.12 ANOVA for overall desirability value

Source	DF	Adj SS	Adj MS	F test	P test	% Significance
A	2	166.79	83.39	44.50	0.000	36.73
B	2	209.60	104.80	55.92	0.000	46.16
C	2	61.30	30.65	16.35	0.000	13.50
D	2	1.40	0.70	0.37	0.695	0.31
E	2	26.24	13.12	7.00	0.008	5.78

Residual Error	14	26.23	1.87			5.78
Total	24	454.05				

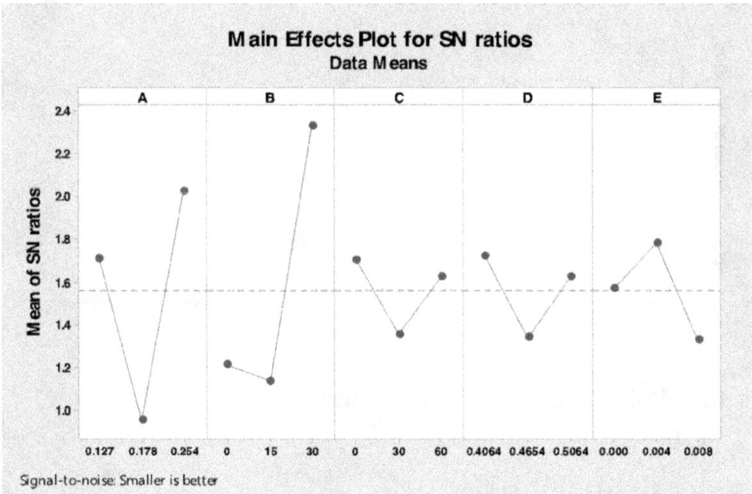

Figure 13.2 Main effcct plot for overall performance coefficient.

Table 13.13 ANOVA for overall performance coefficient

Sourcc	DF	Adj SS	Adj MS	F test	P test	% Significance
A	2	5.44	2.72	2322.10	0.000	34.50
B	2	8.08	4.04	3447.89	0.000	51.22
C	2	0.59	0.29	254.61	0.000	3.78
D	2	0.69	0.34	297.86	0.000	4.43
E	2	0.93	0.46	400.64	0.000	5.95
Residual Error	16	0.01	0.001			0.12
Total	26	15.79				

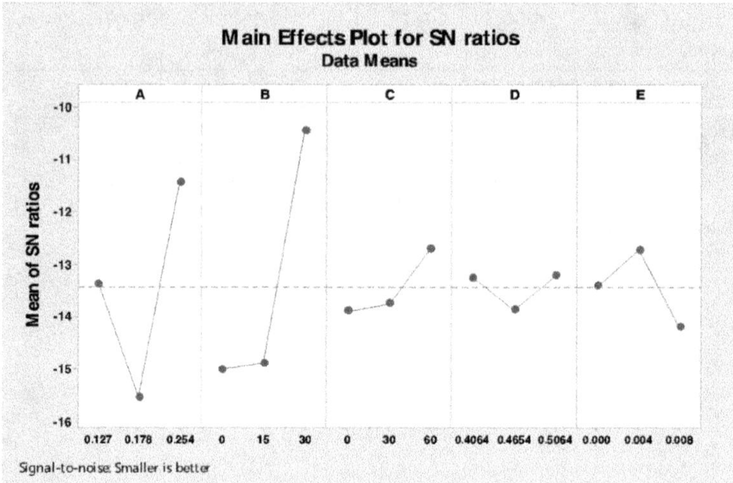

Figure 13.3 Main effect plot for overall utility value.

Table 13.14 ANOVA for overall utility value

Source	DF	Adj SS	Adj MS	F test	P test	% Signifi-cance
A	2	76.05	38.02	104.73	0.00	33.91
B	2	122.35	61.17	168.50	0.00	54.56
C	2	7.62	3.81	10.50	0.00	3.40
D	2	2.47	1.23	3.41	0.05	1.11
E	2	9.95	4.97	13.70	0.00	4.44
Residual Error	16	5.80	0.36			2.59
Total	26	224.27				

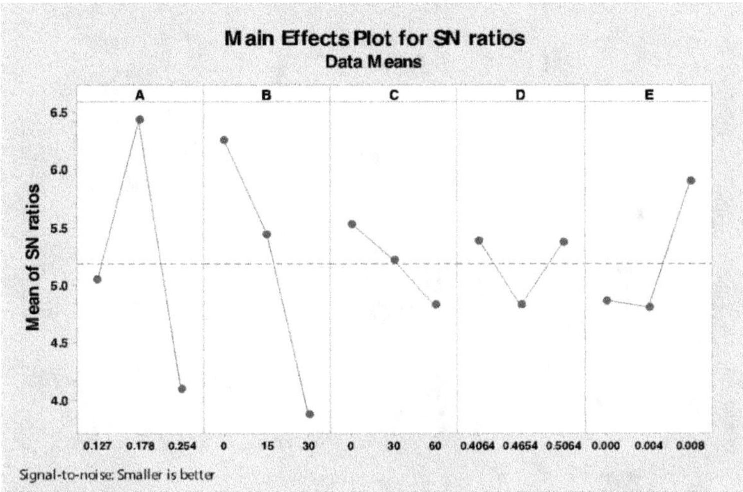

Figure 13.4 Main effect plot for grey relational grade.

Table 13.15 ANOVA for grey relational grade

Source	DF	Adj SS	Adj MS	F test	P test	% Significance
A	2	24.84	12.42	61.59	0.00	38.03
B	2	26.45	13.22	65.59	0.00	40.49
C	2	2.22	1.11	5.52	0.01	3.41
D	2	1.82	0.91	4.52	0.02	2.79
E	2	6.75	3.37	16.75	0.00	10.34
Residual Error	16	3.22	0.20			4.94
Total	26	65.33				

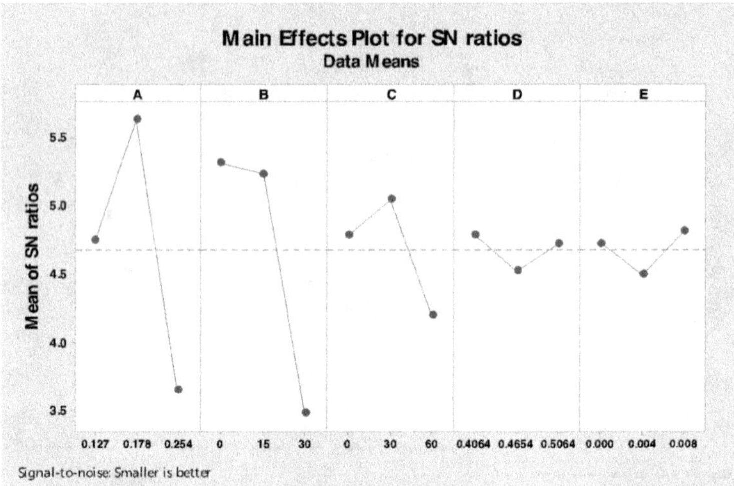

Figure 13.5 Main effect plot for total relative importance OF WASPAS method.

Table 13.16 ANOVA for total relative importance OF WASPAS method

Source	DF	Adj SS	Adj MS	F test	P test	% Significance
A	2	17.86	8.93	152.16	0.00	42.12
B	2	19.38	9.69	165.11	0.00	45.70
C	2	3.42	1.71	29.16	0.00	8.07
D	2	0.33	0.16	2.82	0.08	0.78
E	2	0.47	0.23	4.03	0.03	1.12
Residual Error	16	0.93	0.058			2.21
Total	26	42.41				

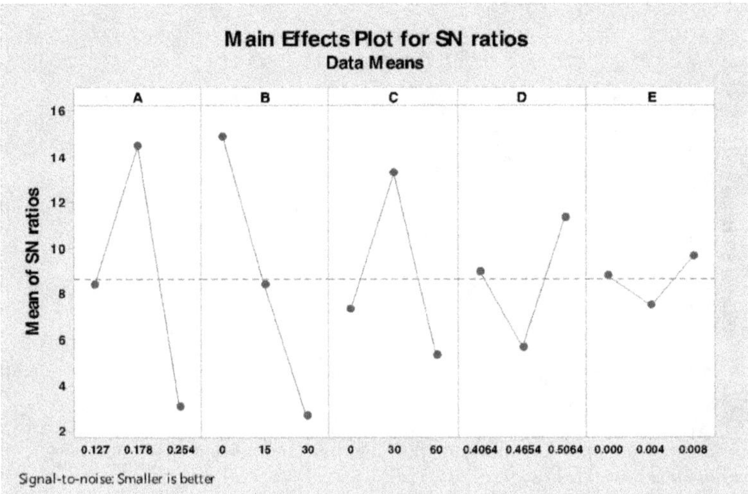

Figure 13.6 Main effect plot for appraisal score of EDAS method.

Table 13.17 ANOVA for appraisal score of EDAS method

Source	DF	Adj SS	Adj MS	F test	P test	% Significance
A	2	587.35	293.67	109.75	0.00	33.09
B	2	672.29	336.14	125.62	0.00	37.88
C	2	307.83	153.91	57.52	0.00	17.34
D	2	143.77	71.88	26.86	0.00	8.10
E	2	20.76	10.37	3.88	0.04	1.17
Residual Error	16	42.81	2.67			2.41
Total	26	1774.81				

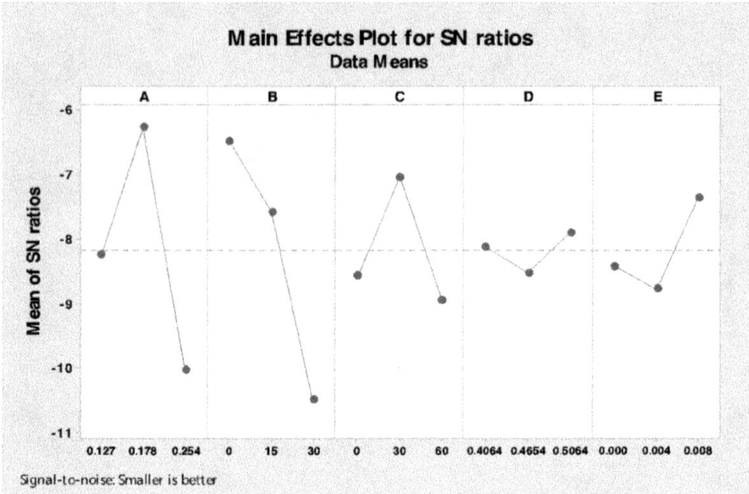

Figure 13.7 Main effect plot for total appraisal score of COCOSO method.

Table 13.18 ANOVA for total appraisal score of COCOSO method

Source	DF	Adj SS	Adj MS	F test	P test	% Significance
A	2	64.06	32.03	51.98	0.00	35.32
B	2	77.53	38.76	62.91	0.00	42.74
C	2	18.41	9.20	14.94	0.00	10.15
D	2	1.79	0.89	1.46	0.26	0.99
E	2	9.74	4.87	7.91	0.00	5.37
Residual Error	16	9.85	0.61			5.43
Total	26	181.40				

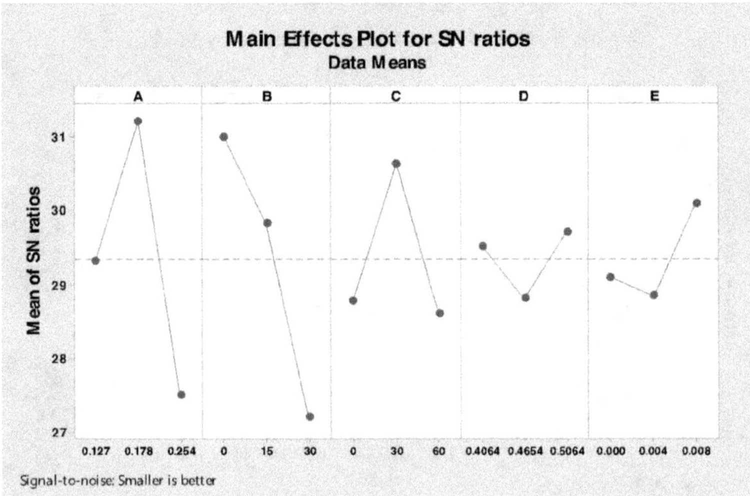

Figure 13.8 Main effect plot for variance of comprehensive scores for MV-COCOSO method.

Table 13.19 ANOVA for variance of comprehensive scores for MV-COCOSO method

Source	DF	Adj SS	Adj MS	F test	P test	% Significance
A	2	62.20	31.10	46.57	0.00	35.44
B	2	68.09	34.04	50.97	0.00	38.80
C	2	22.58	11.29	16.90	0.00	12.87
D	2	4.06	2.03	3.04	0.07	2.32
E	2	7.87	3.93	5.89	0.01	4.49
Residual Error	16	10.68	0.66			6.09
Total	26	175.49				

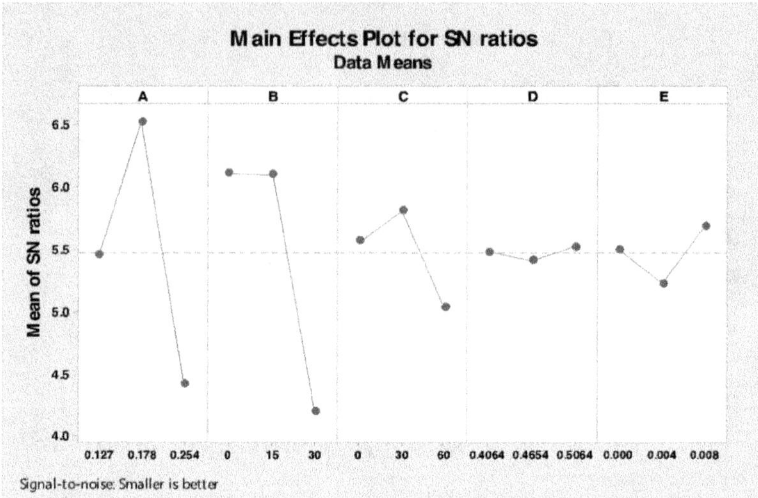

Figure 13.9 Main effect plot for utility function of MARCOS method.

Table 13.20 ANOVA for utility function of MARCOS method

Source	DF	Adj SS	Adj MS	F test	P test	% Significance
A	2	9.98	9.99	108.30	0.00	42.31
B	2	21.93	10.96	118.86	0.00	46.44
C	2	2.83	1.41	15.36	0.00	6.00
D	2	0.05	0.02	0.28	0.75	0.11
E	2	0.95	0.47	5.15	0.01	2.01
Residual Error	16	1.47	0.09			3.13
Total	26	47.23				

It is interesting to notice that factor B i.e. part orientation has been listed as the most significant or dominating parameter for FDM printing part while factor A, i.e. layer thickness is the 2nd most significant or dominating parameter as per the ANOVA results of different optimization methods. Among all methods, residual error resulted in ANOVA of Desirability function analysis, COCOSO method AND MV-COCOSO method are more than 5% which can't be accepted as optimal result. Except these three methods, all the methods have provided optimal results. Among all, TOPSIS method has the better result as the R-squared value is about 99.88%, leaving a residual error of 0.12% only. Hence, TOPSIS method has provided the optimal result for FDM printed part for minimization of surface roughness.

Table 13.21 R-squared value and optimal control parameters obtained from different optimization technique

Method	R-sq value	A	B	C	D	E
Desirability Function Analysis	94.22 %	0.254	30^0	0^0	0.5064	0.004
TOPSIS Method	99.88 %	0.178	15^0	30^0	0.4654	0.008
Utility Concept	97.41 %	0.178	0^0	0^0	0.4654	0.008
GREY Relational Analysis	95.06 %	0.254	30^0	60^0	0.4654	0.004
WASPAS Method	97.79 %	0.254	30^0	60^0	0.4654	0.004
EDAS Method	97.59 %	0.254	30^0	60^0	0.4654	0.004
COCOSO Method	94.57 %	0.254	30^0	60^0	0.4654	0.004
MV-COCOSO Method	93.91 %	0.254	30^0	60^0	0.4654	0.004
MARCOS Method	96.87 %	0.254	30^0	60^0	0.4654	0.004

13.5 Conclusion

The use of the MCDM approach for the optimization of certain modern machining processes has been discussed in this paper. The problem description was to obtain the optimal setting of process parameters for FDM printed parts, and nine different common statistical optimization methods were discussed. Among all methods used, TOPSIS method gave the better results. The obtained findings were compared to those published by previous researchers. The optimum combination of FDM printing variables was found to be the same in Grey Relational Analysis, WASPAS method, EDAS method, COCOSO method, MV-COCOSO method, and MARCOS method. The TOPSIS approach should take into account all of the quality answers and their relative value, leading to the conclusion that it can provide a more precise assessment of the alternatives. As the R-squared value obtained in TOPSIS is more than that of rest optimization methods, it can be recommended as the superior for optimizing FDM parameters in this investigation. Also, the rest of the methods cannot be criticized as these methods have also gave exceptionally optimized results in different engineering problems where TOPSIS method failed.

These methods, on the other hand, are computationally very simple, readable, and stable, and they can be used for a large number of input and output variables. The theoretical method's use in a wide variety of MCDM-based problems in a real-time production context remains a future possibility of the current work.

The selection of a specific machining method is a critical activity in the manufacturing world. The scope of future research will include a comprehensive overview of this complex problem for a given machining application, including material type and necessary geometries, tolerances and surface finish, design specifications, process capabilities, expense, and productivity. In addition, for future studies, the findings of this analysis can be compared to those of other MCDM approaches.

References

1. Madić, M., Gecevska, V., Radovanović, M., & Petković, D. (2014) Multi-criteria economic analysis of machining processes using the WASPAS method. Journal of Production Engineering, 17(2), 1-6.
2. Stević, Ž. E. L. J. K. O., Tanackov, I., Vasiljević, M., & Vesković, S. (2016) Evaluation in logistics using combined AHP and EDAS method. In Proceedings of the XLIII International Symposium on Operational Research, Belgrade, Serbia, 20-23.
3. Ghorabaee, K., M., Zavadskas, E. K., Olfat, L., & Turskis, Z. (2015) Multi-criteria inventory classification using a new method of evaluation based on distance from average solution (EDAS). Informatica, 26(3), 435-451.
4. Chakraborty, S., Zavadskas, E. K., & Antucheviciene, J. (2015) Applications of WASPAS method as a multi-criteria decision-making tool. Economic Computation and Economic Cybernetics Studies and Research, 49(1), 5-22.
5. Yazdani, M., Zarate, P., Zavadskas, E. K., & Turskis, Z. (2019) A Combined Compromise Solution (CoCoSo) method for multi-criteria de-

cision-making problems. Management Decision. 57(9).

6. Lai, H., Liao, H., Wen, Z., Zavadskas, E. K., & Al-Barakati, A. (2020) An Improved CoCoSo Method with a Maximum Variance Optimization Model for Cloud Service Provider Selection. Engineering Economics, 31(4), 411-424.
7. Stević, Ž., Pamučar, D., Puška, A., & Chatterjee, P. (2020) Sustainable supplier selection in healthcare industries using a new MCDM method: Measurement of alternatives and ranking according to Compromise solution (MARCOS). Computers & Industrial Engineering, 140, 106231.
8. Bagal, D. K., Parida, B., Barua, A., Jeet, S., & Sahoo, B. B. (2019) Multi-parametric optimization in CNC dry turning of chromoly steel using taguchi coupled desirability function analysis and utility concept. Int. J. Appl. Eng. Res, 14(13), 21-26.
9. Barua, A., Jeet, S., Cherkia, H., Bagal, D. K., & Sahoo, B. B. (2019) Parametric optimization of FDM processed part for improving surface finish using MOORA technique and desirability function analysis. International Journal of Applied Engineering Research, 14.
10. Bagal, D. K., Parida, B., Barua, A., Jeet, S., & Sahoo, B. B. (2019) Multi-parametric optimization in CNC dry turning of Chromoly steel using Taguchi coupled desirability function analysis and utility concept. Int. J. Appl. Eng. Res, 14(13), 21-26.
11. Panda, S. N., Bagal, D. K., Pattanaik, A. K., Patnaik, D., Barua, A., Jeet, S., Parida B. & Naik, B. (2020) Comparative Evaluation for Studying the Parametric Influences on Quality of Electrode Using Taguchi Method Coupled with MOORA, DFA, and TOPSIS Method for Electrochemical Machining. In Recent Advances in Mechanical Infrastructure, Springer, Singapore, 115 129.
12. Jeet, S., Barua, A., Bagal, D. K., Pattanaik, A. K., Agrawal, P. K., & Panda, S. N. (2019) Multi-parametric optimization during drilling of aerospace alloy (UNS A97068) using hybrid RSM-GRA, GA and SA. International Journal of Management, Technology and Engineering, 9.
13. Barua, A., Jeet, S., Parida, B., Sahoo, B. B., Bagal, D. K., & Samantray, A. (2018) Virtual Optimization of Motorcycle Sprocket Material by Using FEA and Taguchi Coupled TOPSIS-GA-SA. International Journal of Advanced Scientific Research and Management, 3, 54-63.
14. Jeet, S., Barua, A., Parida, B., Sahoo, B. B., & Bagal, D. K. (2018) Multi-objective optimization of welding parameters in GMAW for stainless steel and low carbon steel using hybrid RSM-TOPSIS-GA-SA approach. Int J Tech Innov Mod Eng Sci, 4, 683-692.
15. Bagal, D. K., Barua, A., Jeet, S., Satapathy, P., & Patnaik, D. (2019) MCDM optimization of parameters for wire-EDM machined stainless steel using hybrid RSM-TOPSIS, genetic algorithm and simulated annealing. International Journal of Engineering and Advanced

Technology, 9, 366-71.

16. Khan, A., & Maity, K. P. (2016) Application of MCDM-based TOPSIS method for the optimization of multi quality characteristics of modern manufacturing processes. In International Journal of Engineering Research in Africa, 23, 33-51.
17. Bagal, D. K., Barua, A., Pattanaik, A. K., Jeet, S., & Patnaik, D. (2020) Parametric Optimization Based on Mechanical Characterization of Fused Deposition Modelling Fabricated Part Using Utility Concept. In Advances in Materials Processing, Springer, Singapore 313-325.
18. Bagal, D. K., Parida, B., Barua, A., Jeet, S., & Sahoo, B. B. (2019) Multi-parametric optimization in CNC dry turning of chromoly steel using Taguchi coupled desirability function analysis and utility concept. Int. J. Appl. Eng. Res, 14(13), 21-26.
19. Barua, A., Jeet, S., Bagal, D. K., Bagal, P., & Pattanaik, A. K. (2019) Comparative analysis based on MCDM optimization of printing parameters affecting compressive and tensile strength of fused deposition modelling processed parts. International Journal of Technical Innovation in Modern Engineering Science, 5, 383-92.
20. Parida, B., Barua, A., Jeet, S., & Bagal, D. K. (2018) Fabrication and mechanical characterization of jute-glass-silk fiber polymer composites based on hybrid RSM-GRA-FIS and RSM-TOPSIS approach. International Journal for Research in Engineering Application & Management, 4(5), 25-33.
21. Sahoo, B., Barua, A., Jeet, S., & Bagal, D. K. (2018) Multi Objective Optimization of WEDM Process Parameters Using Hybrid RSM-GRA-FIS, GA and SA Approach. International Journal of Research in Advent Technology, 6(7), 1752-1761.
22. Jeet, S., Barua, A., Cherkia, H., & Bagal, D. K. (2019) Comparative Investigation Based on MOORA, GRA and TOPSIS Method of Turning of Nickel-Chromium-Molybdenum Steel under the influence of Low Cost Oil Mist Lubrication System. International Journal of Applied Engineering Research, 14(13), 8-20.
23. Naik, B., Paul, S., Barua, A., Jeet, S., & Bagal, D. K. (2019) Fabrication and strength analysis of hybrid jute-glass-silk fiber polymer composites based on hybrid Taguchi-WASPAS method. International Journal of Management, Technology Engineering Economics, 9, 3472-9.
24. Jeet, S., Barua, A., Bagal, D. K., Pradhan, S., Patnaik, D., & Pattanaik, A. K. (2021) Comparative Investigation of CNC Turning of Nickel-Chromoly Steel under Different Cutting Environment with a Fabricated Portable Mist Lubricator: A Super Hybrid Taguchi-WASPAS-GA-SA-PSO Approach. In Advanced Manufacturing Systems and Innovative Product Design: Select Proceedings of IPDIMS 2020, Springer Singapore 515-531.
25. Bagal, D. K., Mahapatra, S. K., Barua, A., Jeet, S., Pattanaik, A. K., & Patnaik, D. (2021) Multi-parametric Optimization of Wire-EDM of

Inconel 718 Super Alloy Using Taguchi-Coupled WASPAS Method. In Advances in Mechanical Processing and Design, Springer, Singapore, 459-467.

26. Bagal, D. K., Patra, A. K., Jeet, S., Barua, A., Pattanaik, A. K., Patnaik, D. (2021) MCDM Optimization of Karanja Biodiesel Powered CI Engine to Improve Performance Characteristics Using Super Hybrid Taguchi-Coupled WASPAS-GA, SA, PSO Method. In Next Generation Materials and Processing Technologies: Select Proceedings of RDMPMC 2020. Springer Singapore. DOI: 10.1007/978-981-16-0182-8_37
27. Bagal, D. K., Giri, A., Pattanaik, A. K., Jeet, S., Barua, A., Panda, S. N. (2021) MCDM Optimization of Characteristics in Resistance Spot Welding for Dissimilar Materials Utilizing Advanced Hybrid Taguchi Method-Coupled CoCoSo, EDAS and WASPAS Method. In Next Generation Materials and Processing Technologies: Select Proceedings of RDMPMC 2020. Springer Singapore. DOI: 10.1007/978-981-16-0182-8_36
28. Naik, B., Paul, S., Mishra, S. P., Rout, S. P., Barua, A., & Bagal, D. K. (2019) Performance Analysis of M40 Grade Concrete by Partial Replacement of Portland Pozzolana Cement with Marble Powder and Fly Ash Using Taguchi-EDAS Method. JASC: Journal of Applied Science Computations, 6, 733-43.
29. Bagal, D. K., Panda, S. K., Barua, A., Jeet, S., Pattanaik, A. K., & Patnaik, D. (2021) Parametric Appraisal of CNC Micro-drilling of Aerospace Material (PMMA) Using Taguchi-Based EDAS Method. In Advances in Mechanical Processing and Design, Springer, Singapore, 449-458.
30. Barua, A., Jeet, S., Bagal, D. K., Satapathy, P., & Agrawal, P. K. (2019) Evaluation of mechanical behavior of hybrid natural fiber reinforced nano sic particles composite using hybrid Taguchi-CoCoSo method. International Journal of Innovative Technology Exploring Engineering, 8, 3341-5.
31. Bagal, D. K., Naik, B., Parida, B., Barua, A., Jeet, S., Singh, S. K., & Pattanaik, A. K. (2020). Comparative Mechanical Characterization of M30 Concrete Grade by Fractional Replacement of Portland Pozzolana Cement with Industrial Waste Using CoCoSo and CODAS Methods. In IOP Conference Series: Materials Science and Engineering, 970(1), 012015.

Printed in the USA
CPSIA information can be obtained
at www.ICGtesting.com
LVHW021530291023
762483LV00046B/741

9 781922 617187